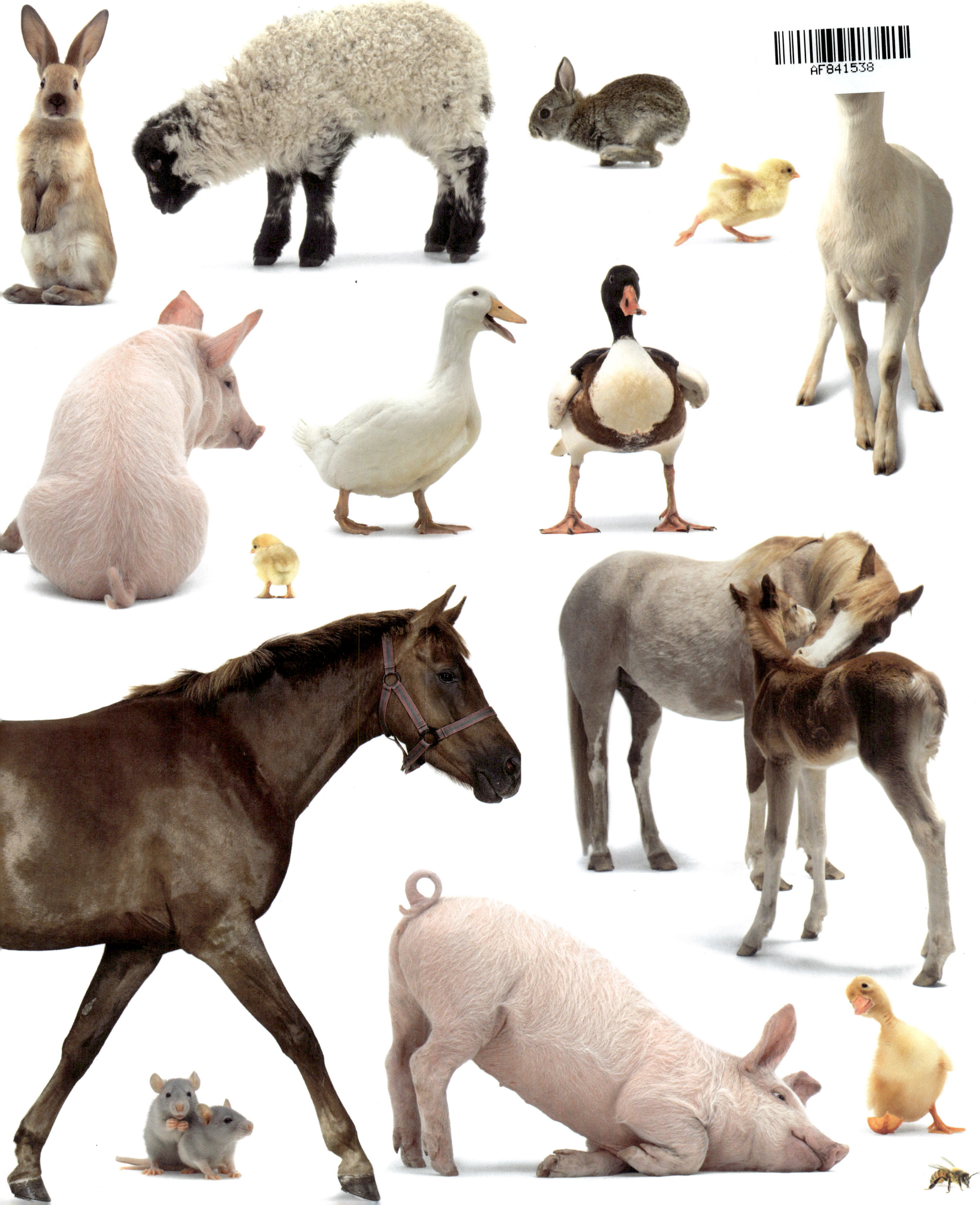

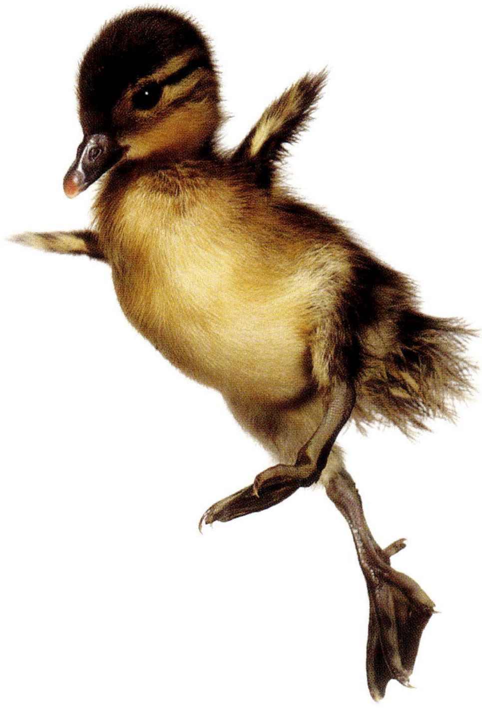

Fur and Feathers...

...An Unusual Farm

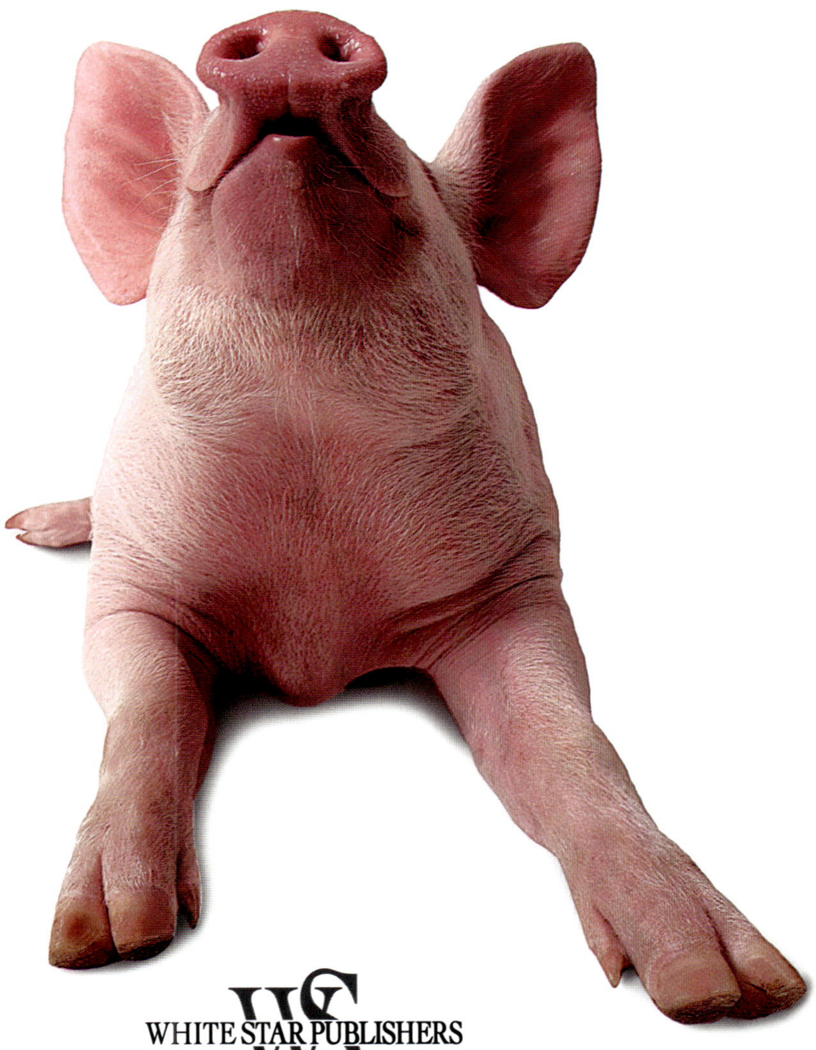

WHITE STAR PUBLISHERS

Editorial Project: Valeria Manferto De Fabianis - Graphic Design: Clara Zanotti

Preface: Giorgio Ferrero - Editorial Assistant: Giorgio Ferrero

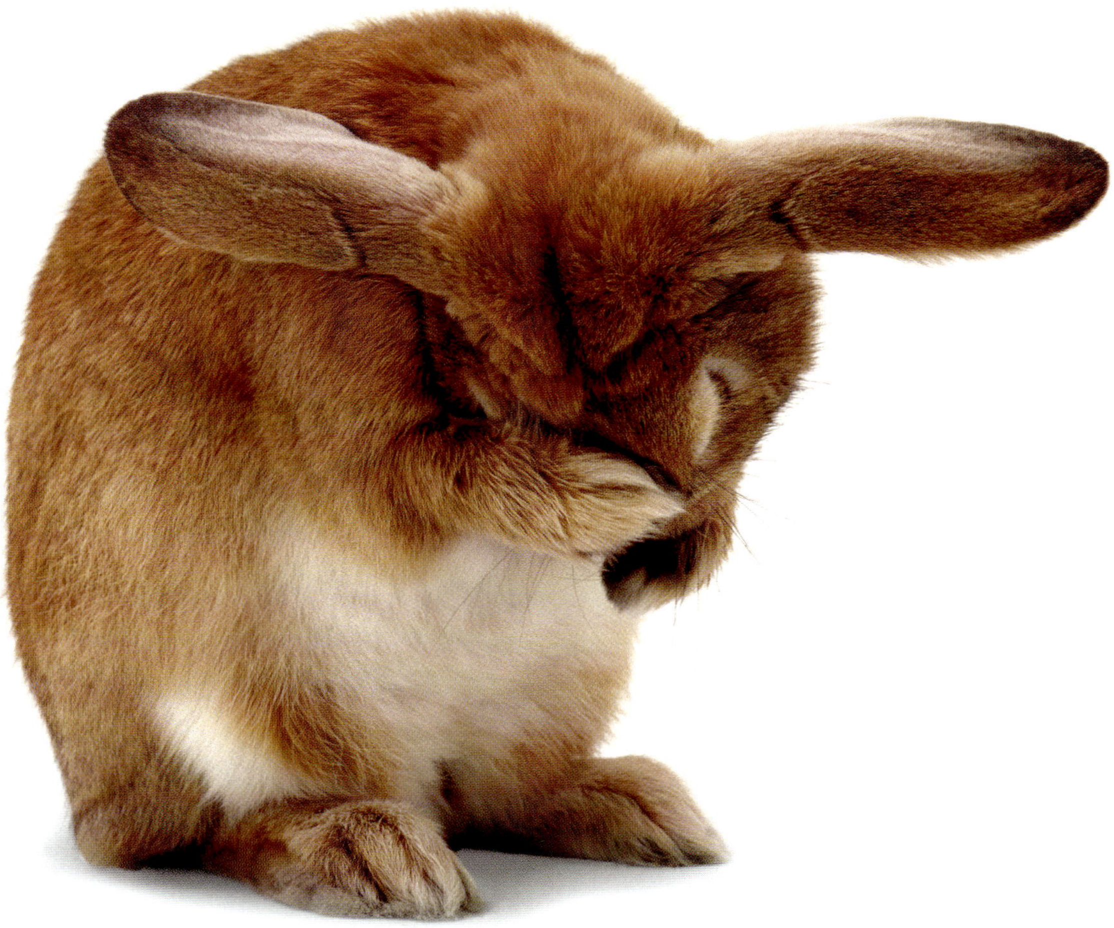

Fur and Feathers… …An Unusual Farm

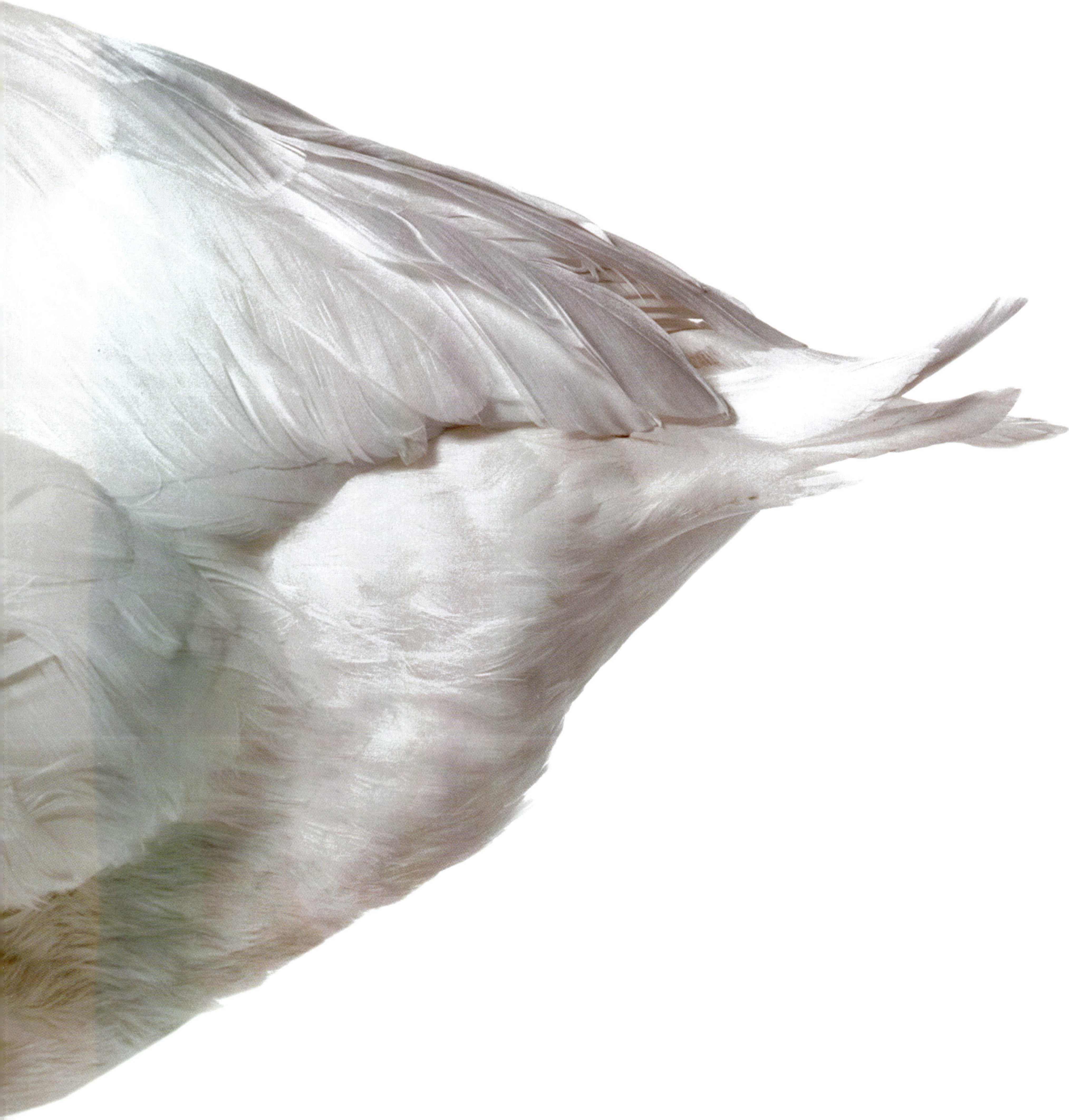

“All animals are equal
but some animals are more equal
than others.”

(George Orwell)

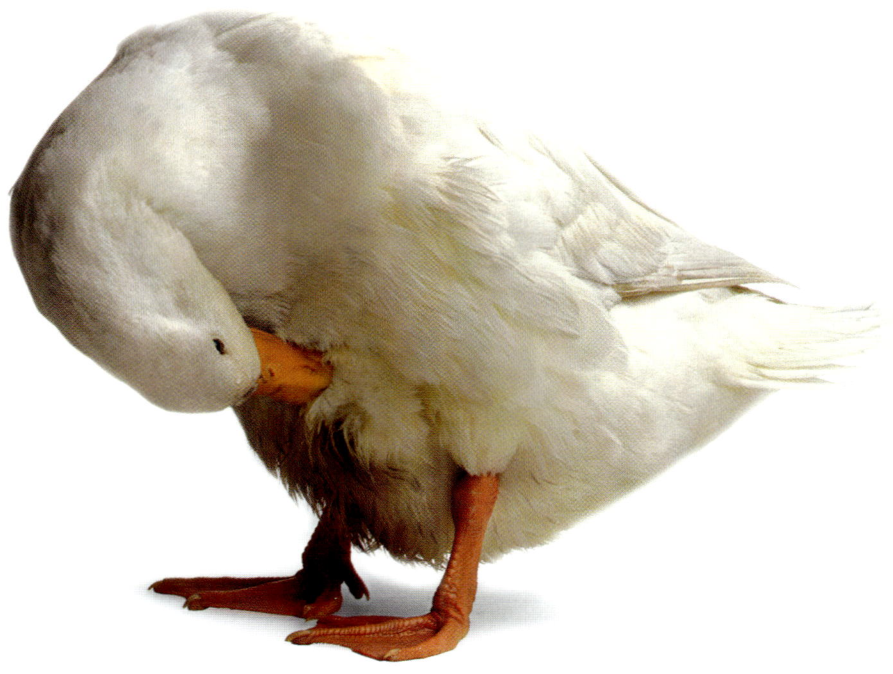

Preface by
GIORGIO FERRERO

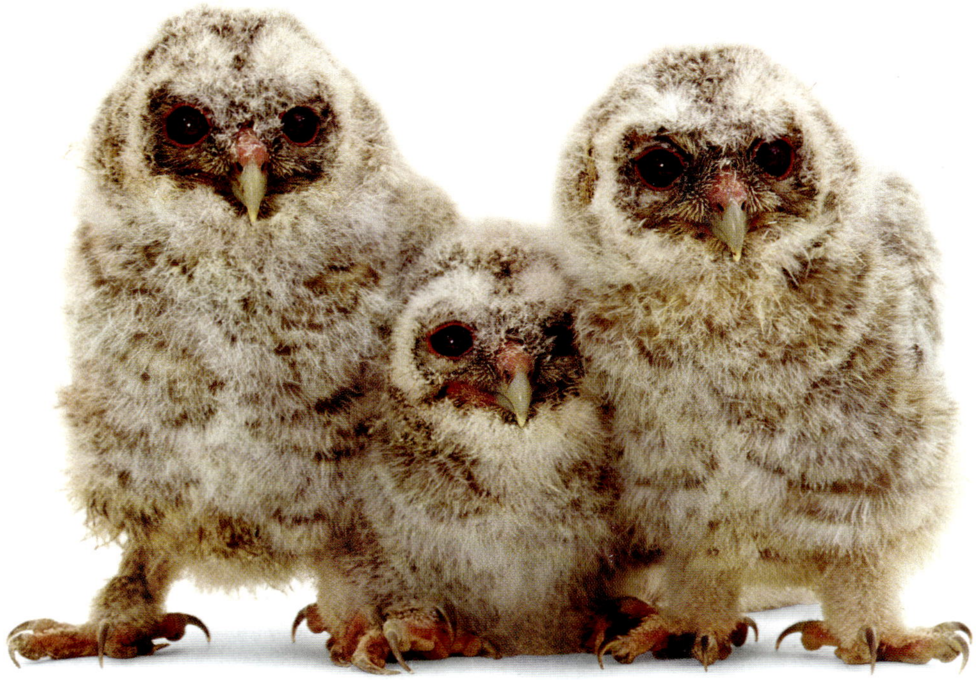

From Aesop to Orwell, from Phaedrus to La Fontaine, passing through the medieval "fabliaux," farm animals keep on living in the pages of the books as well as in the countryside's farmyards and stables. Cattle and sheep, goats and pigs, geese and rabbits together form a unique and variegated microcosm, sometimes chaotic but certainly never obviously regulated. Animal society is almost the mirror image of human society: different and conflicting relationships and connections develop among animals as well as among men, and are just as powerful.

“Beasts of England,
beasts of Ireland, beasts of every land
and clime, hearken to my joyful
tidings of the golden future time.”

(George Orwell)

Nowadays it’s rarer and rarer to enter a true farm environment unless we’re talking with farmers.

But we are nonetheless familiar with farm animals - and generally think well of them. We unthinkingly accept the horse’s elegance, the pig’s ungainliness, the hedgehog’s suspicion, the mouse’s curiosity, the cock’s aggressiveness, the cow’s tranquillity. They are all animals - but they remind us of ourselves.

Some of them live in stables, others spend their days ruminating in the fields; some, like foxes, are predatory hunters or are fleet-footed thieves. Some fly only by night, others sing at daybreak. Hens look after their broods, peacocks proudly spread their tails hoping to attract a partner. A young goose learns how to swim in a small pond, a sheep observes with curiosity a grasshopper about to spring forward.

Such are the scenes of everyday life, the usual farm activities.

“All animals are equal but some animals are more equal than others,” was the motto of the tyrant-pig in *Animal Farm*. All animals, of course, differ from each another: each animal has its own story to tell.

In this lavish book of numerous finely conceived photographs, the world of farm animals reveals itself top the reader's eye. Behaviors, poses, expressions, movements - all have their secret (or less secret) motivation; an animal world that is strangely human. . . .

“It is a blind goose that cometh to the fox’s sermon.” (John Lyly)

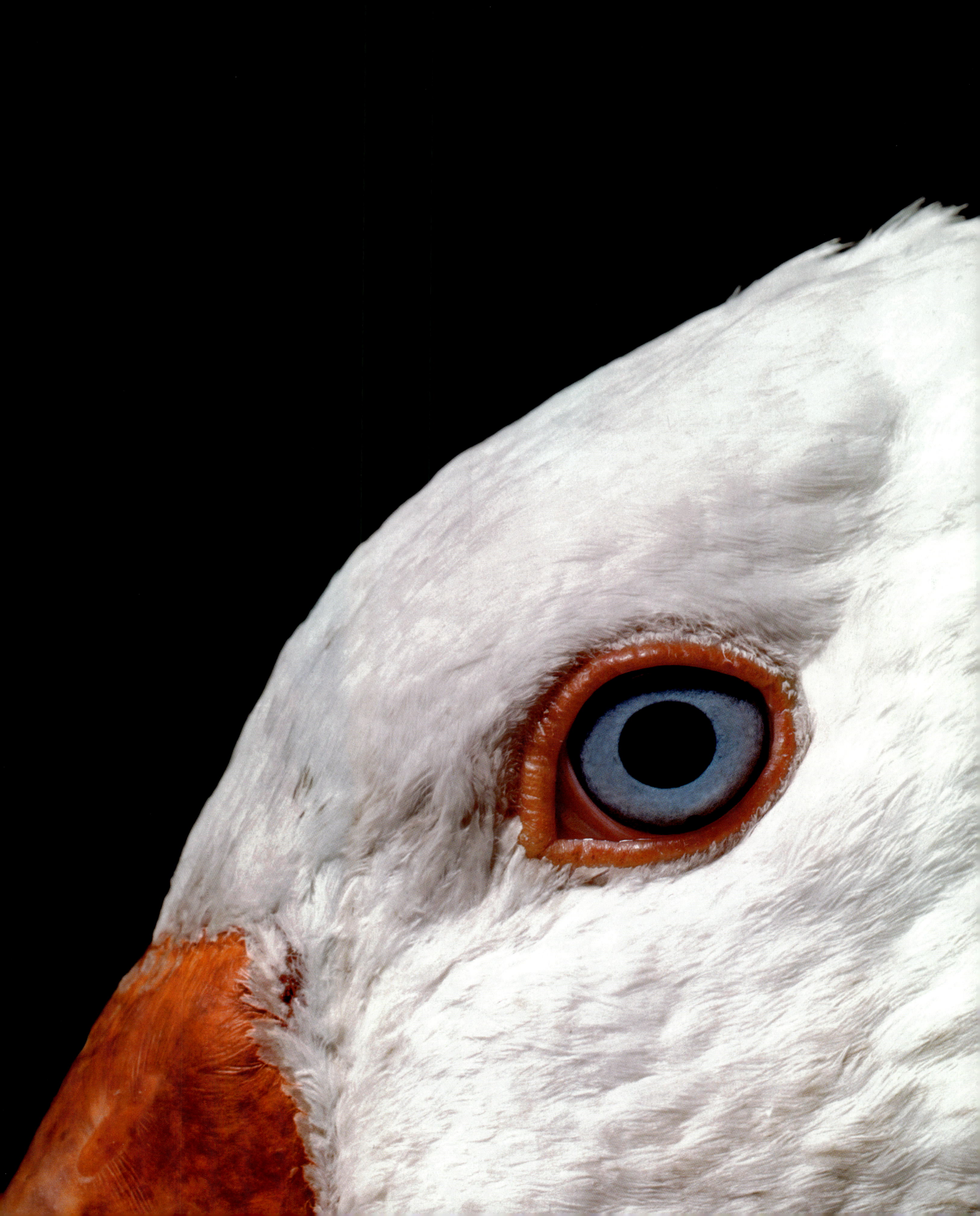

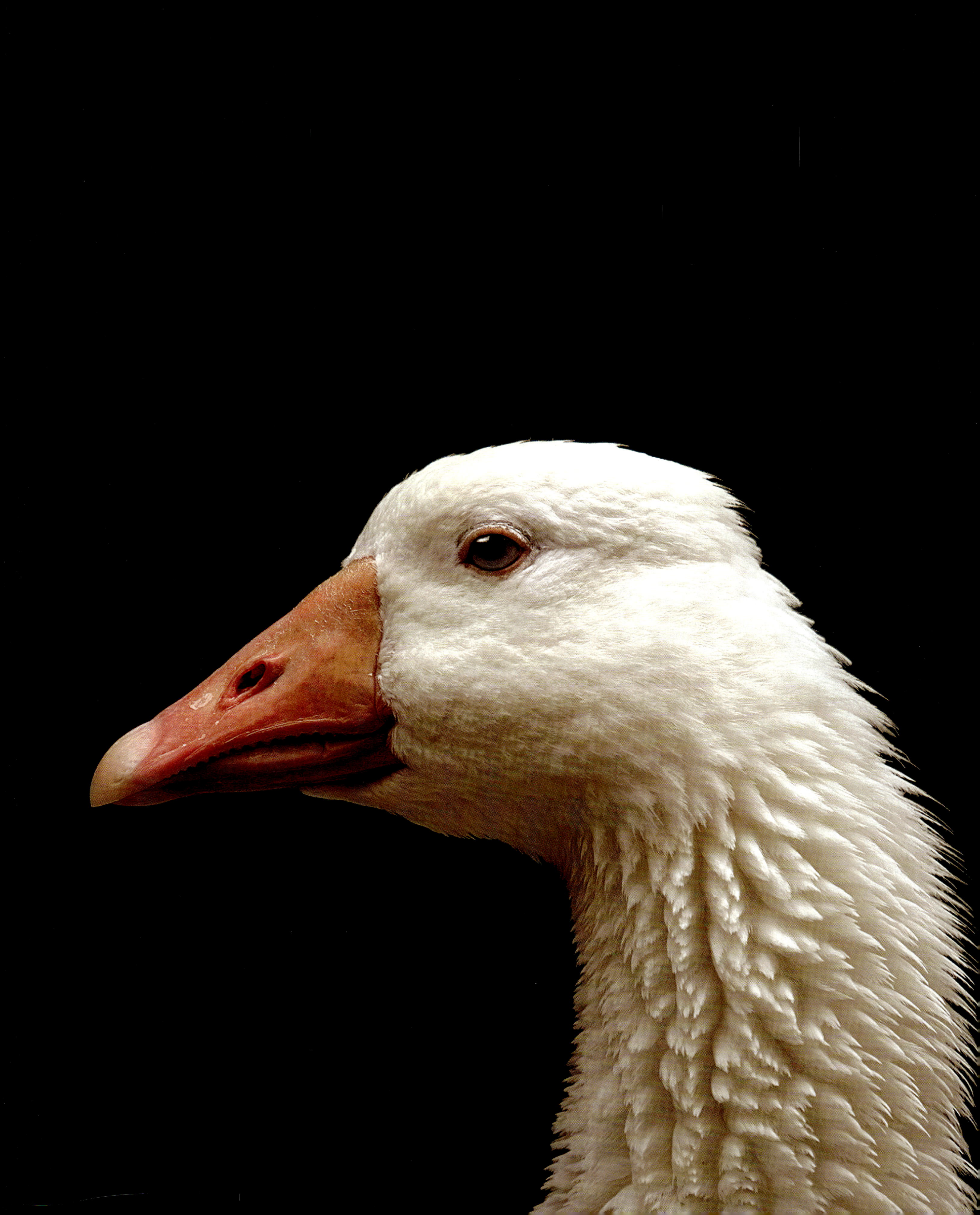

"As far as the goose cranes its neck, it will never become a swan." (Italian proverb)

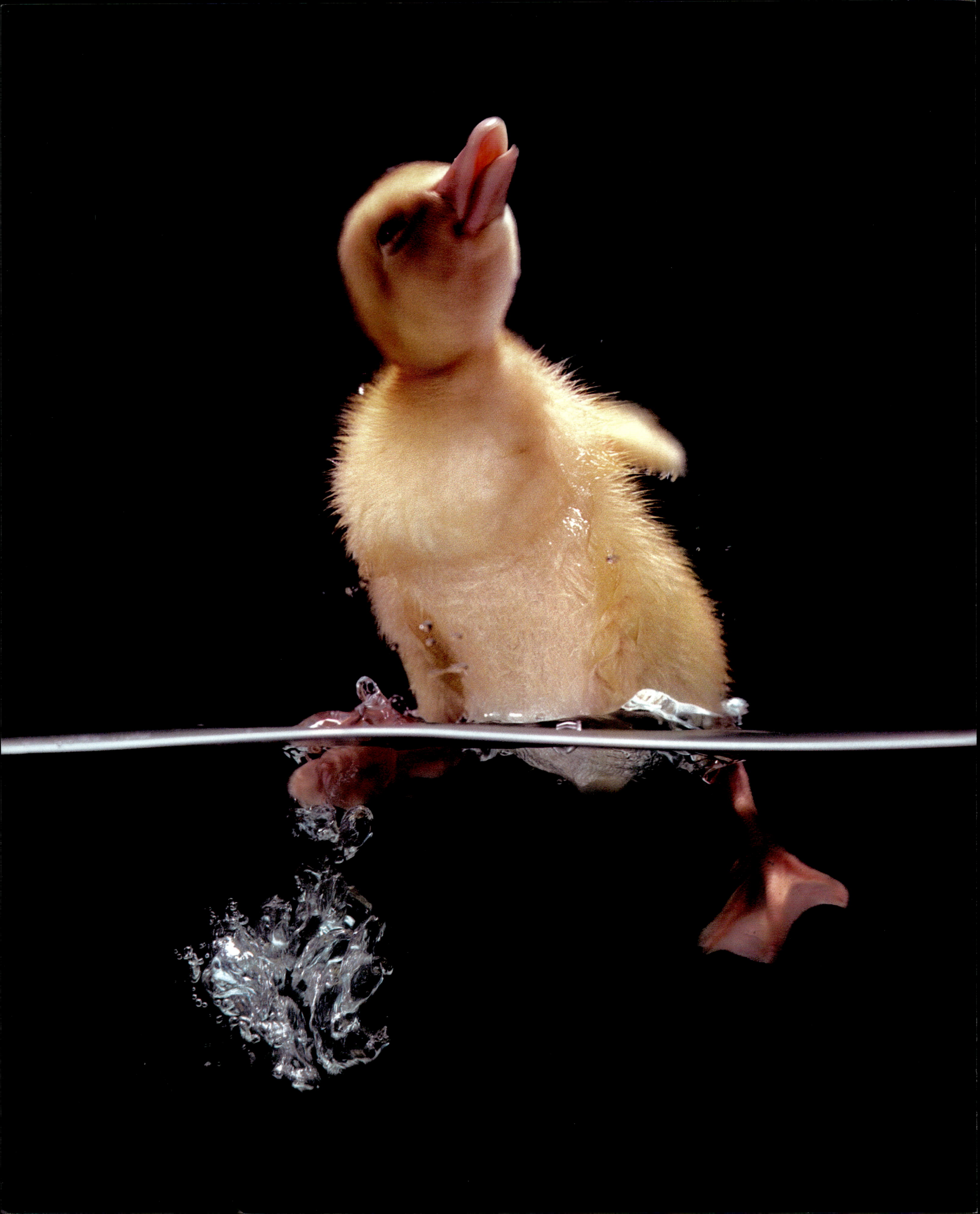

SWIMMING LESSON

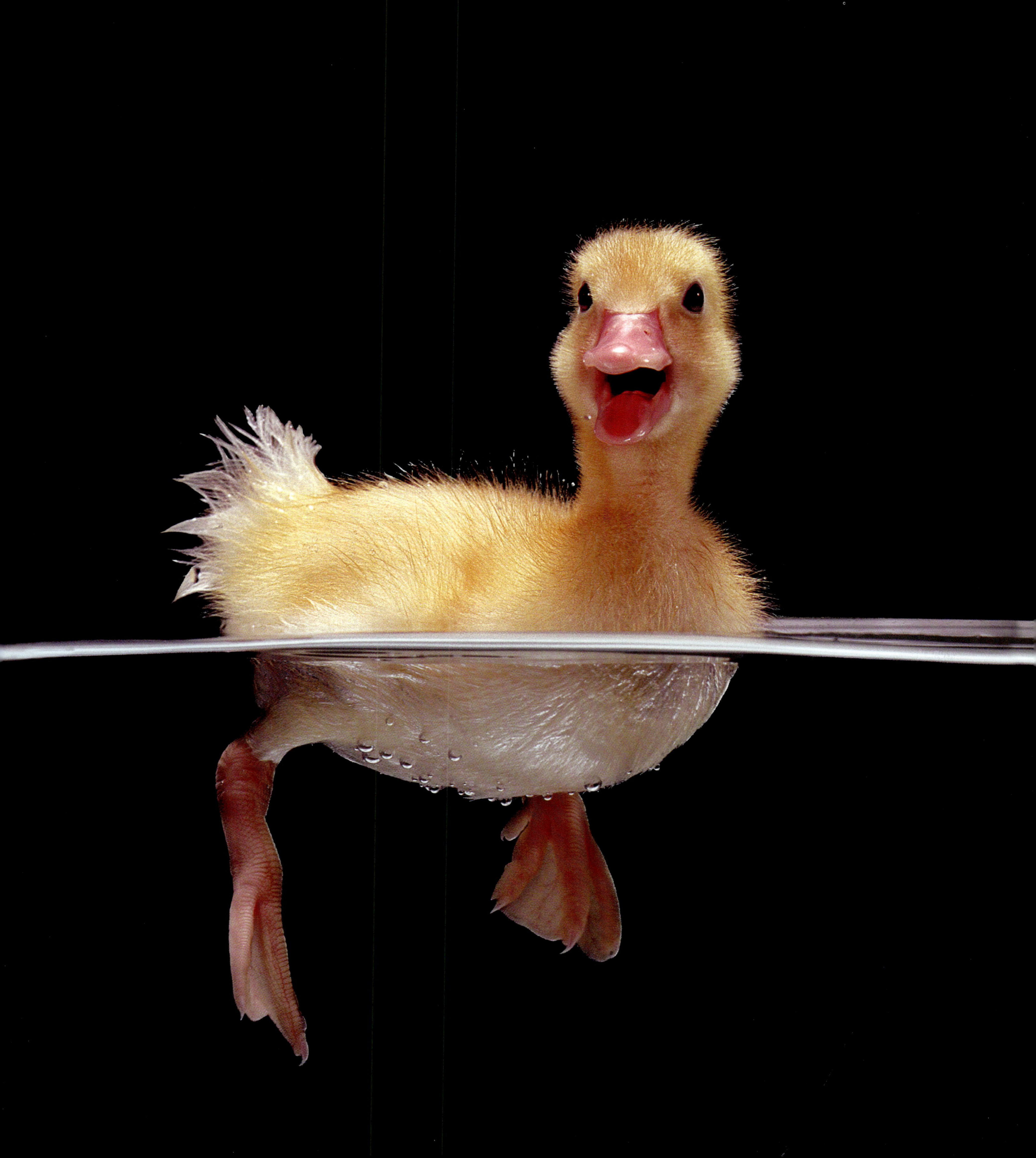

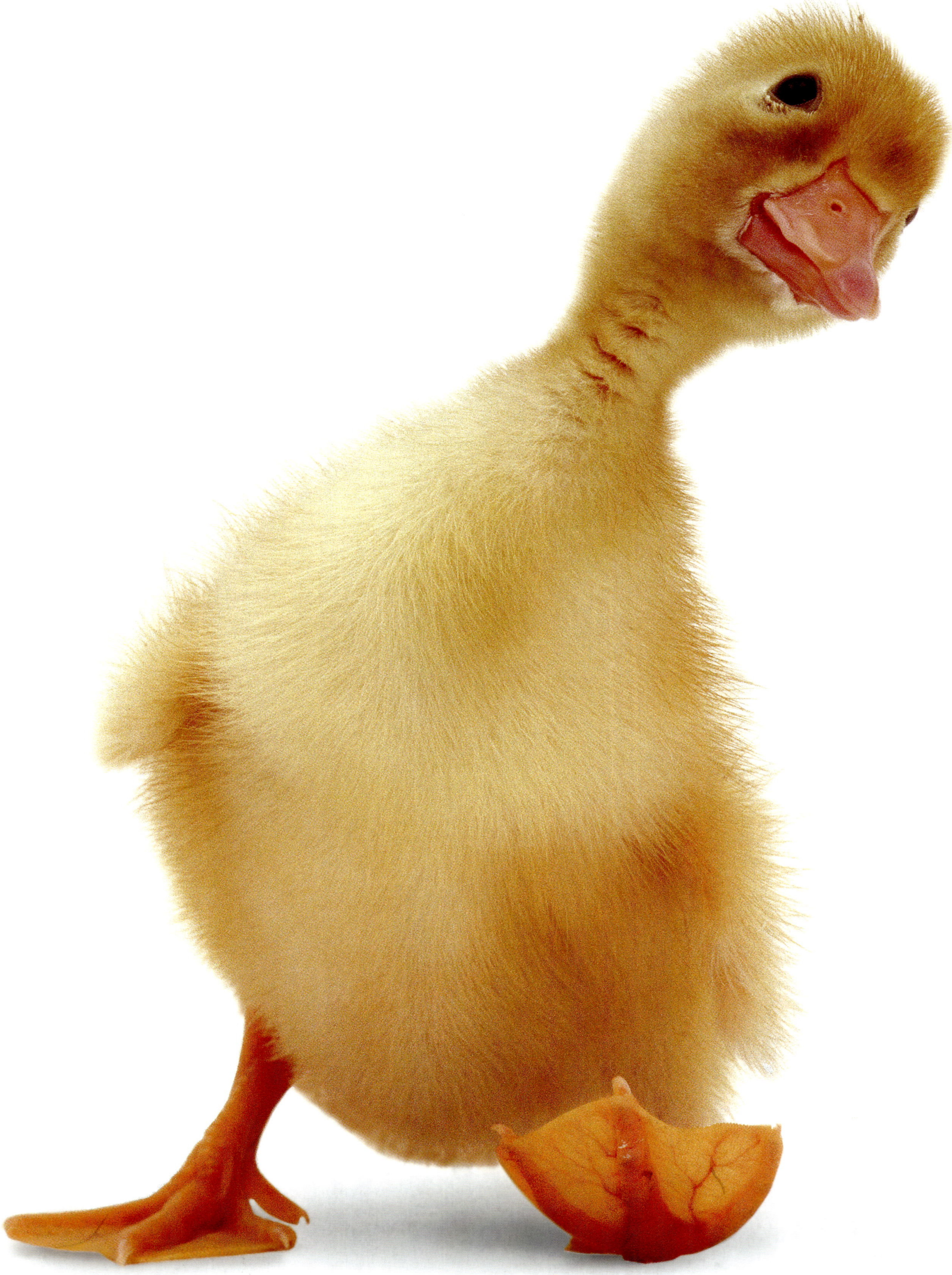

SEARCHING FOR BALANCE

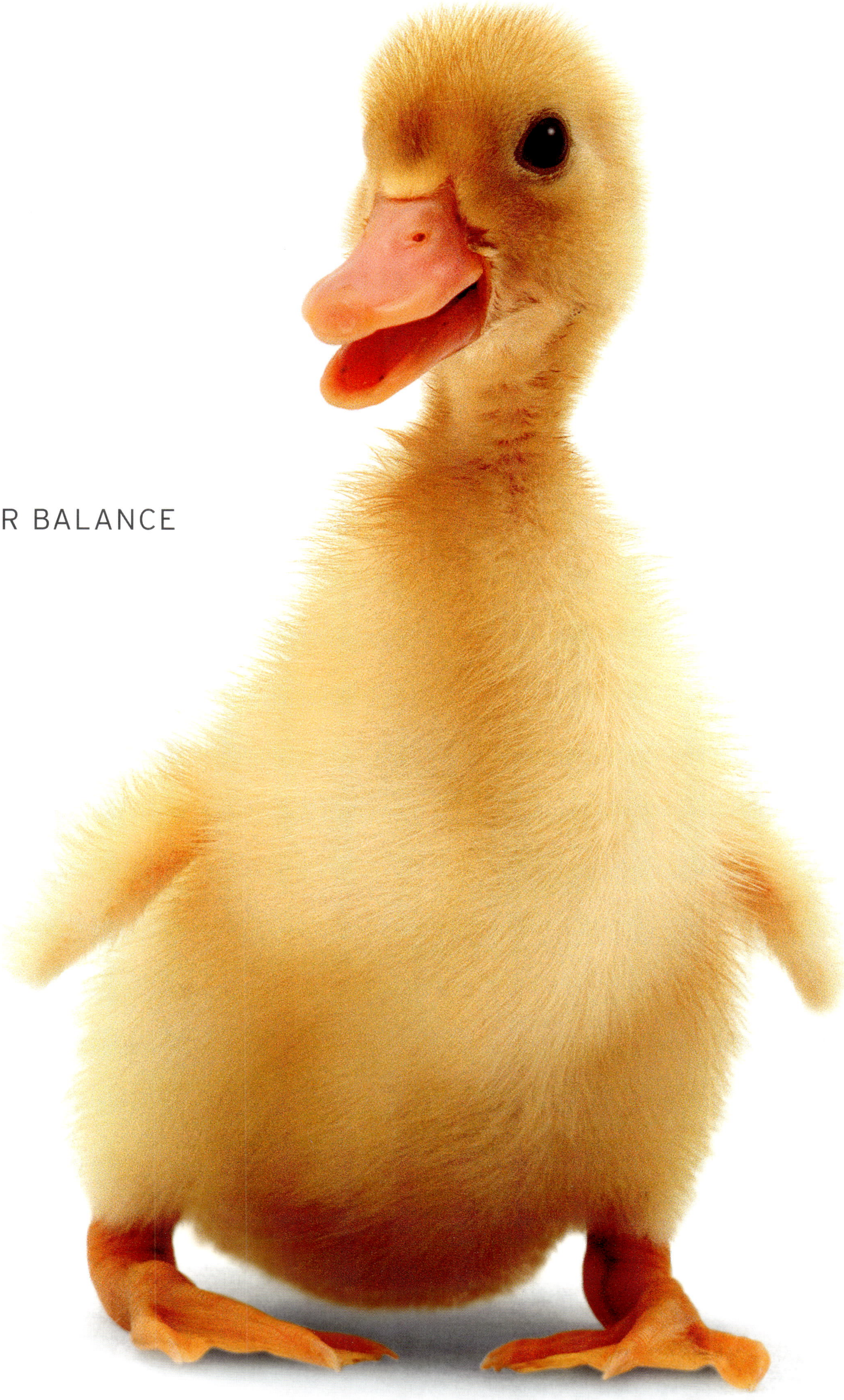

BORN TO BE A LEADER

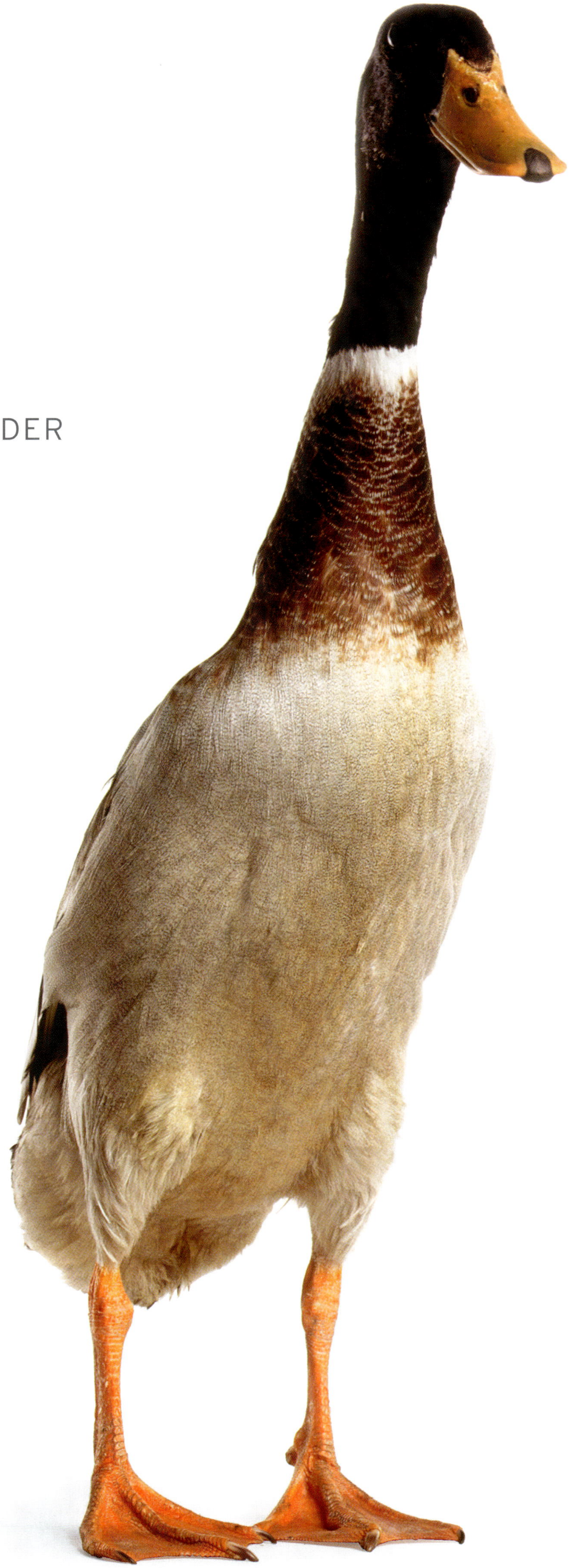

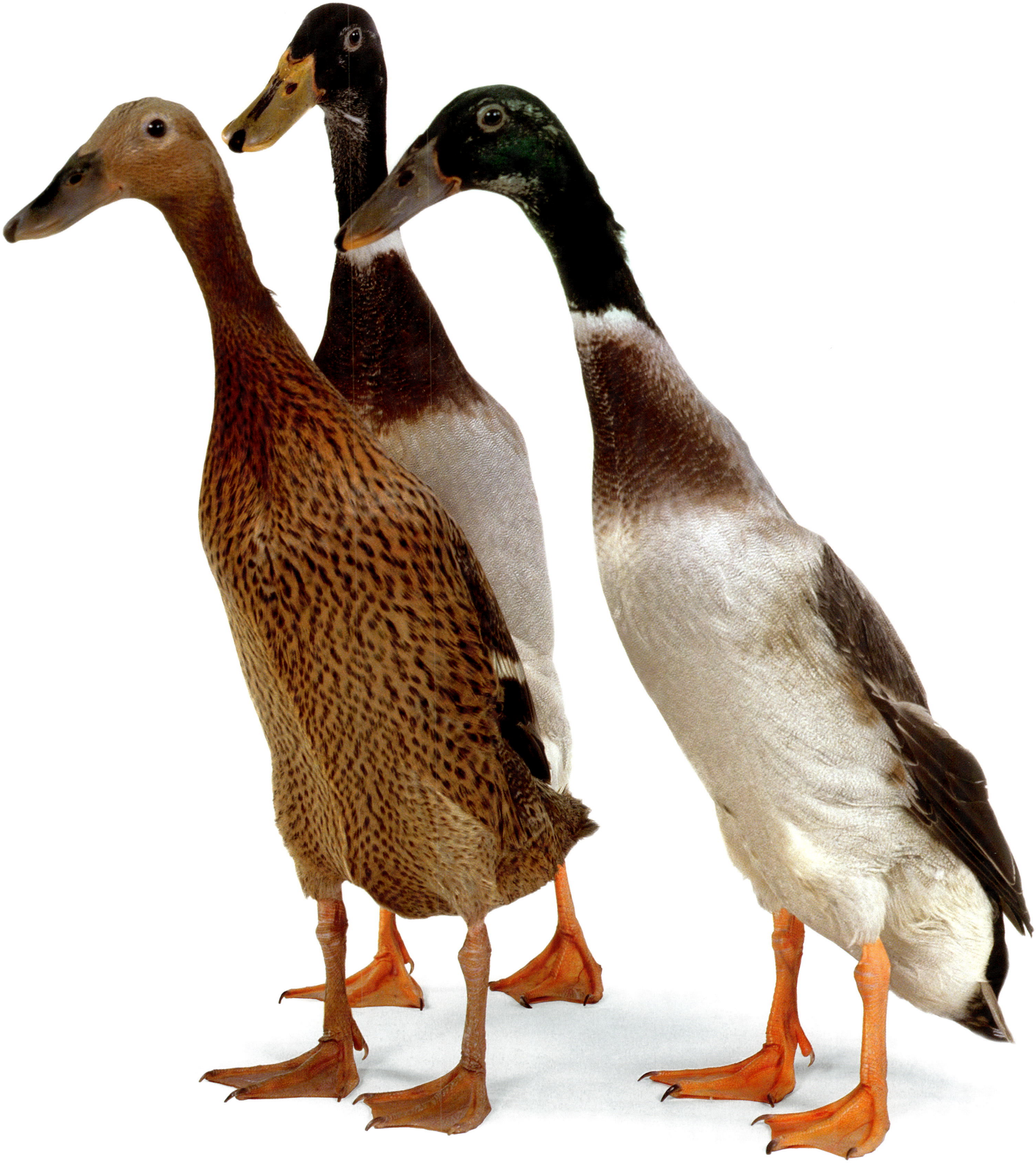

THE WRIGHT BROTHERS

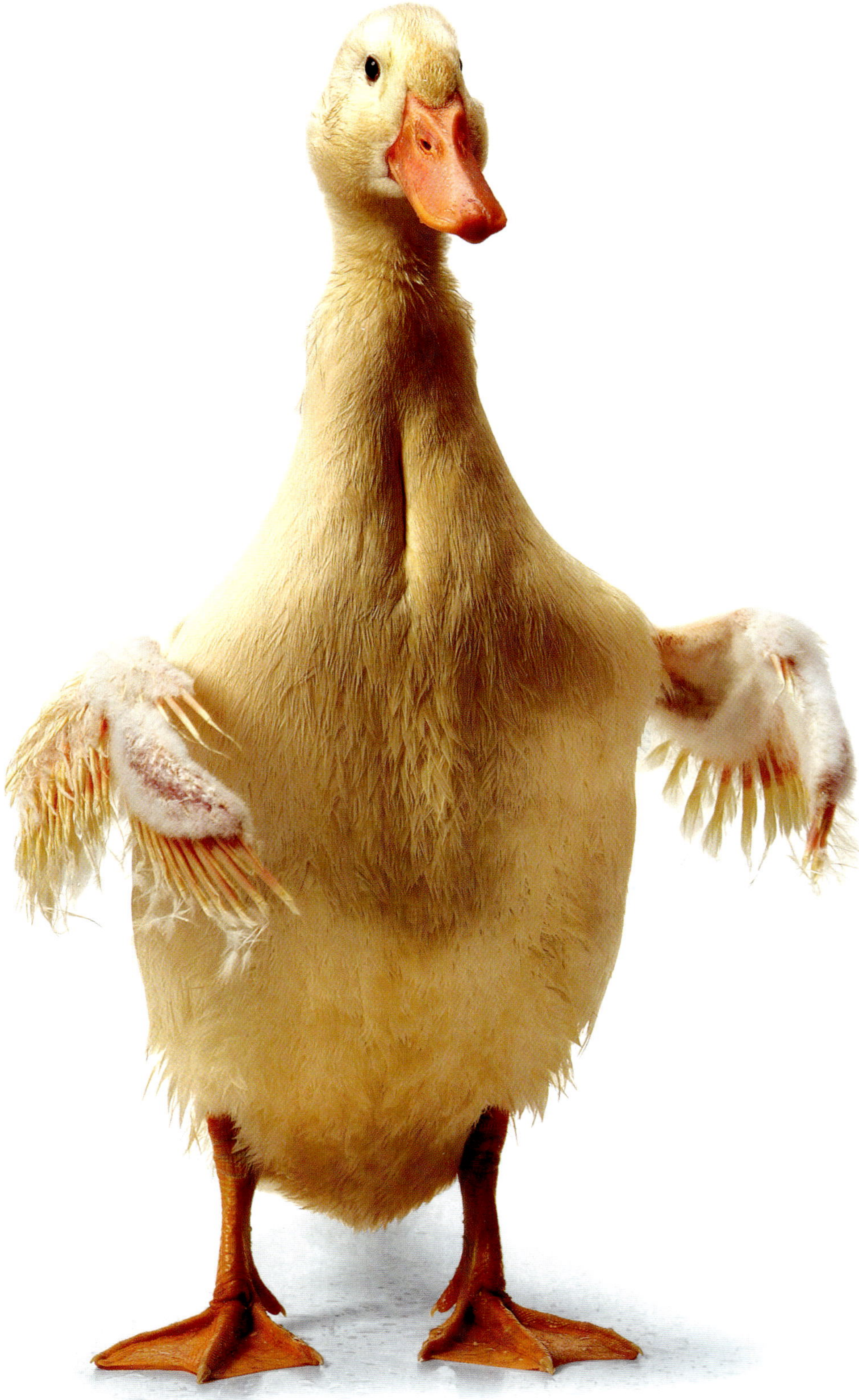

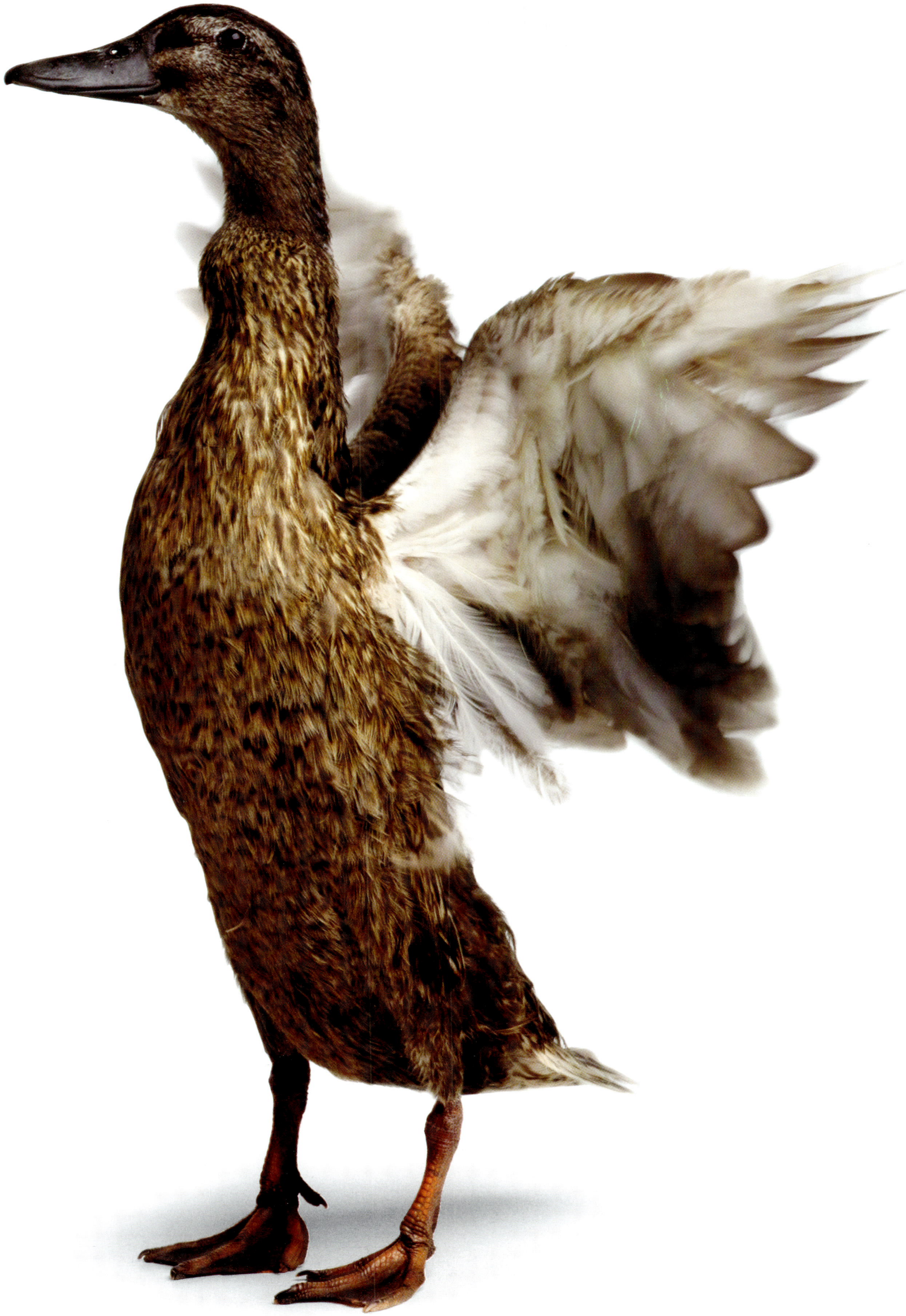

ATTENTION! . . . STAND AT EASE

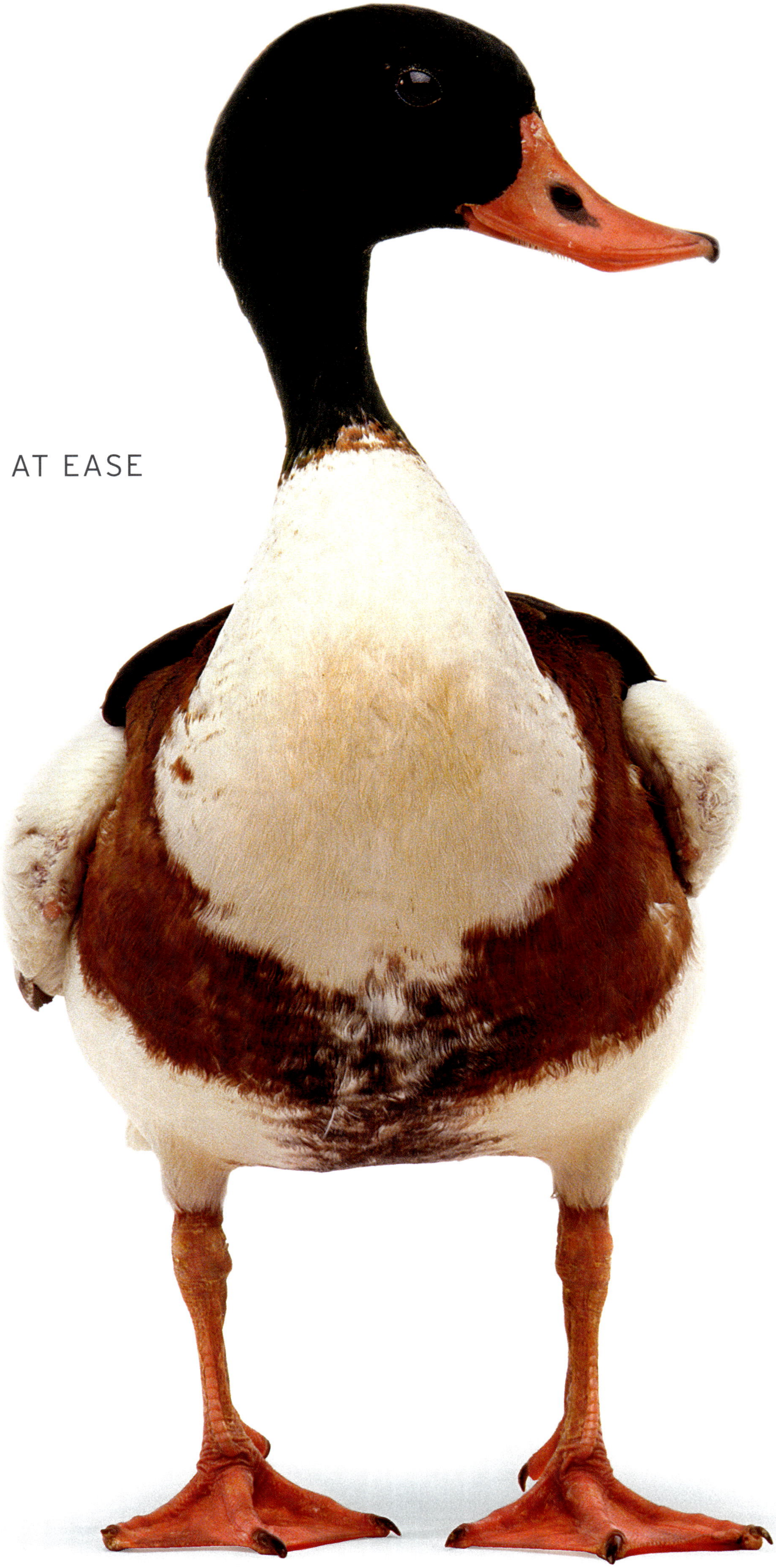

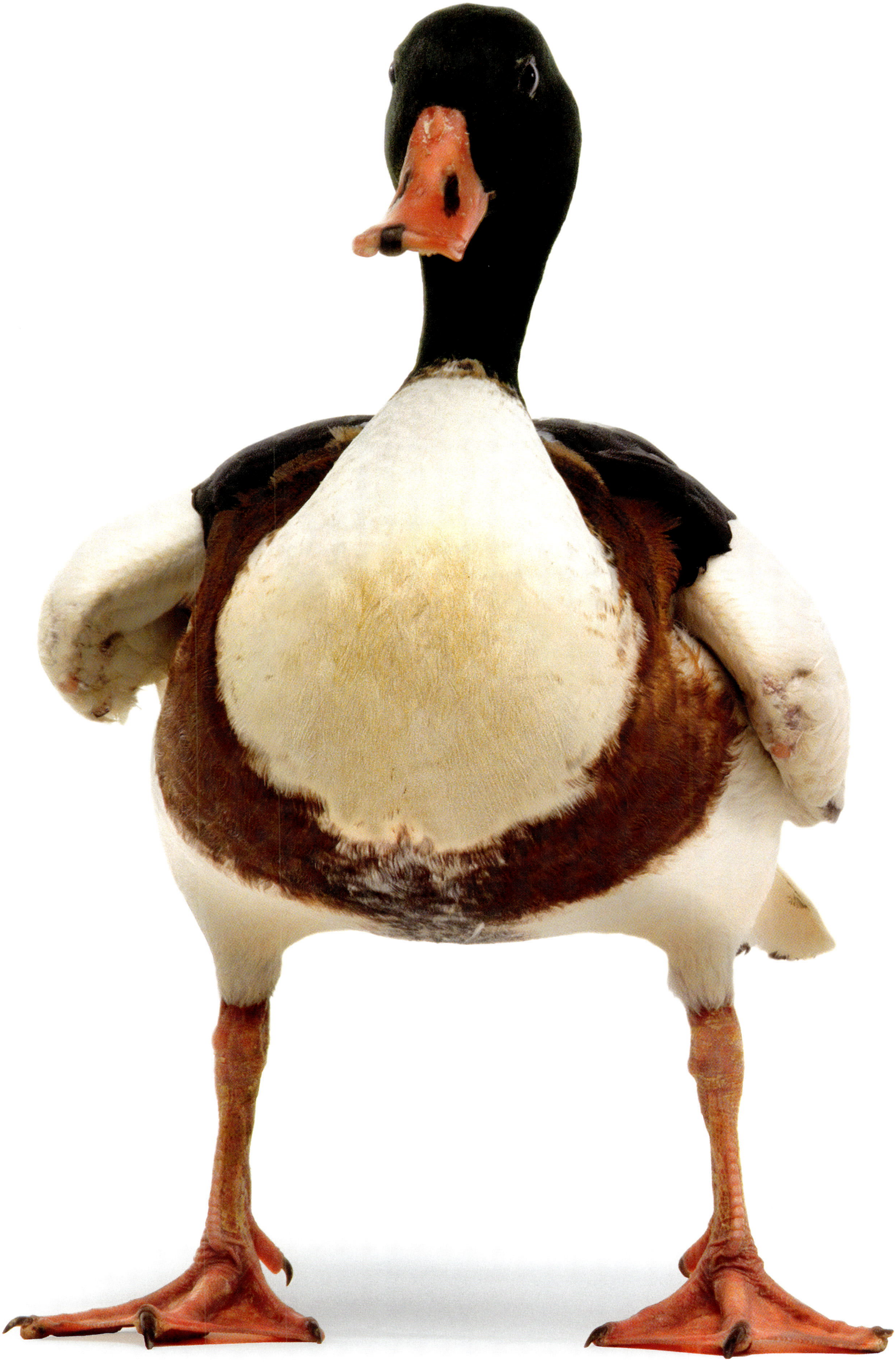

DROWSE

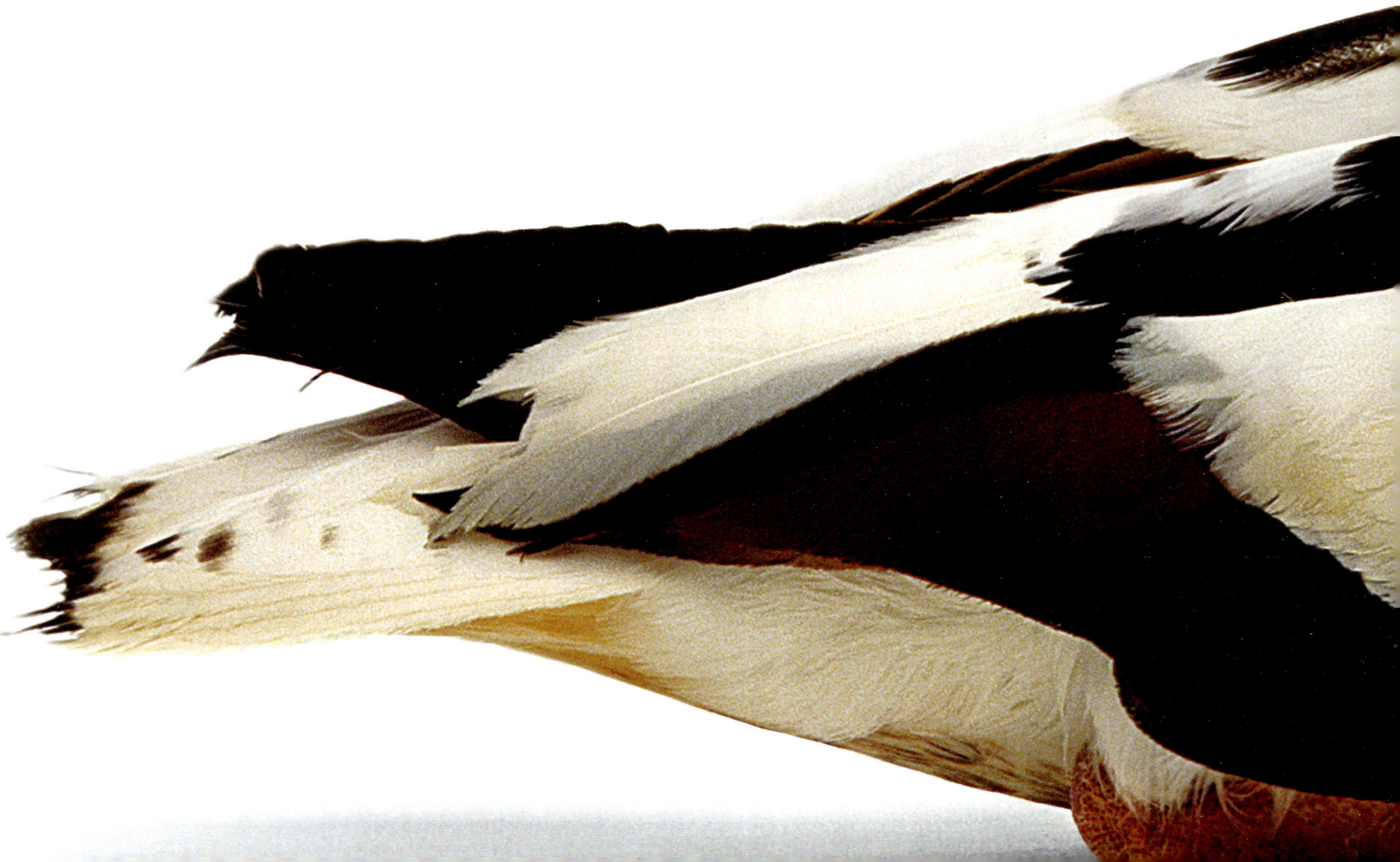

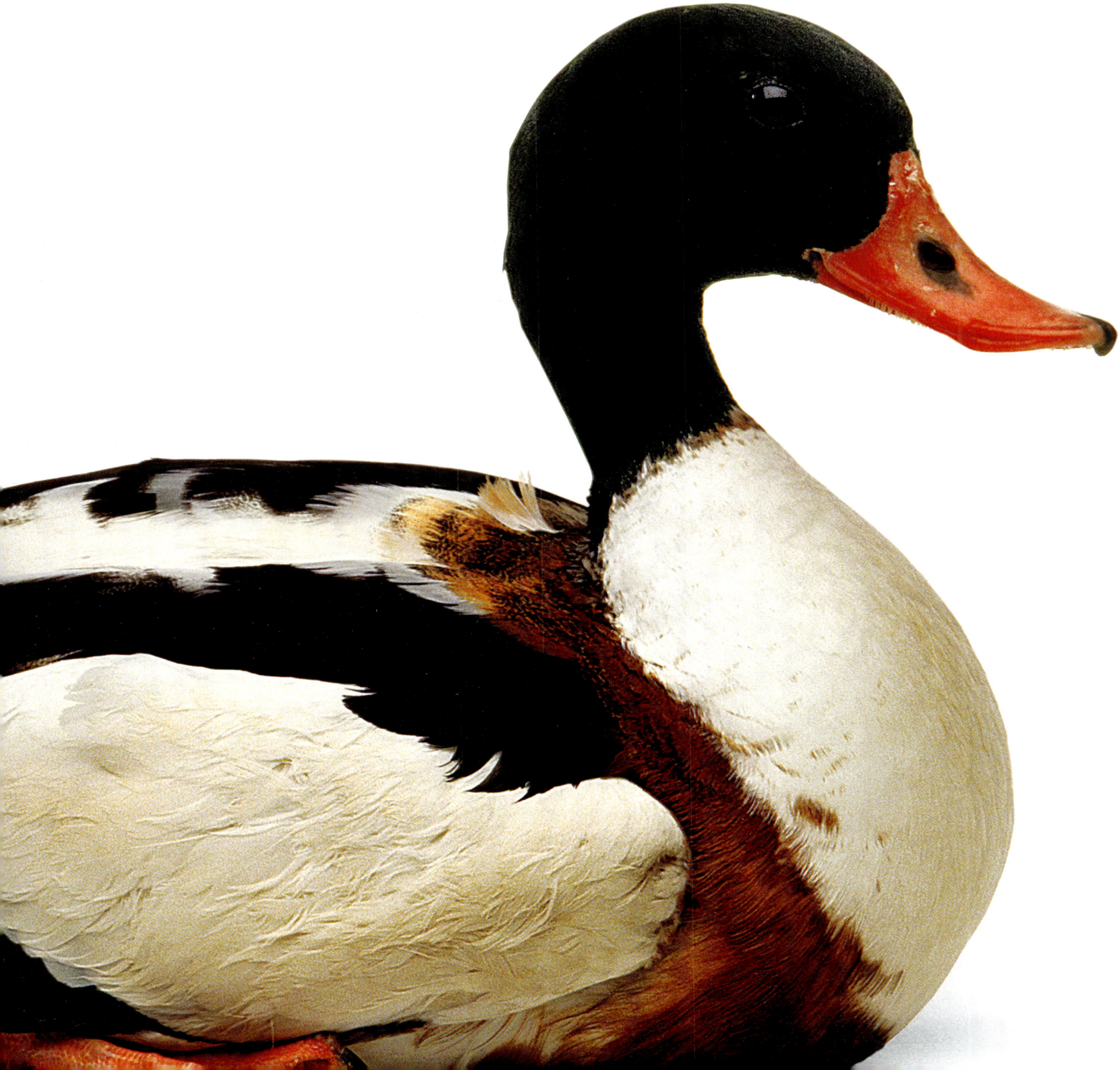

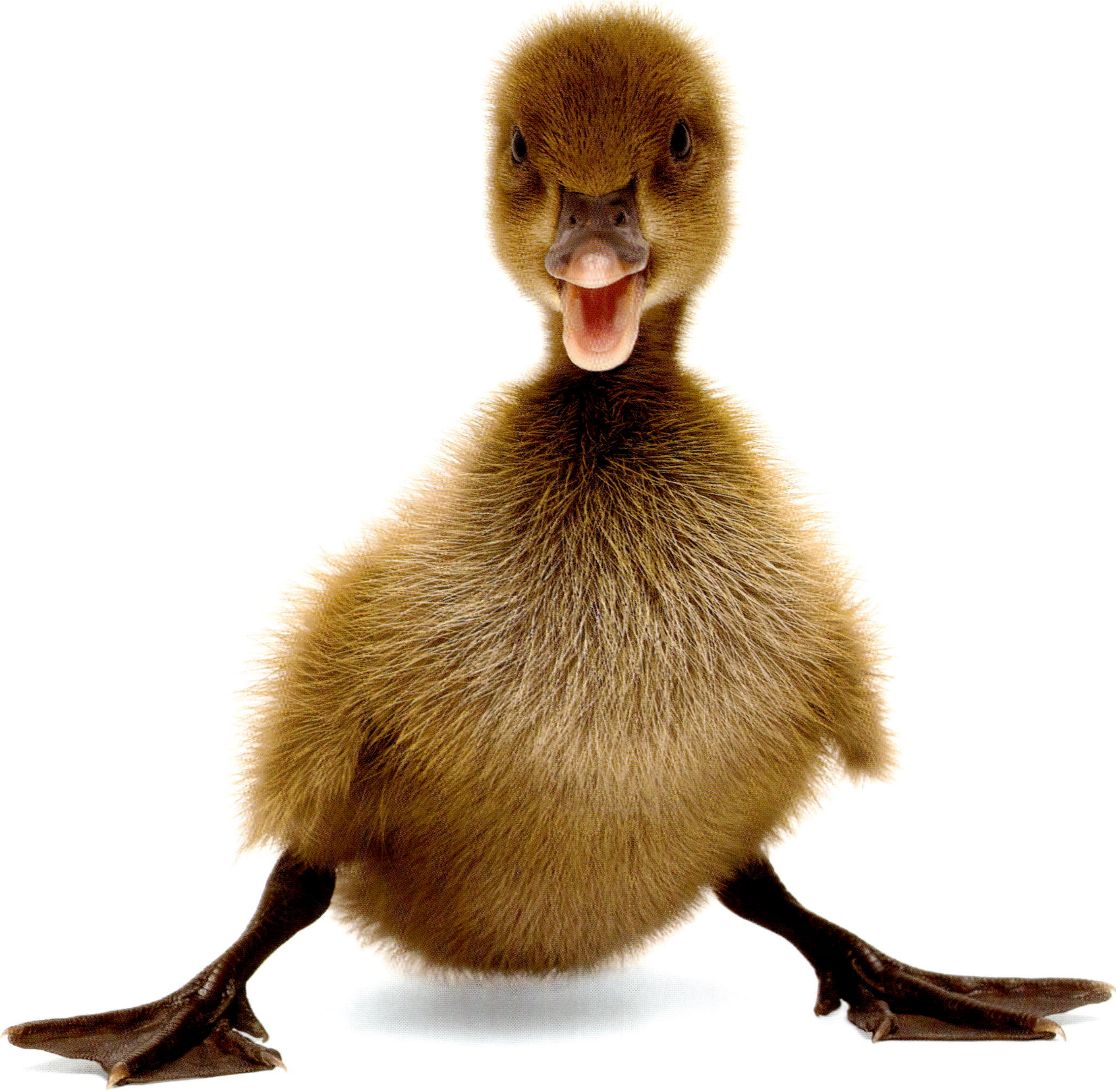

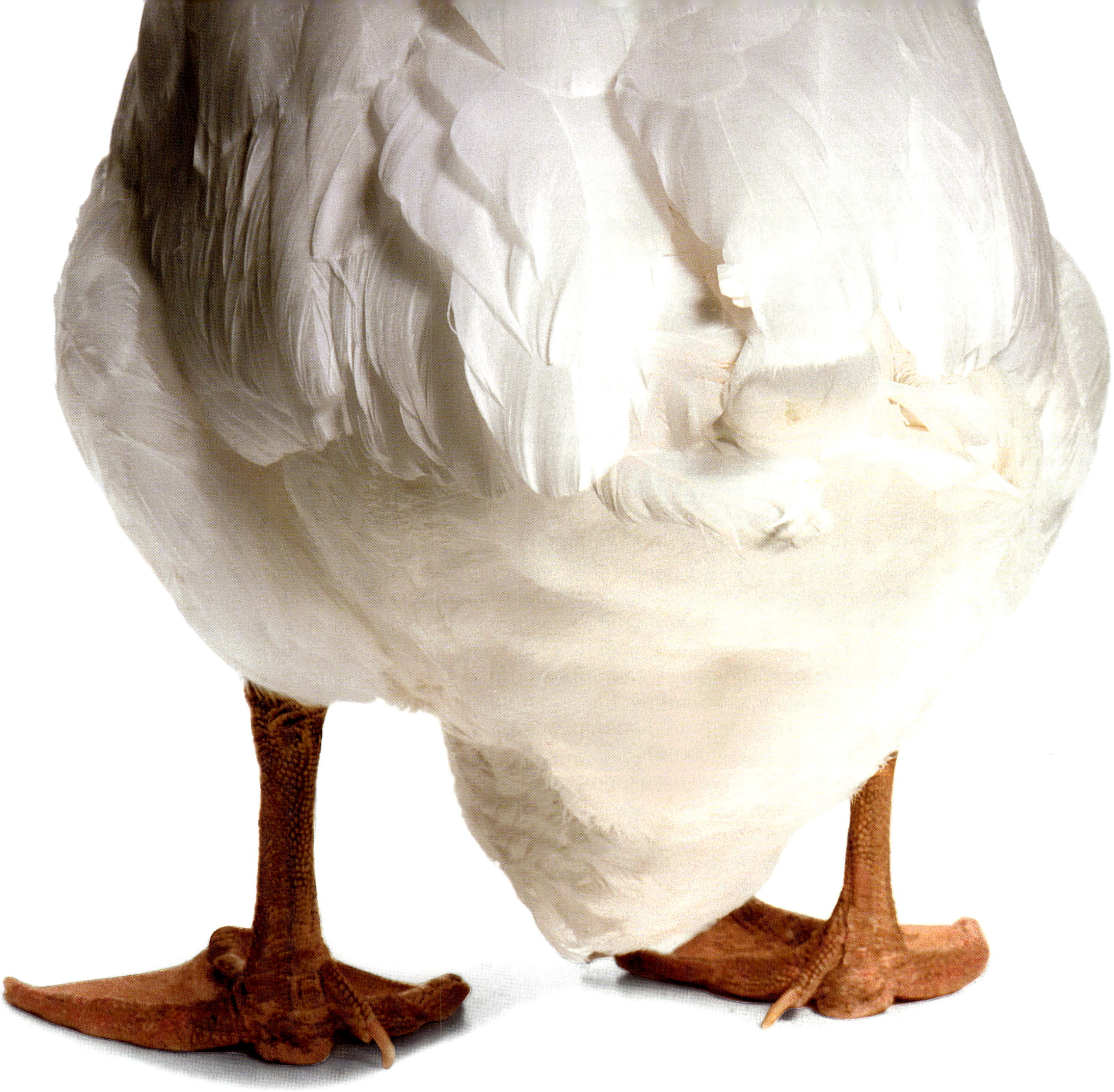

“The distance is nothing;
it is only the first step that is difficult.”
(Marie du Deffand)

COUNTERMELODY

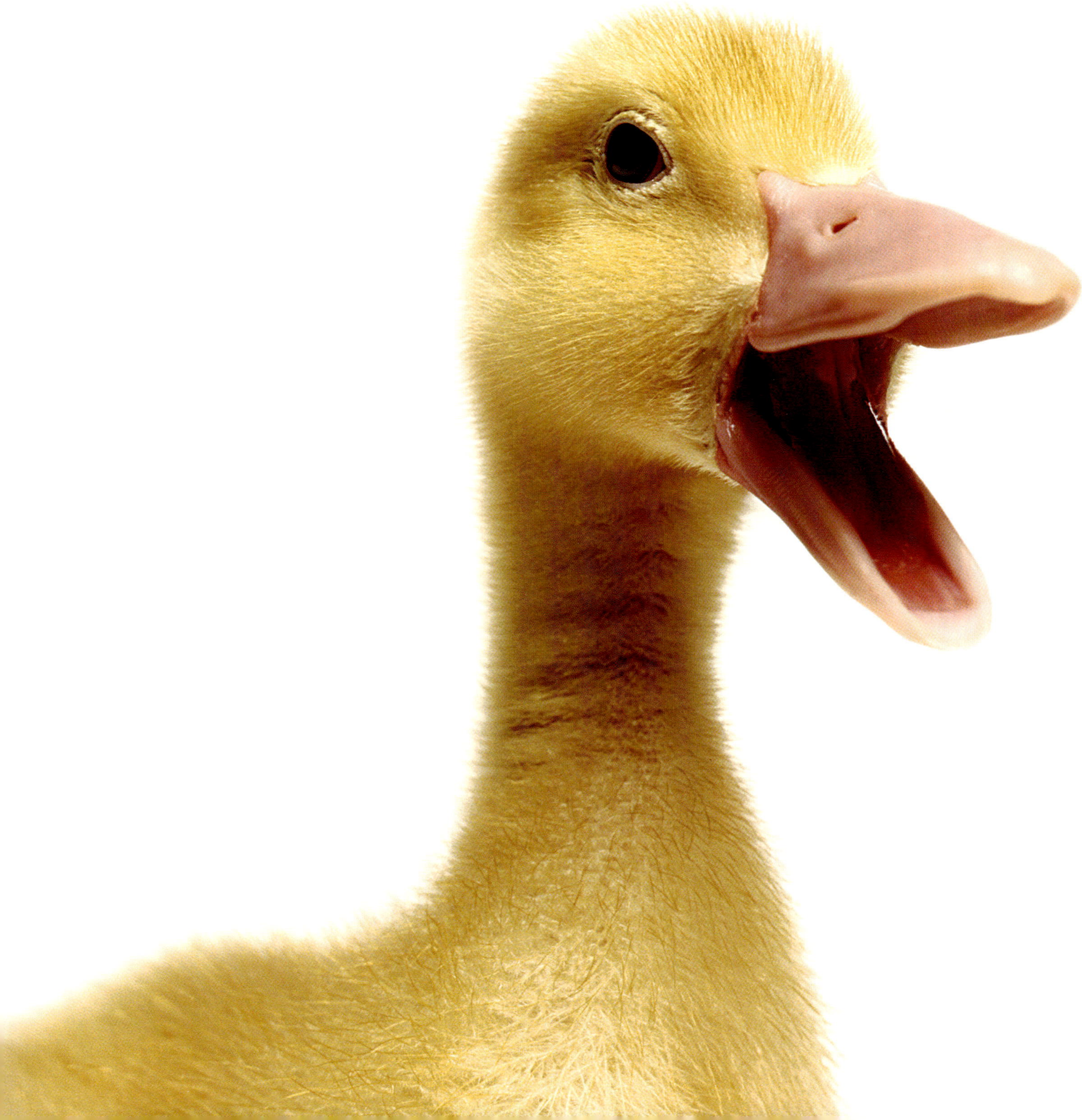

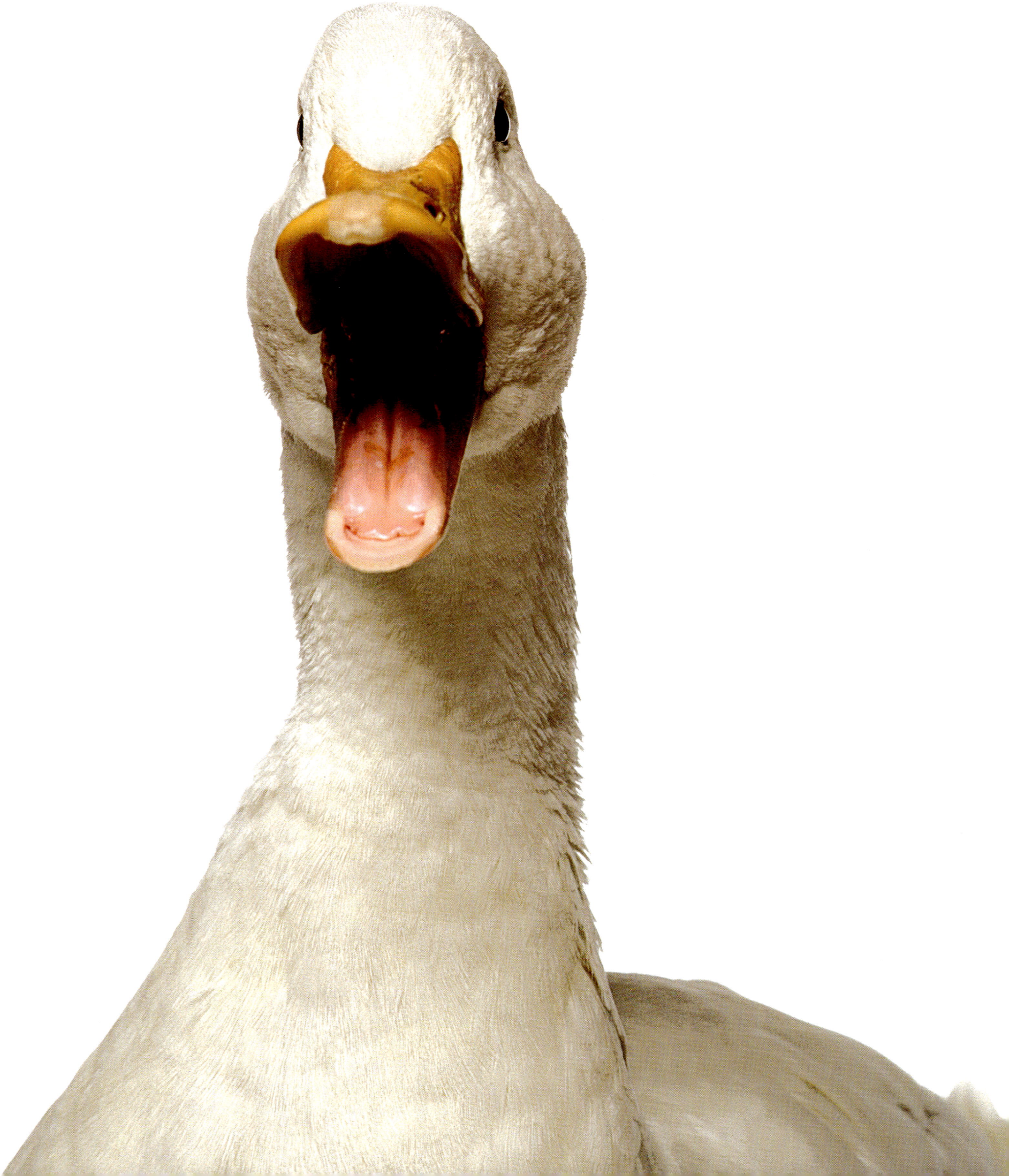

DOUBLE-FACED

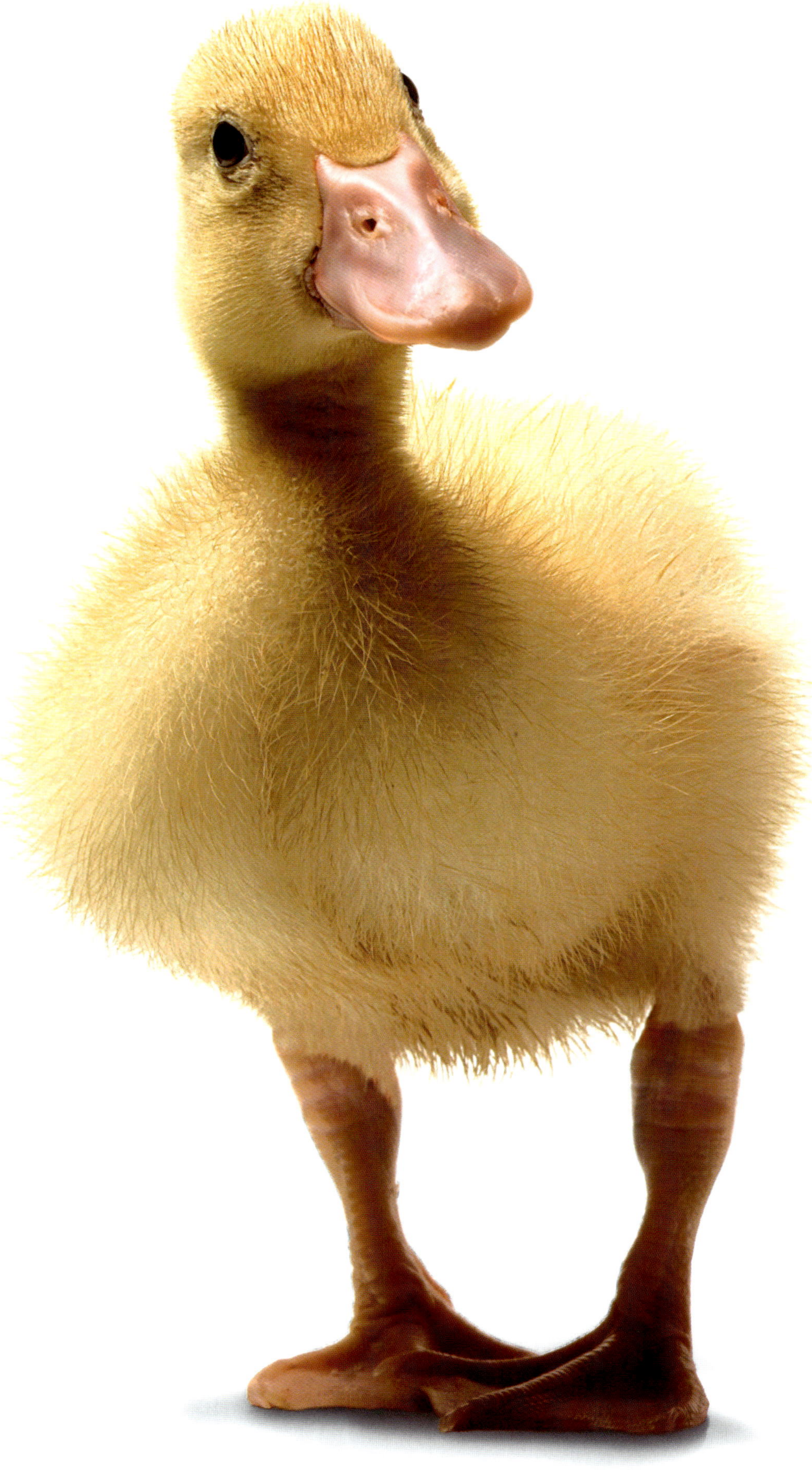

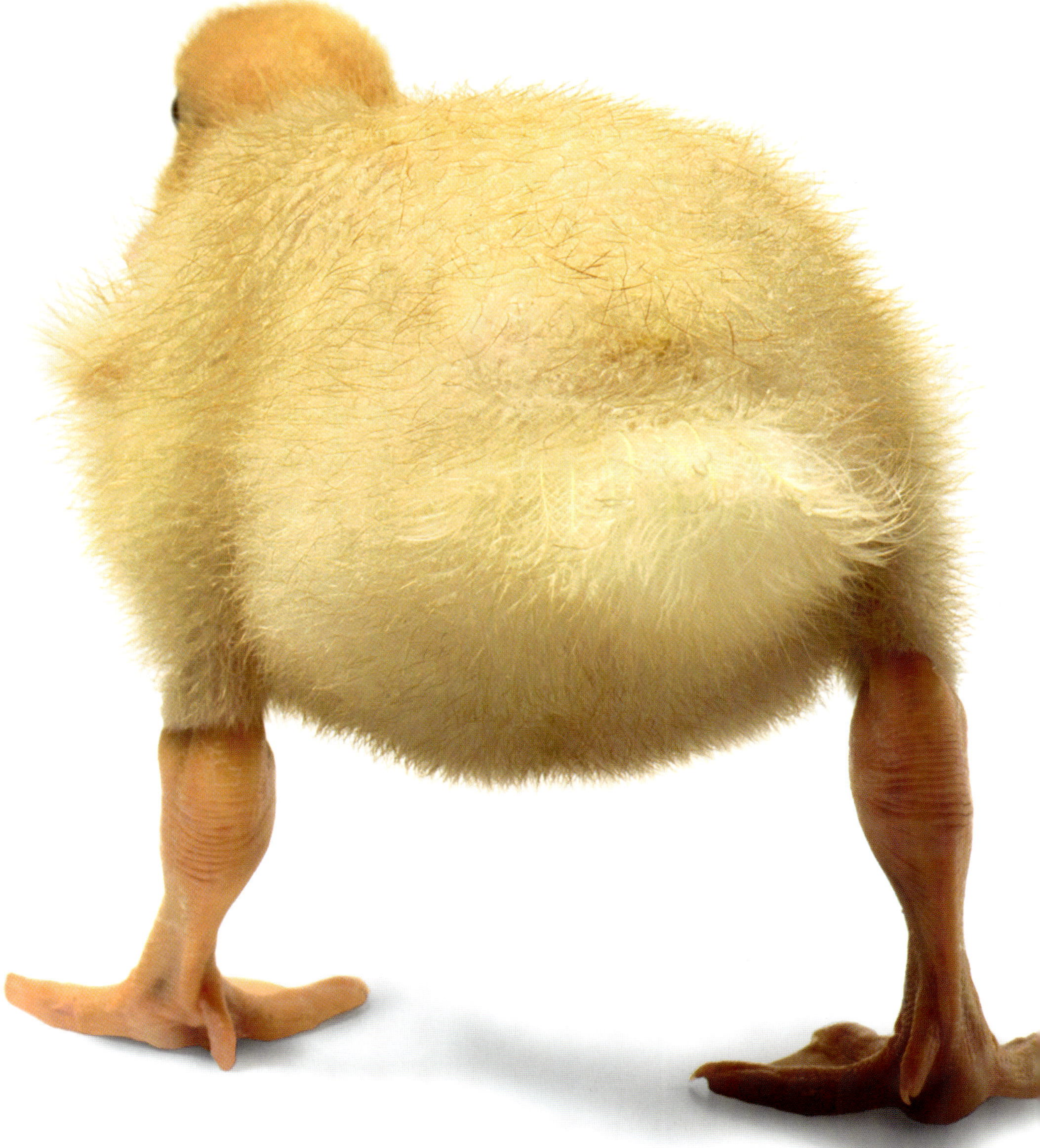

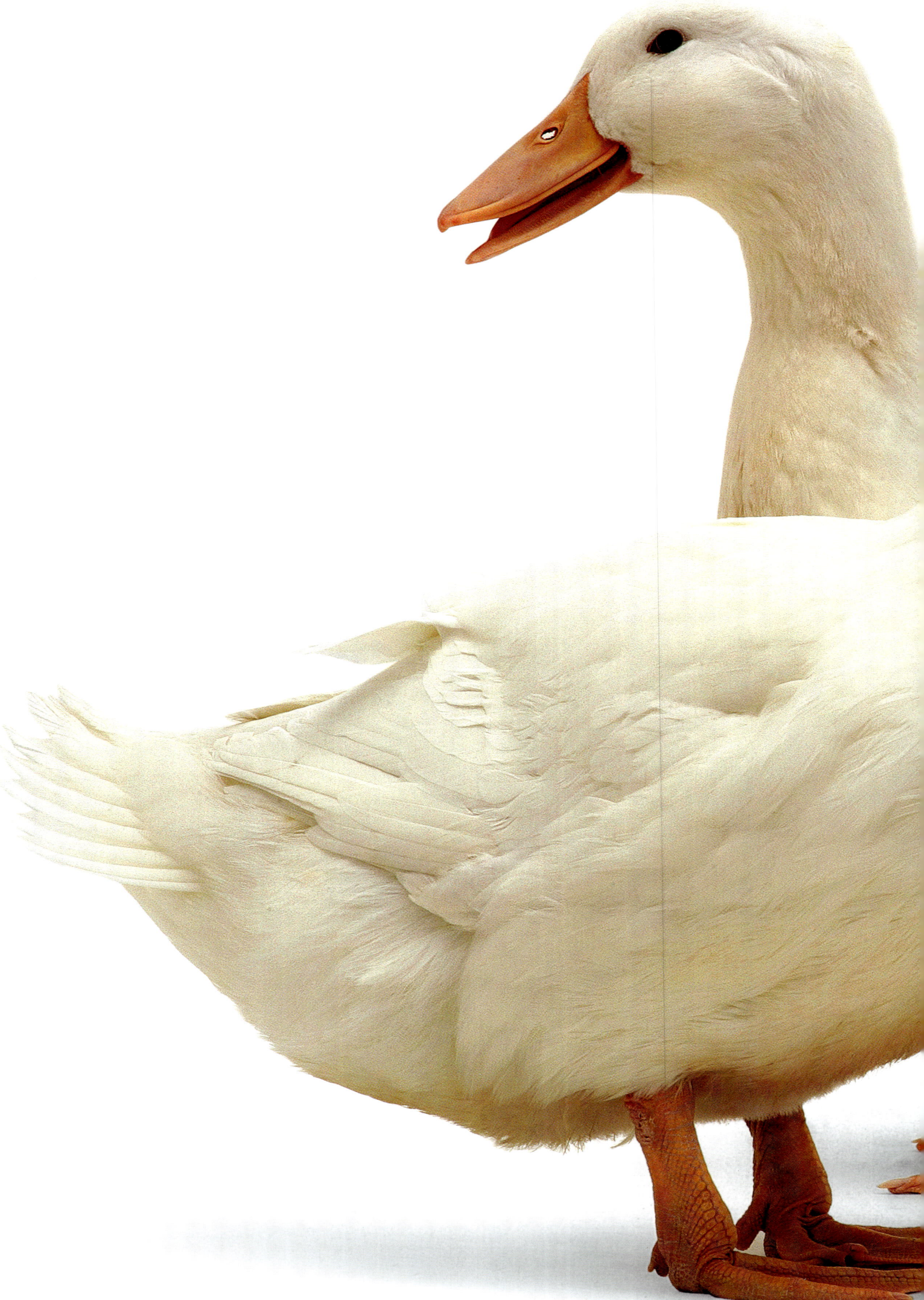

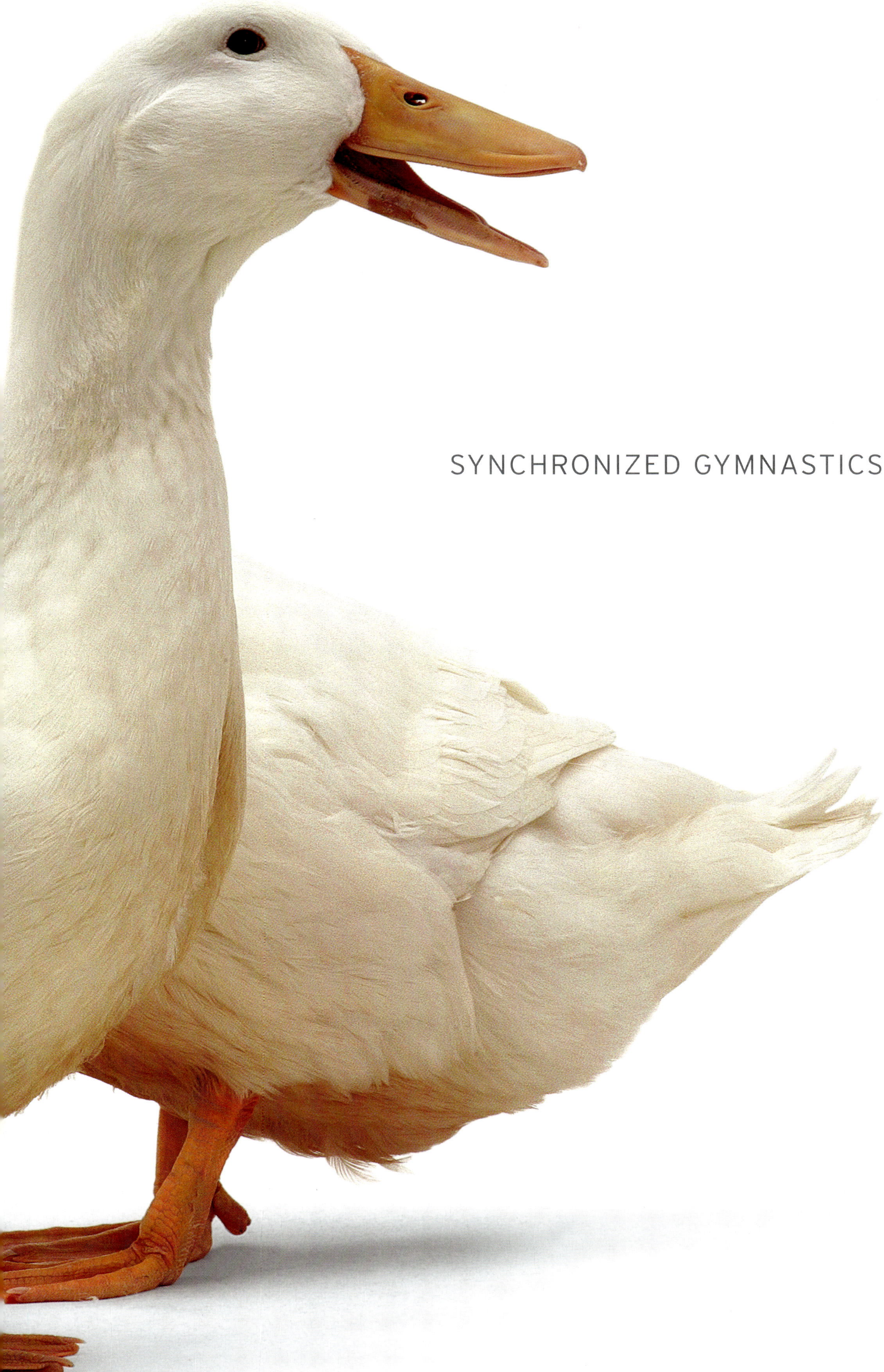

SYNCHRONIZED GYMNASTICS

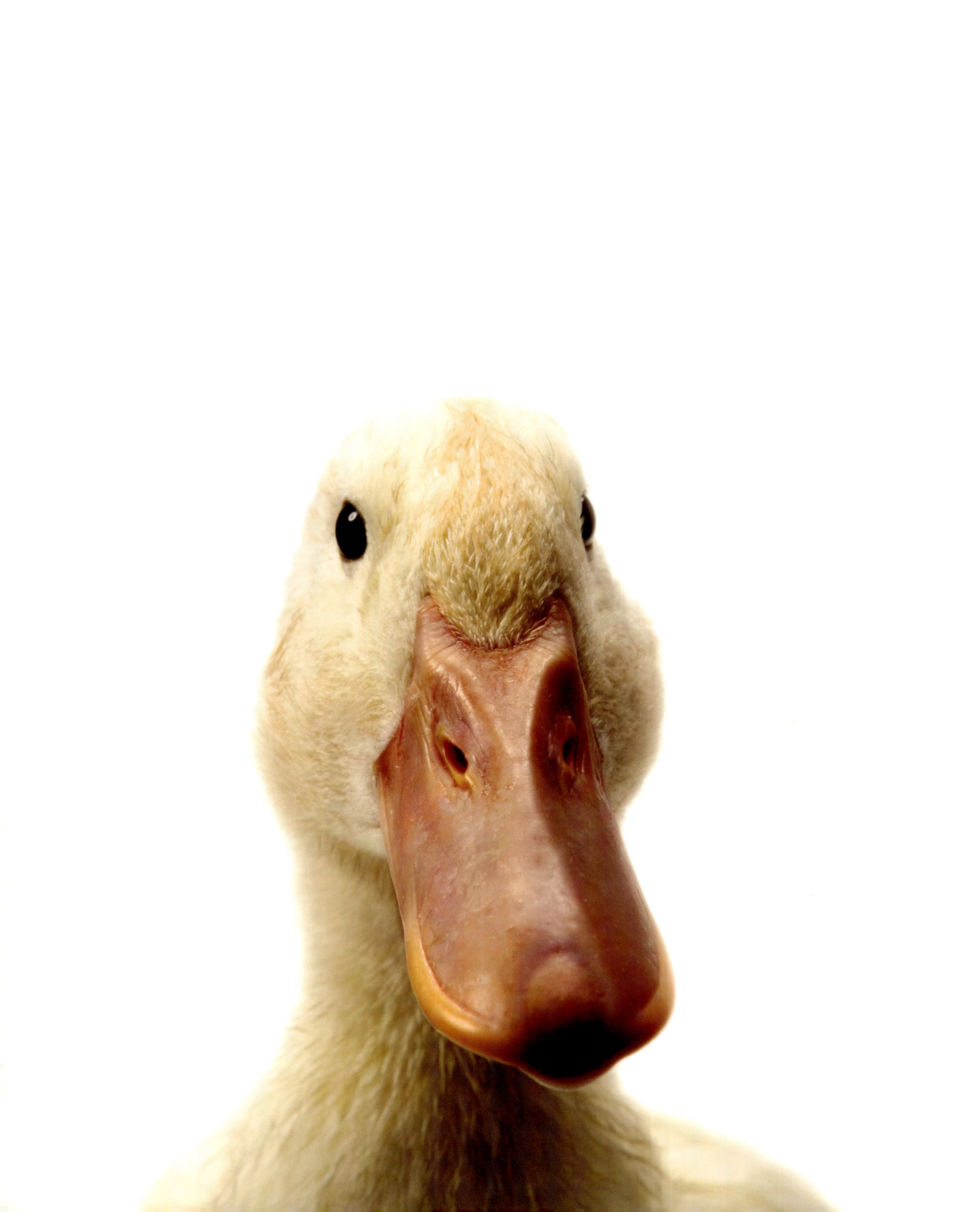

IDENTIFICATION PHOTO

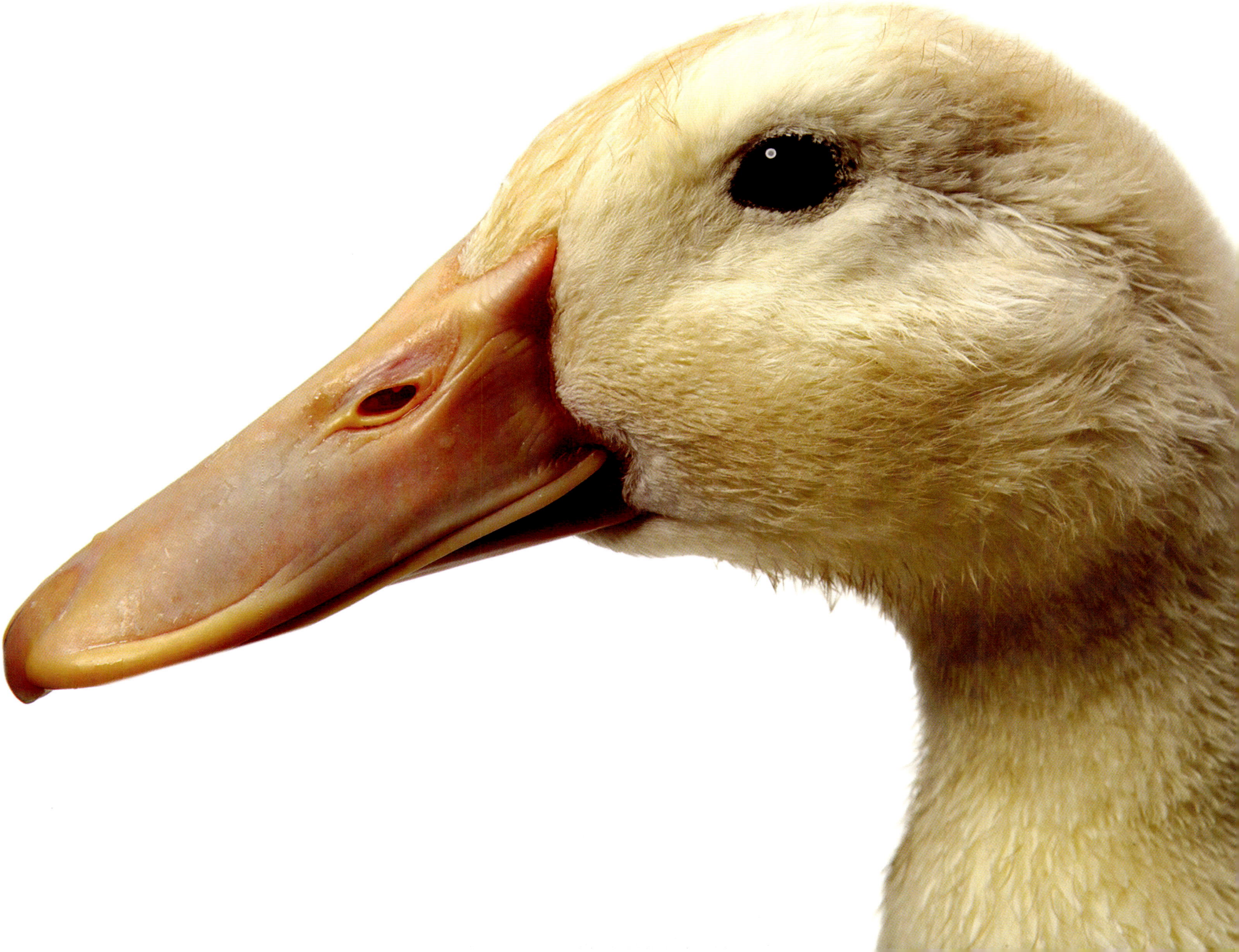

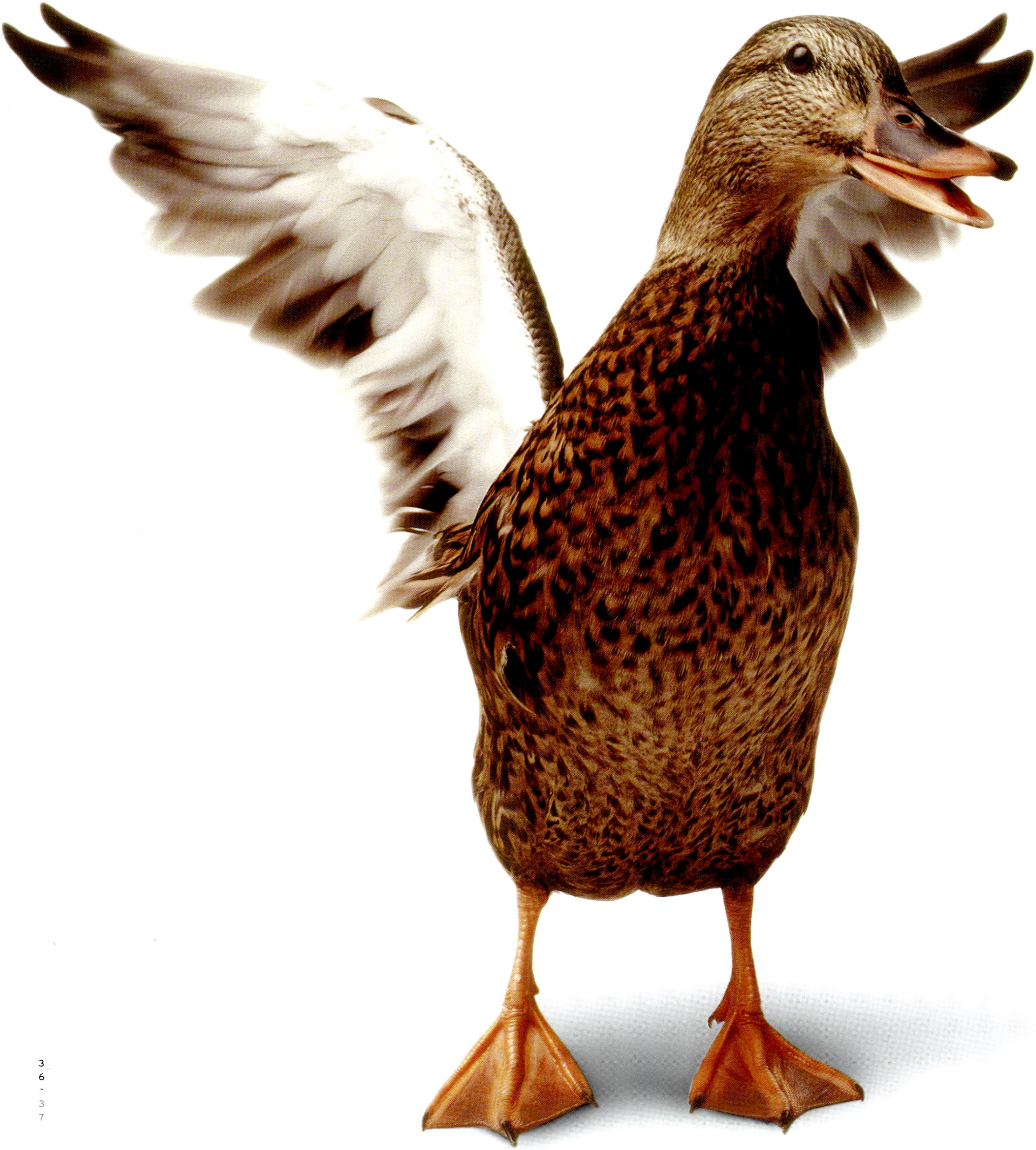

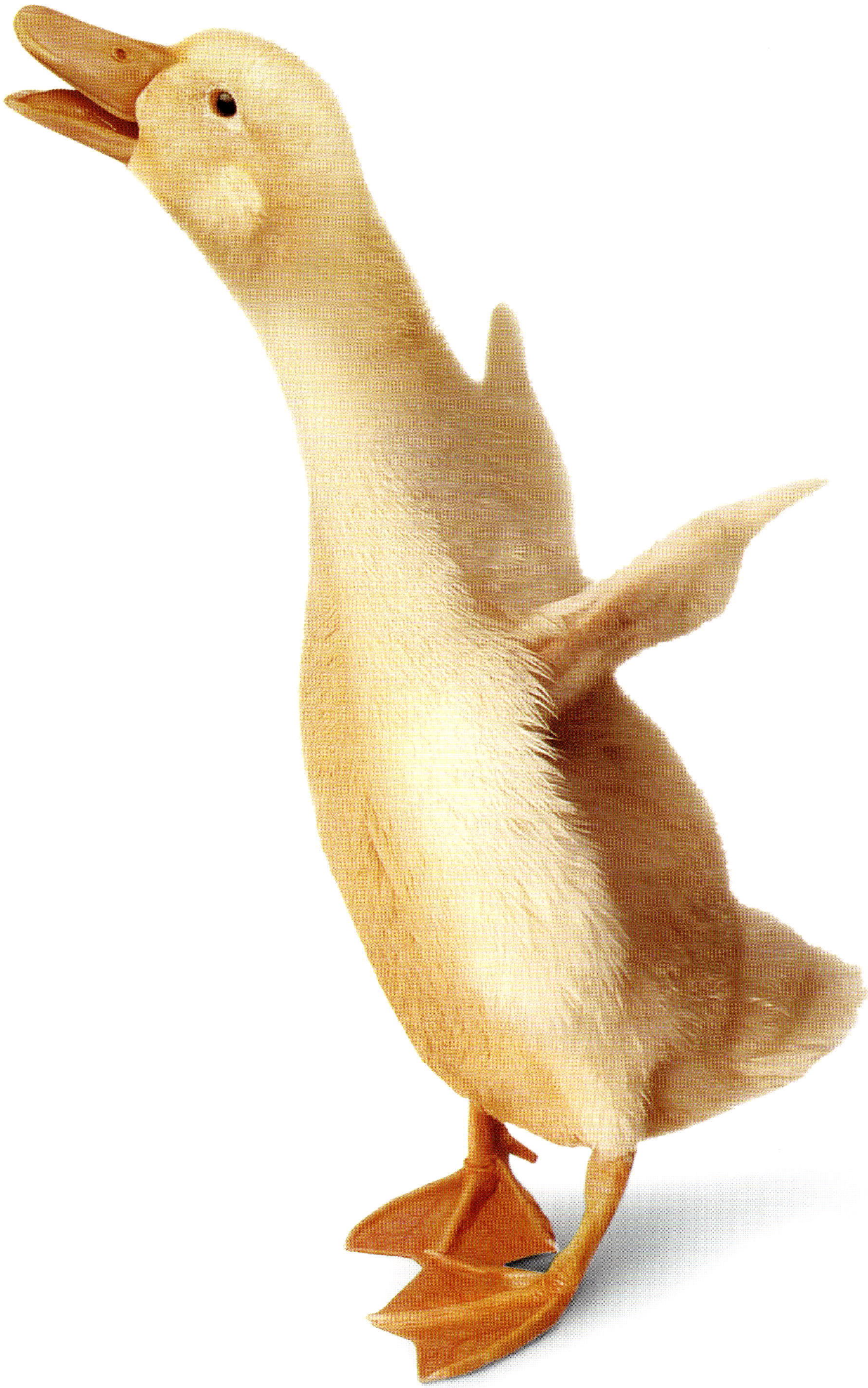

SQUABBLE

A FEATHER FOR EVERY SEASON

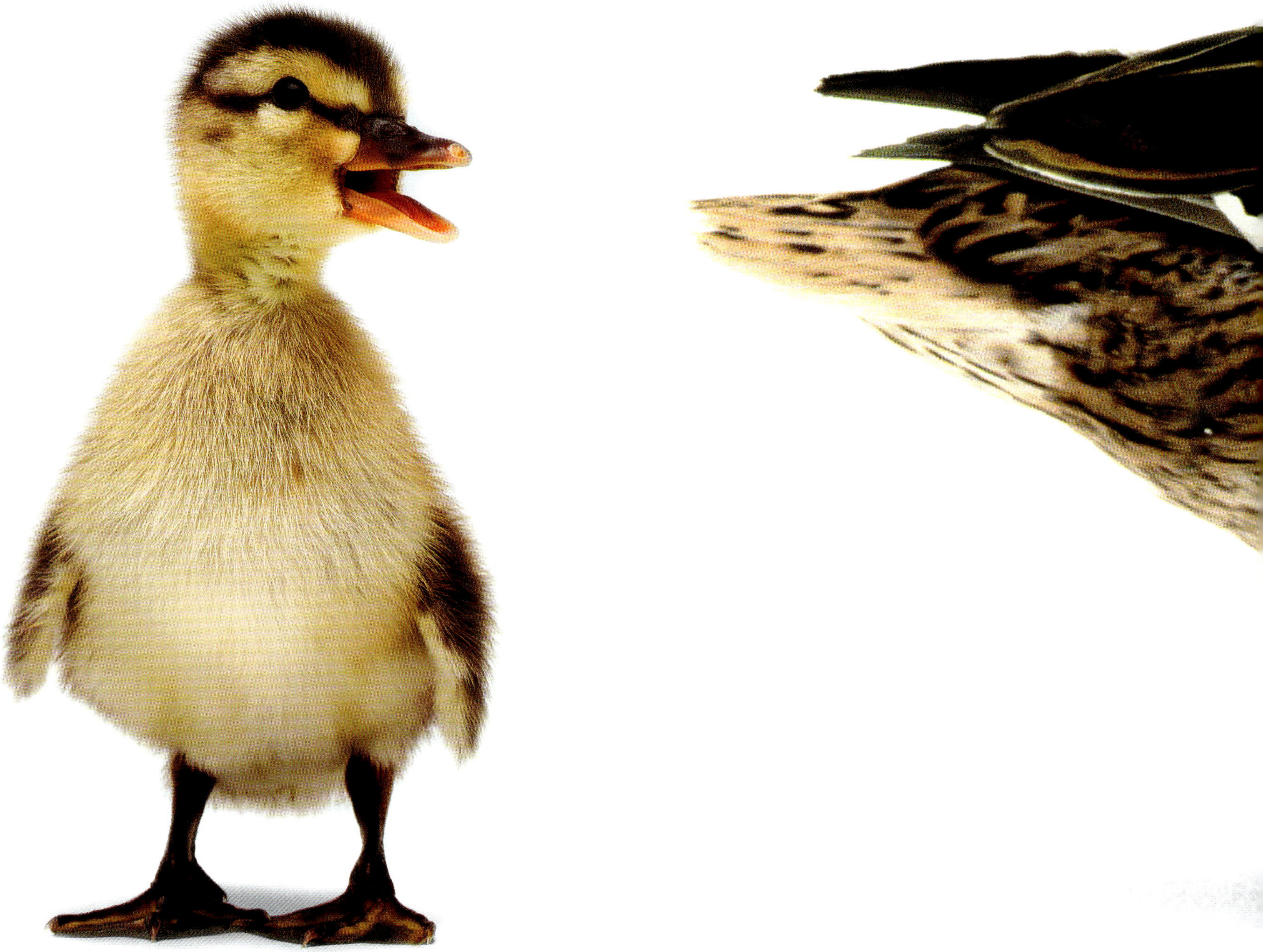

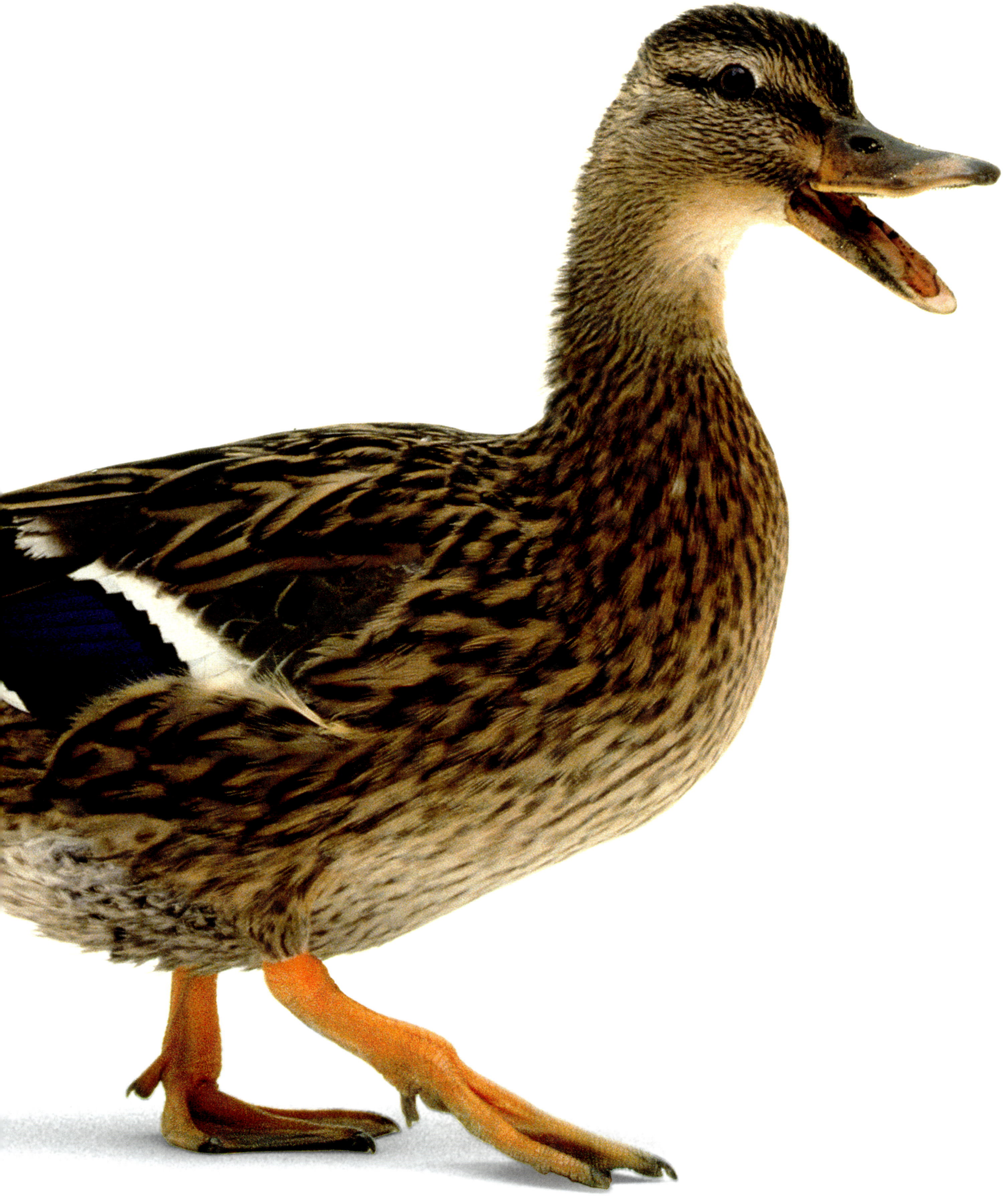

IMPRINTING POWER

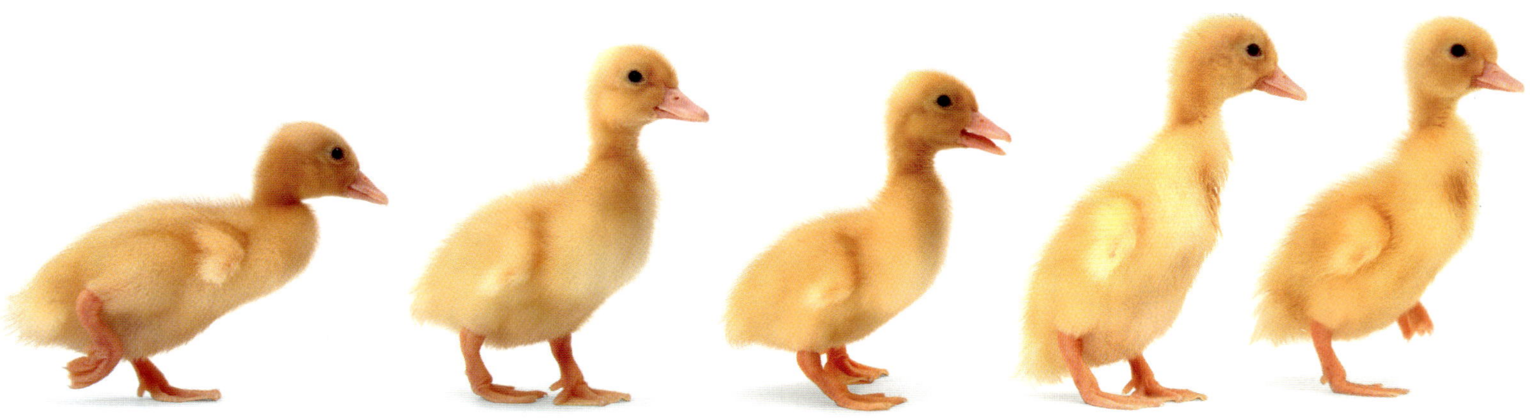

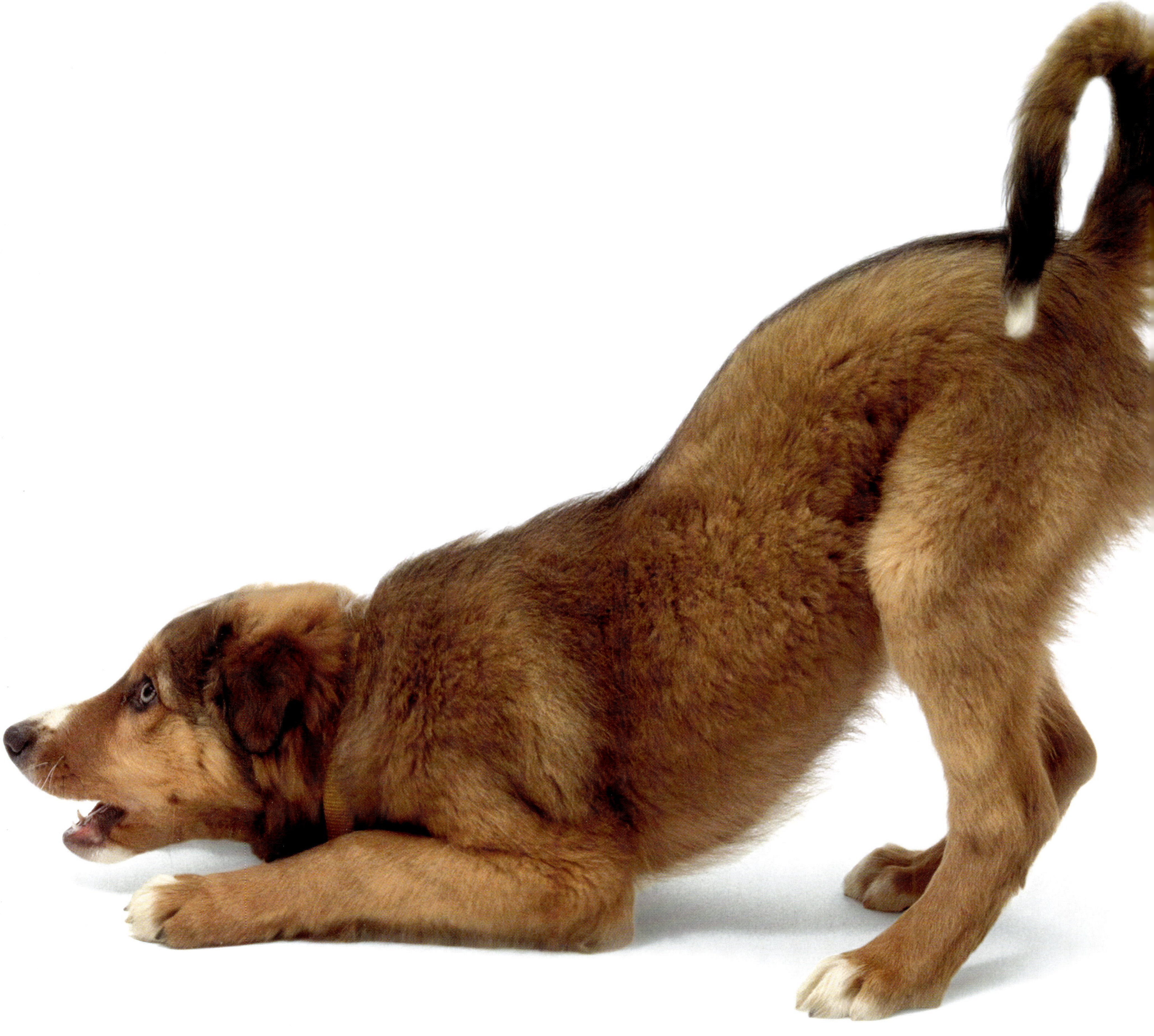

“It isn’t by size that you win or fail –
be the best of whatever you are.”
(Douglas Malloch)

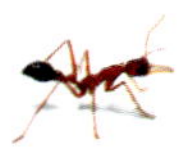

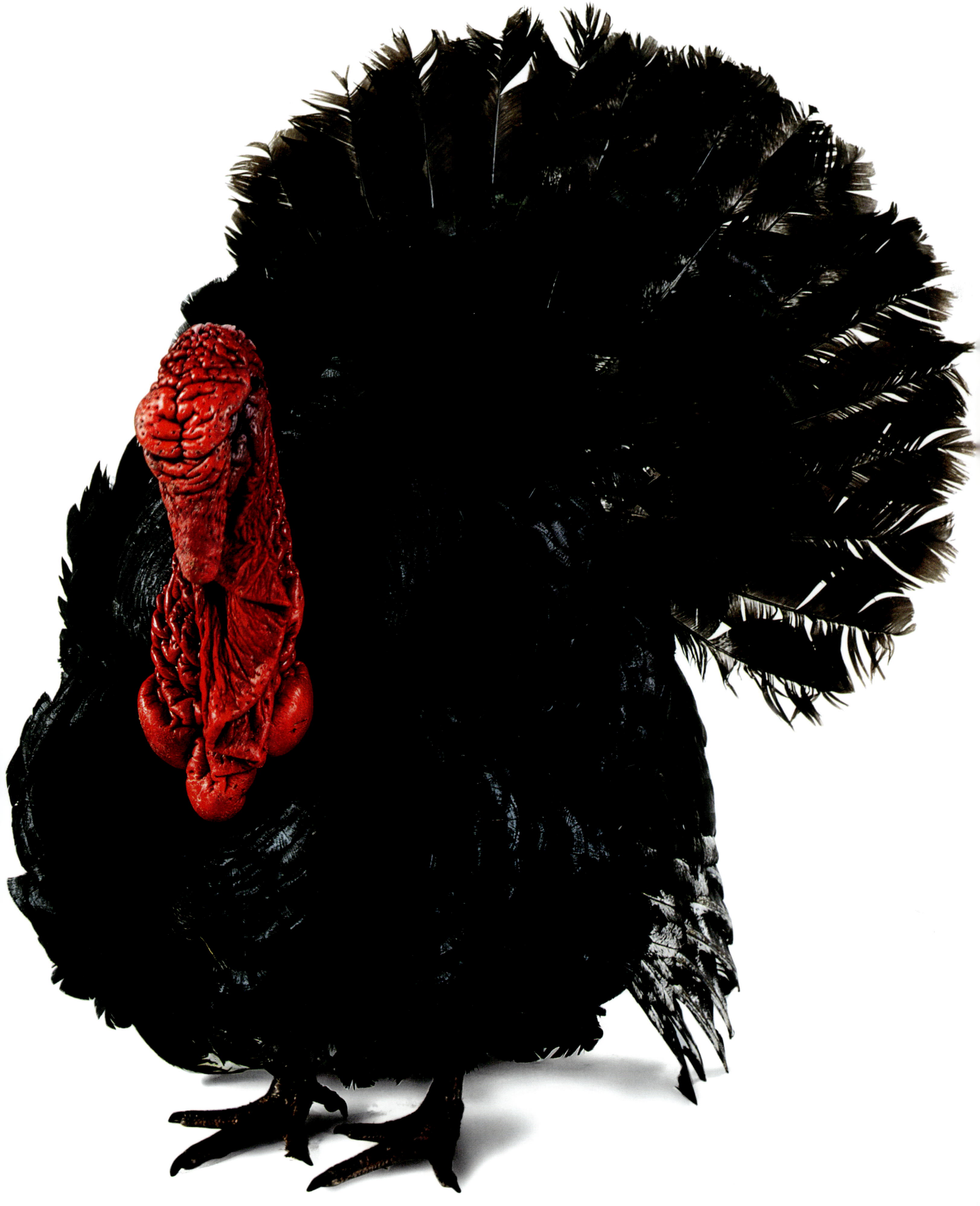

COMPOSITE HEAD

Giuseppe Arcimboldo

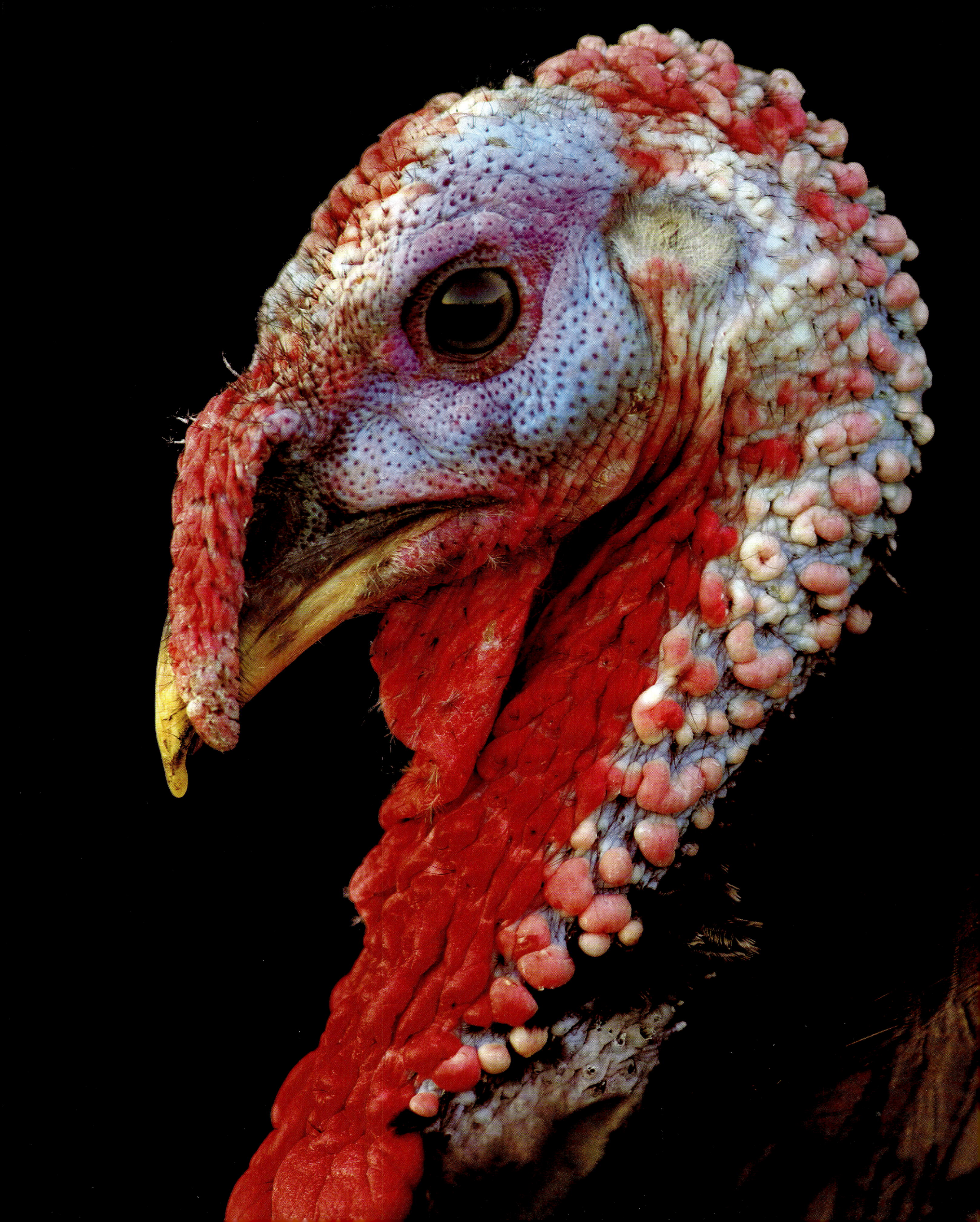

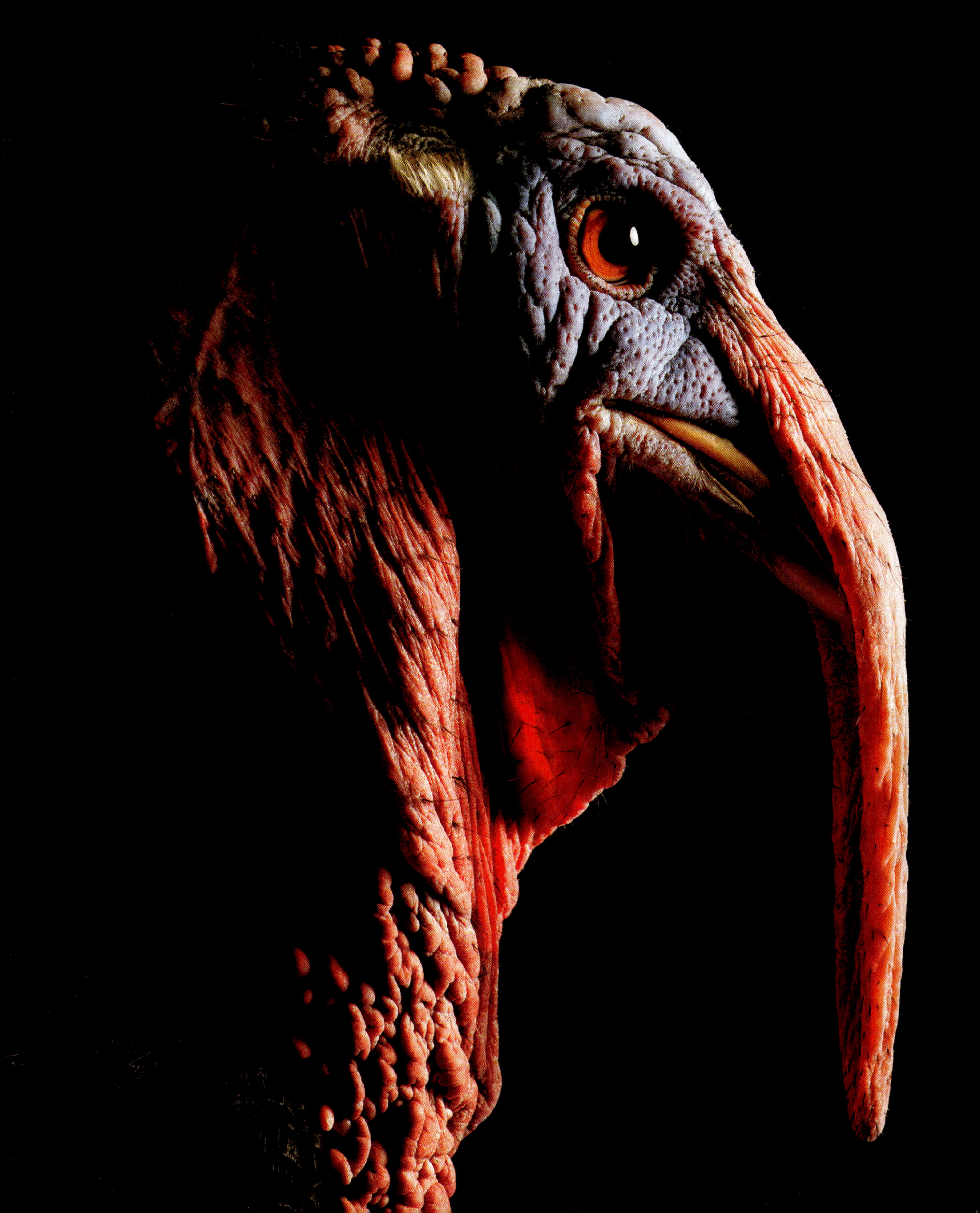

PORTRAIT OF AN AUTHOR

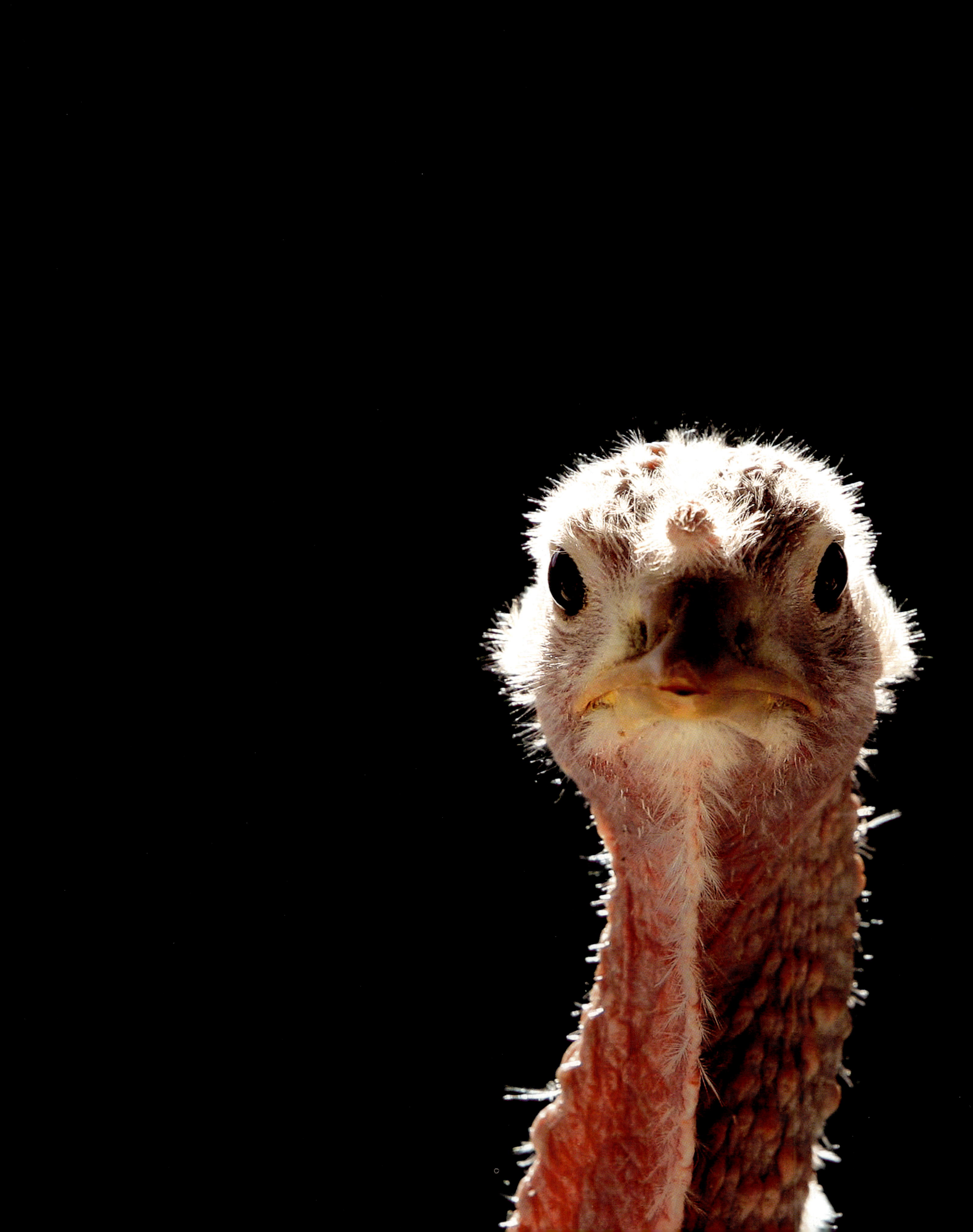

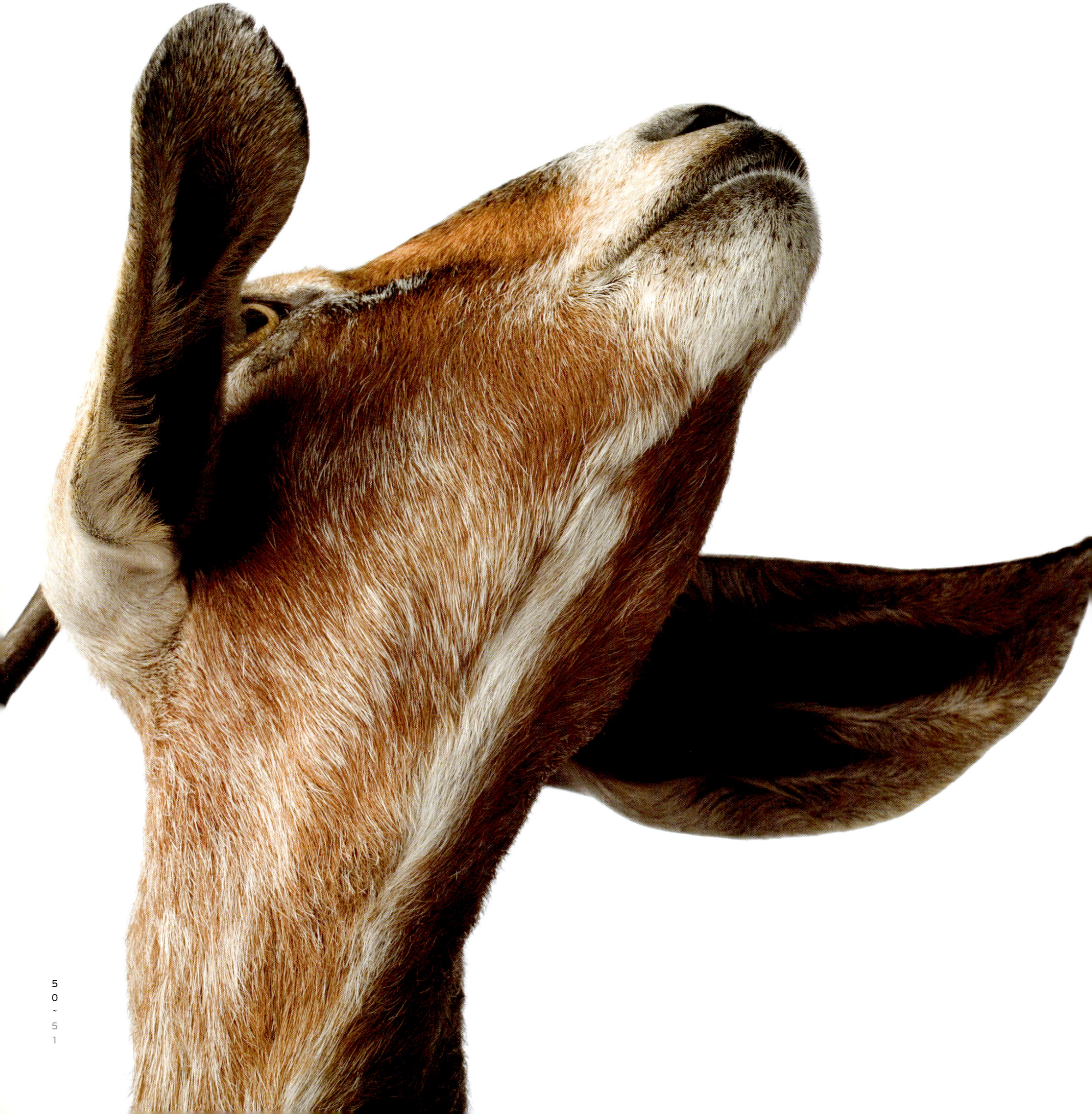

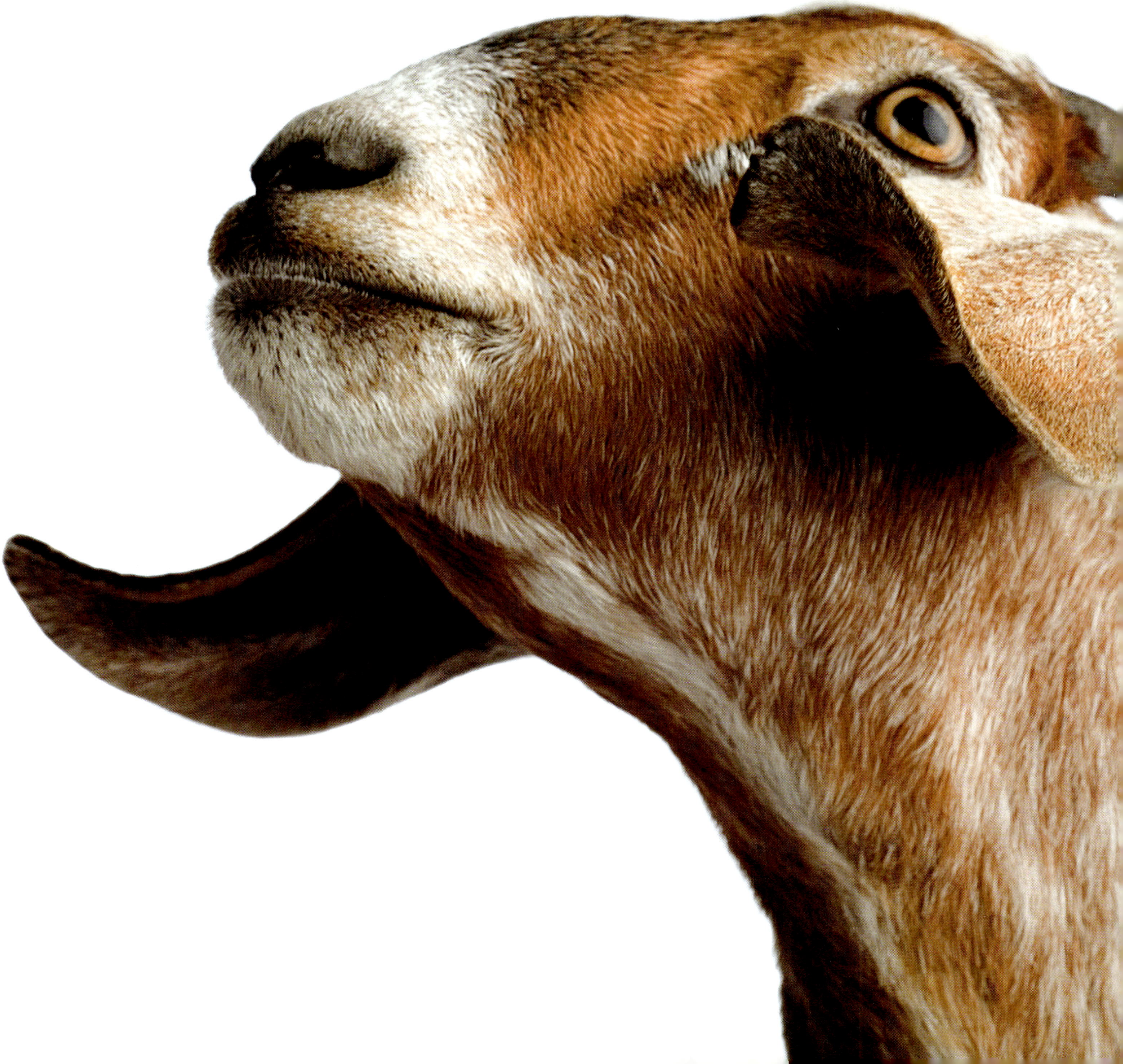

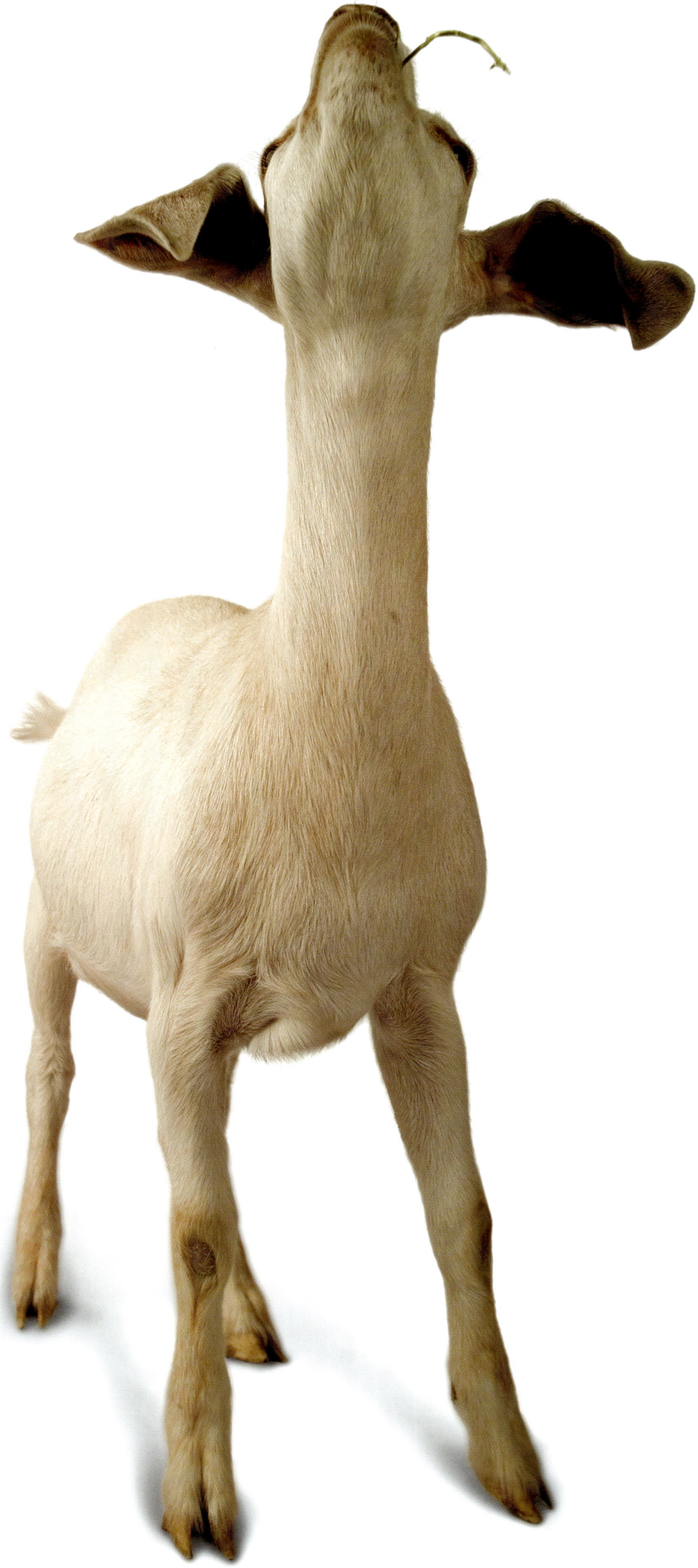

“A hungry stomach has no ears.”
(Jean de La Fontaine)

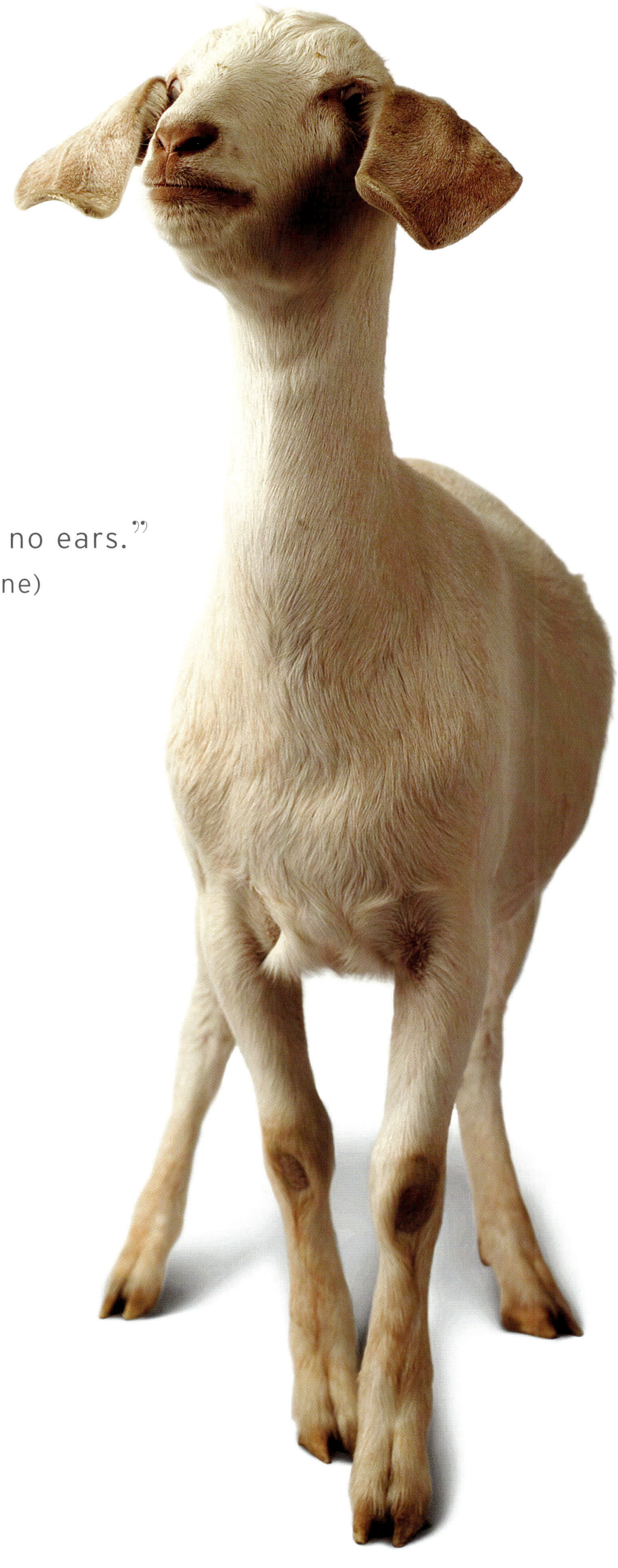

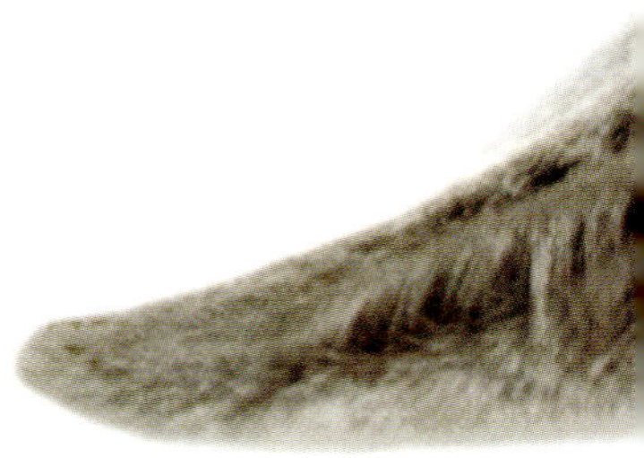

“A prudent man does not make
the goat his gardener.”

(English proverb)

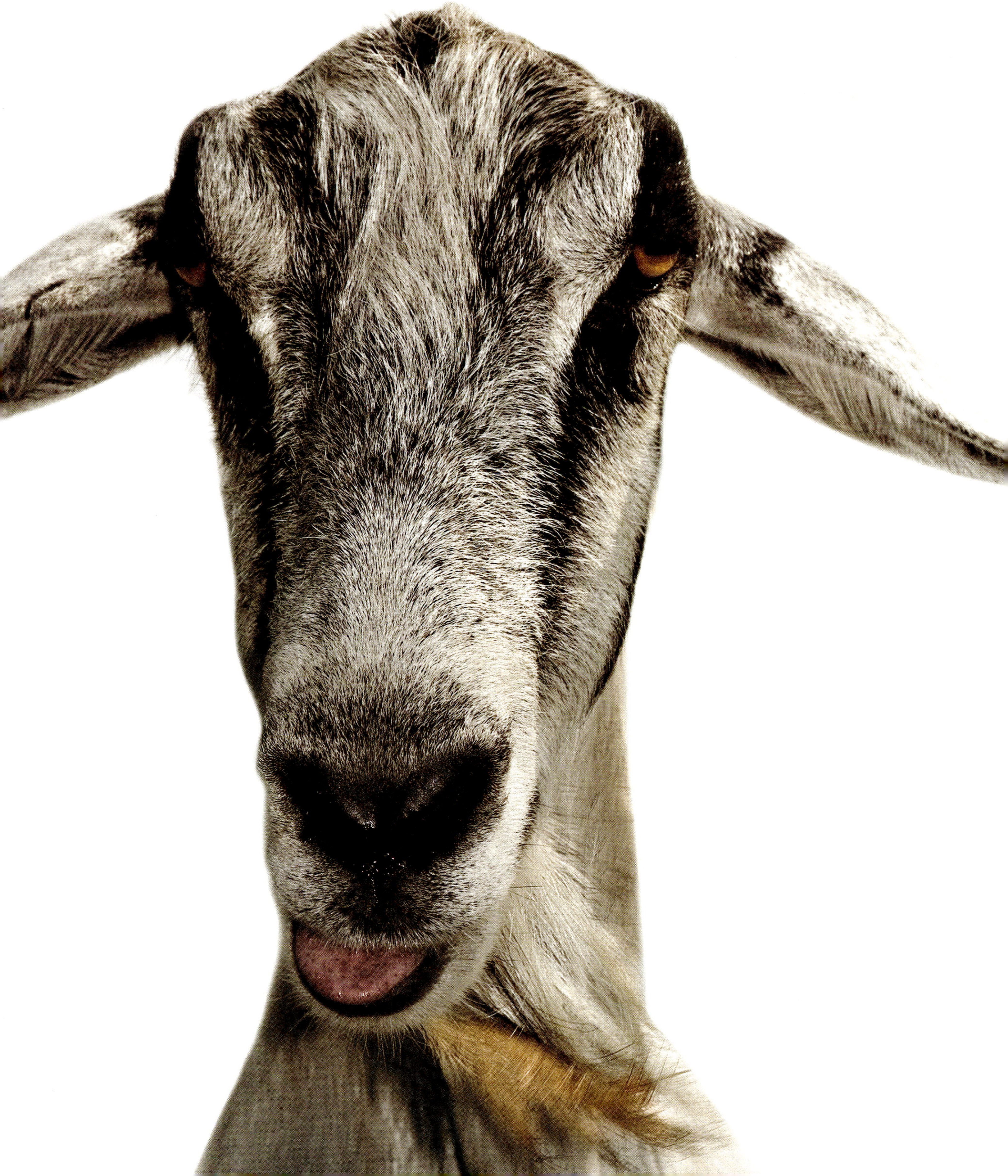

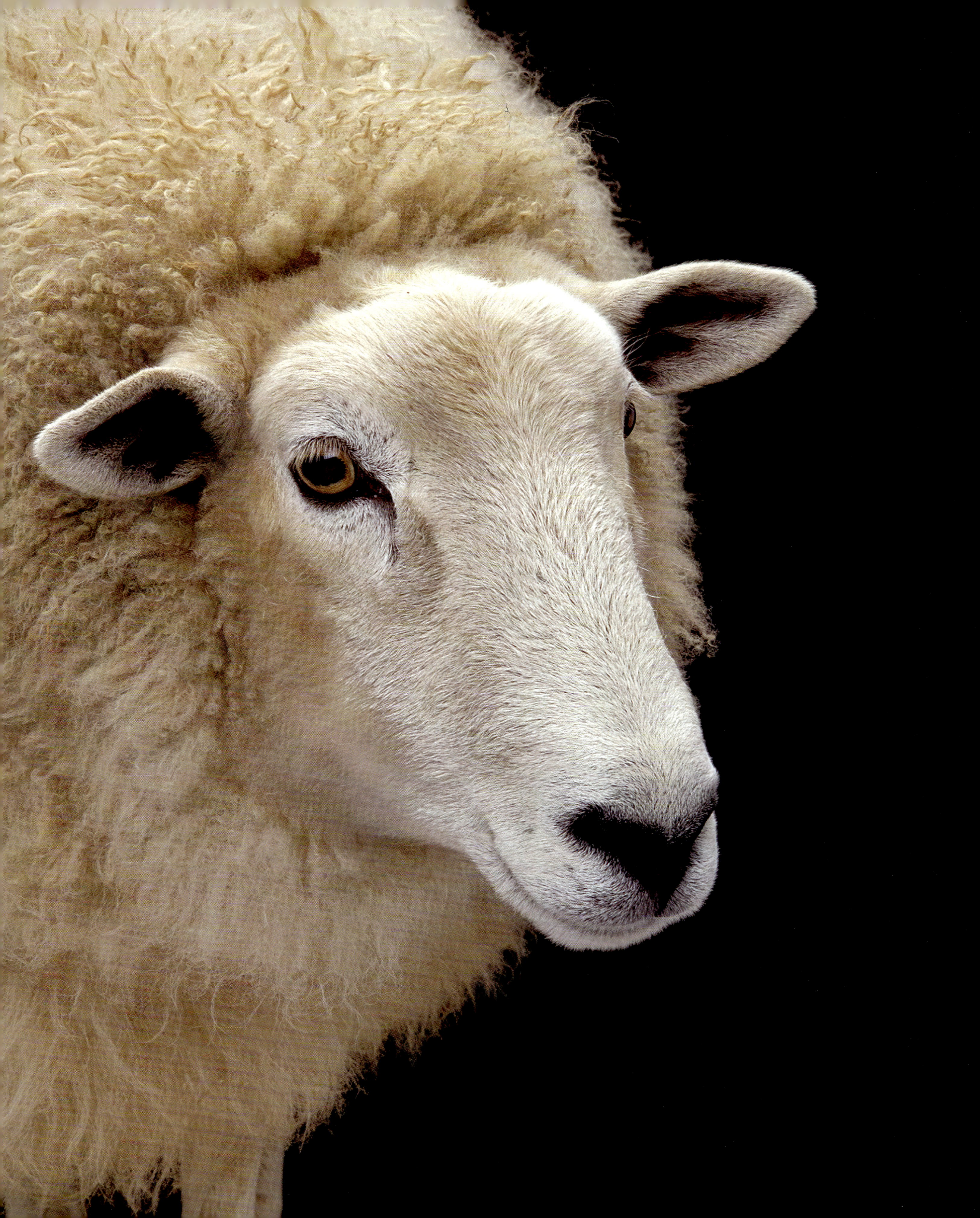

“It’s better to be a lion for a day than a sheep all your life.”

(Elizabeth Kenny)

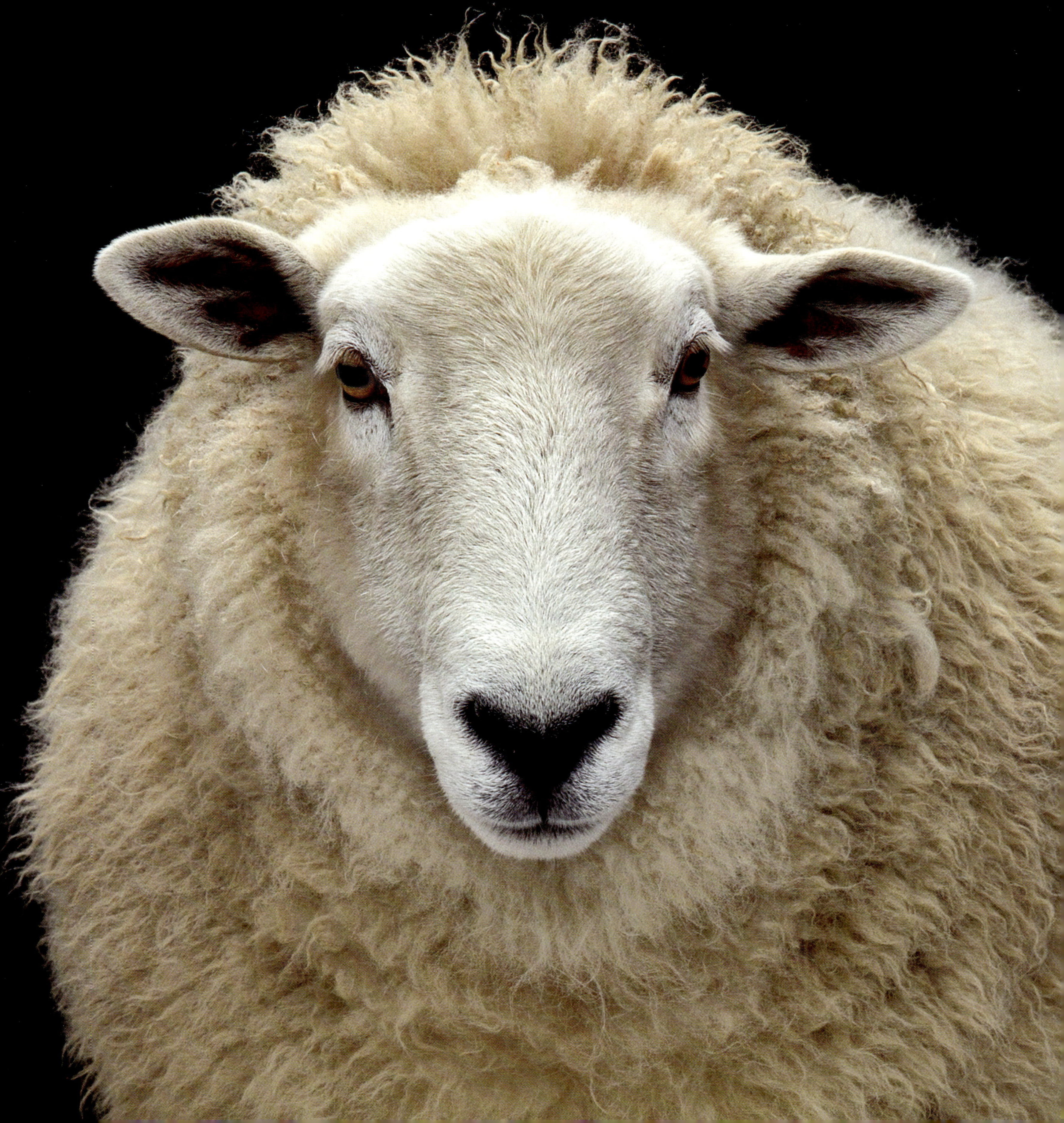

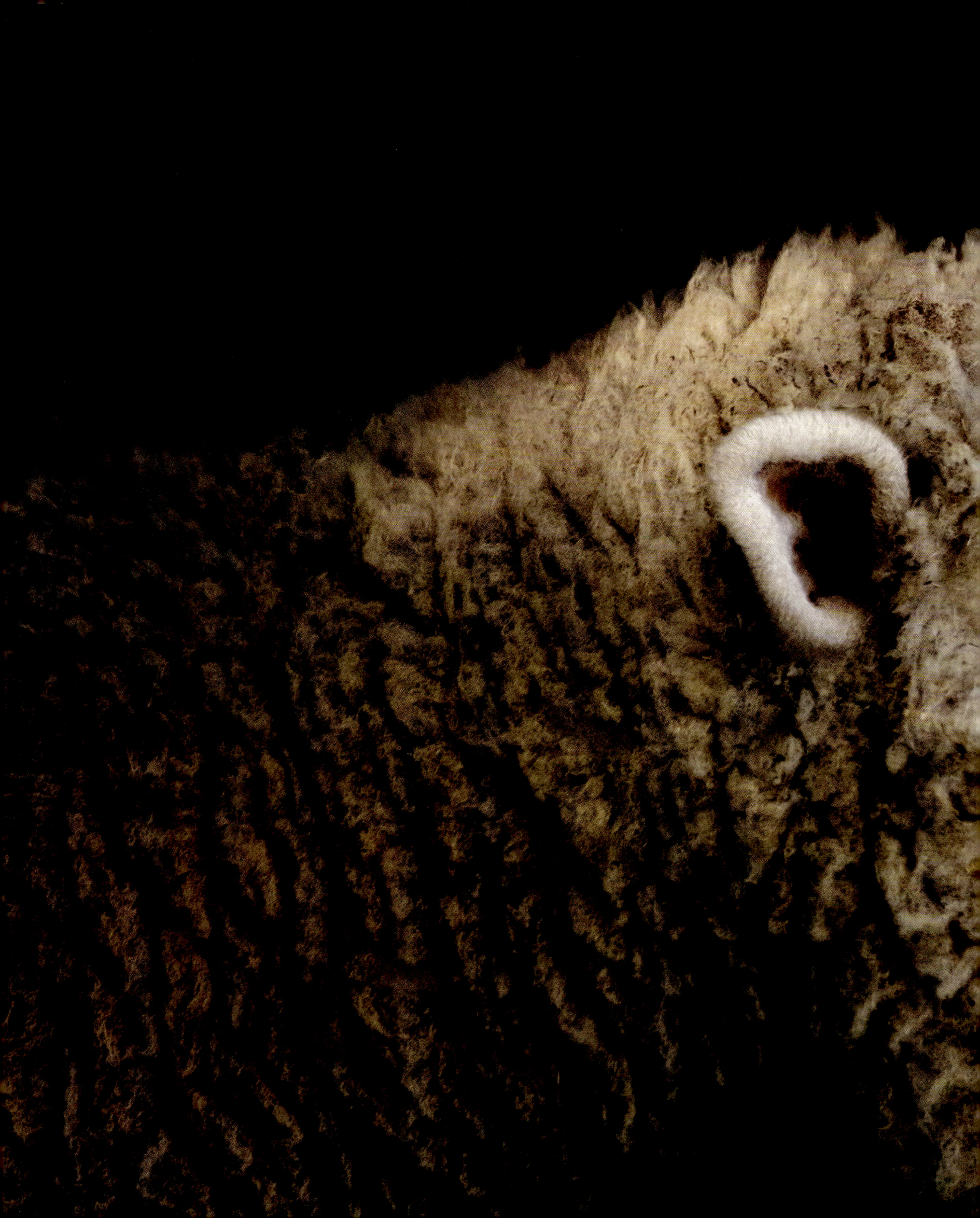

“In order to be an immaculate member of a flock of sheep,
one must above all be a sheep oneself.”

(Albert Einstein)

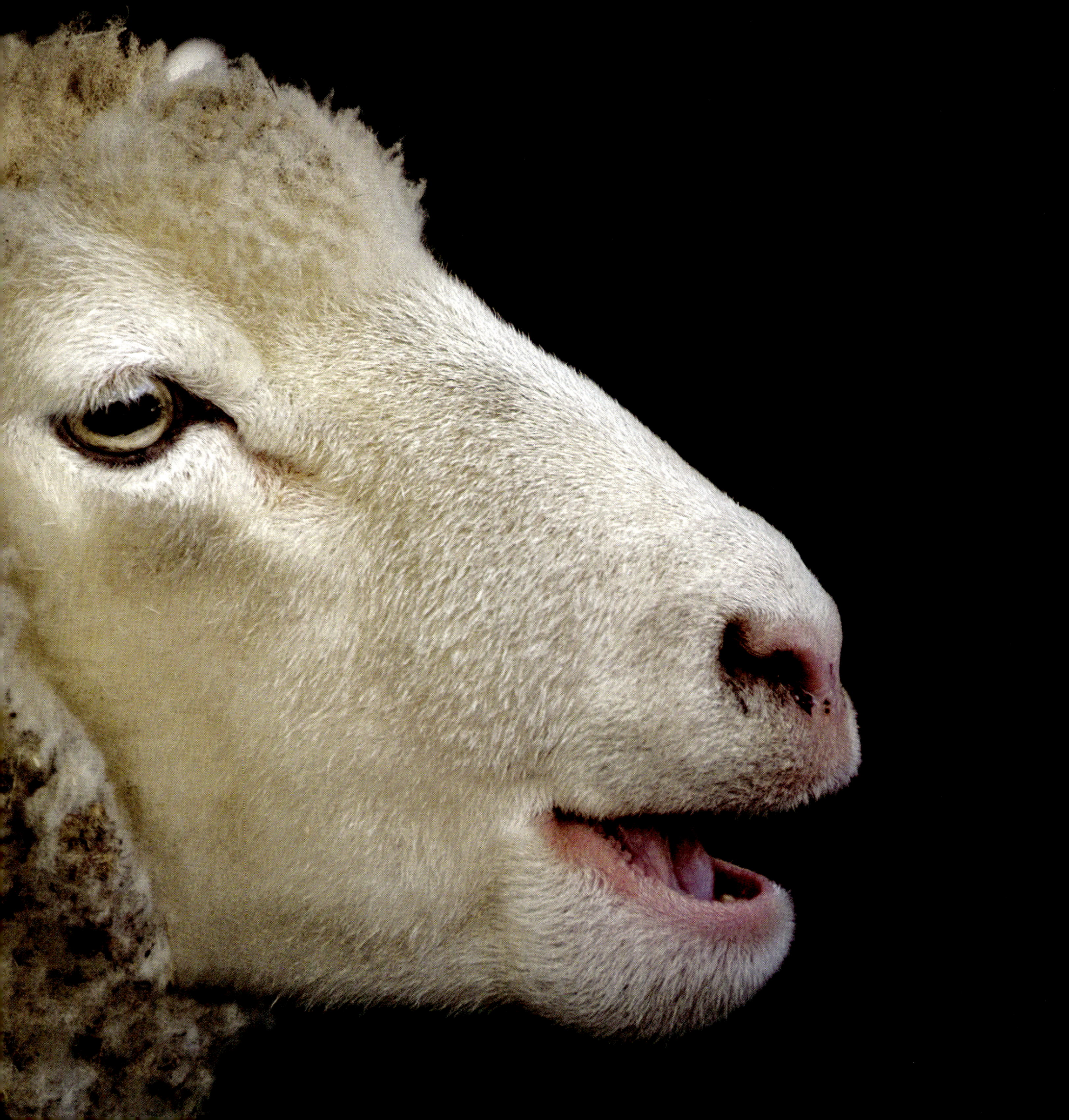

"Nature does not make jumps."

(Carl Linnaeus)

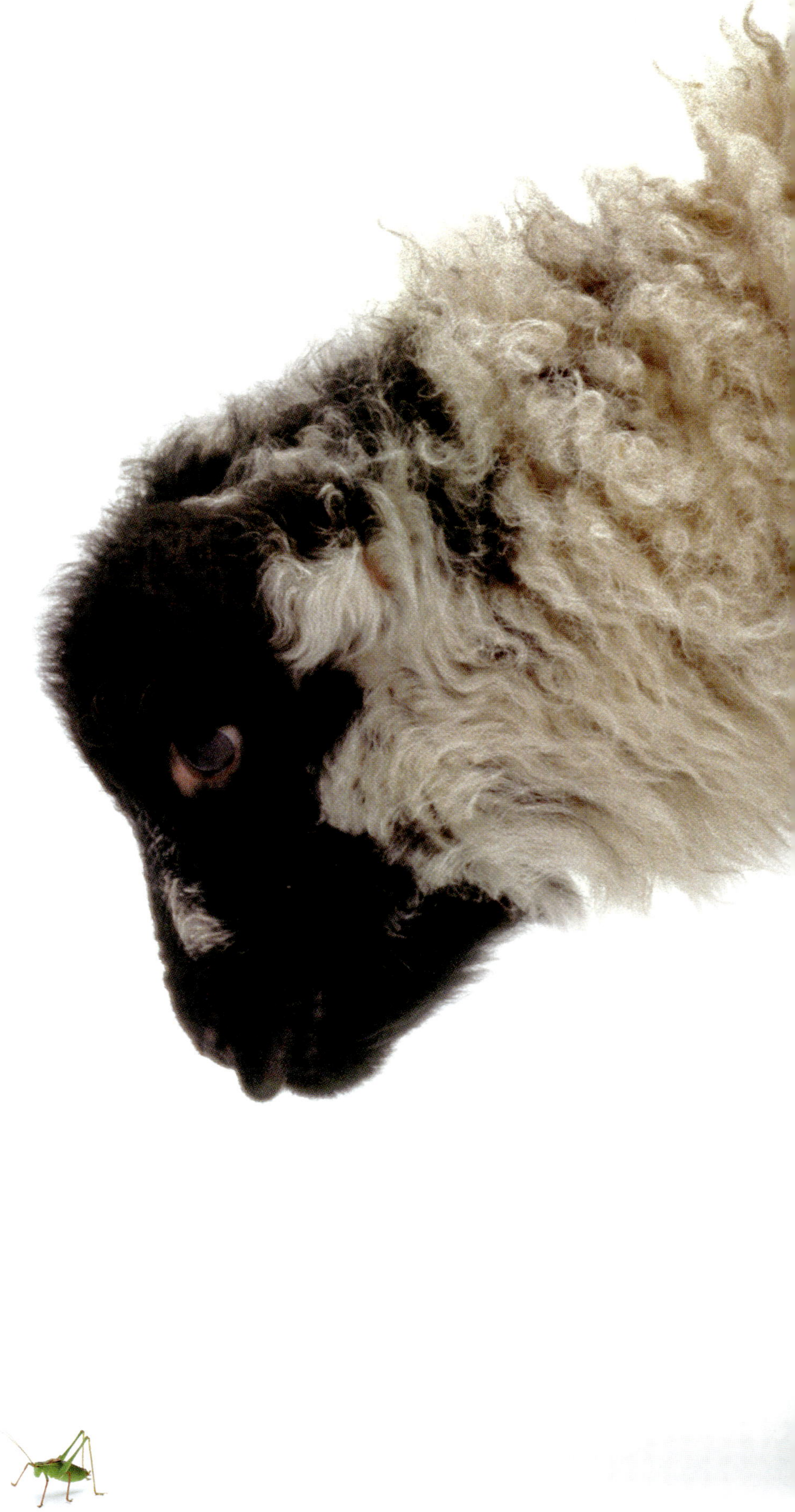

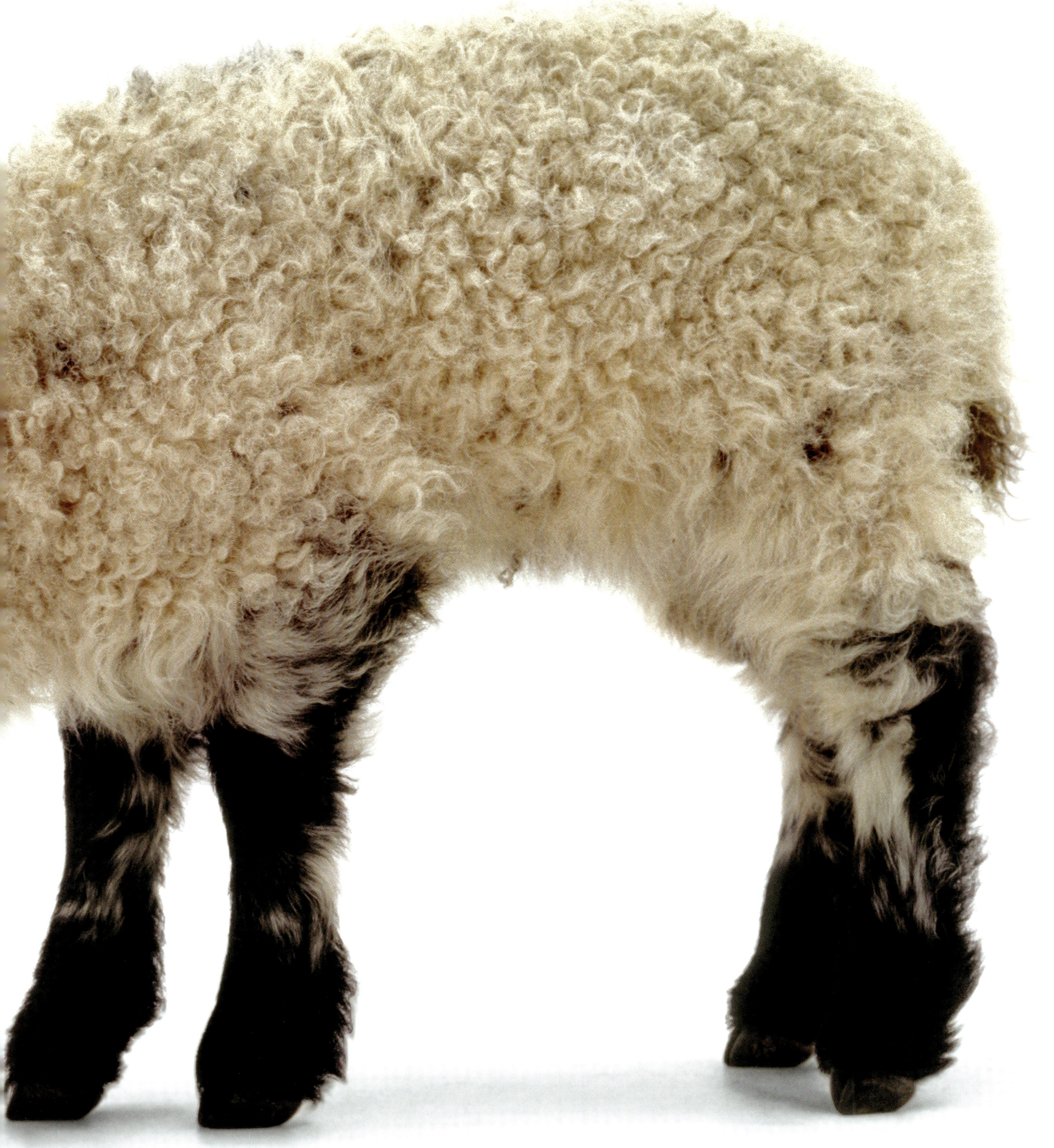

“People, like sheep, tend to follow a leader – occasionally in the right direction.”

(Alexander Chase)

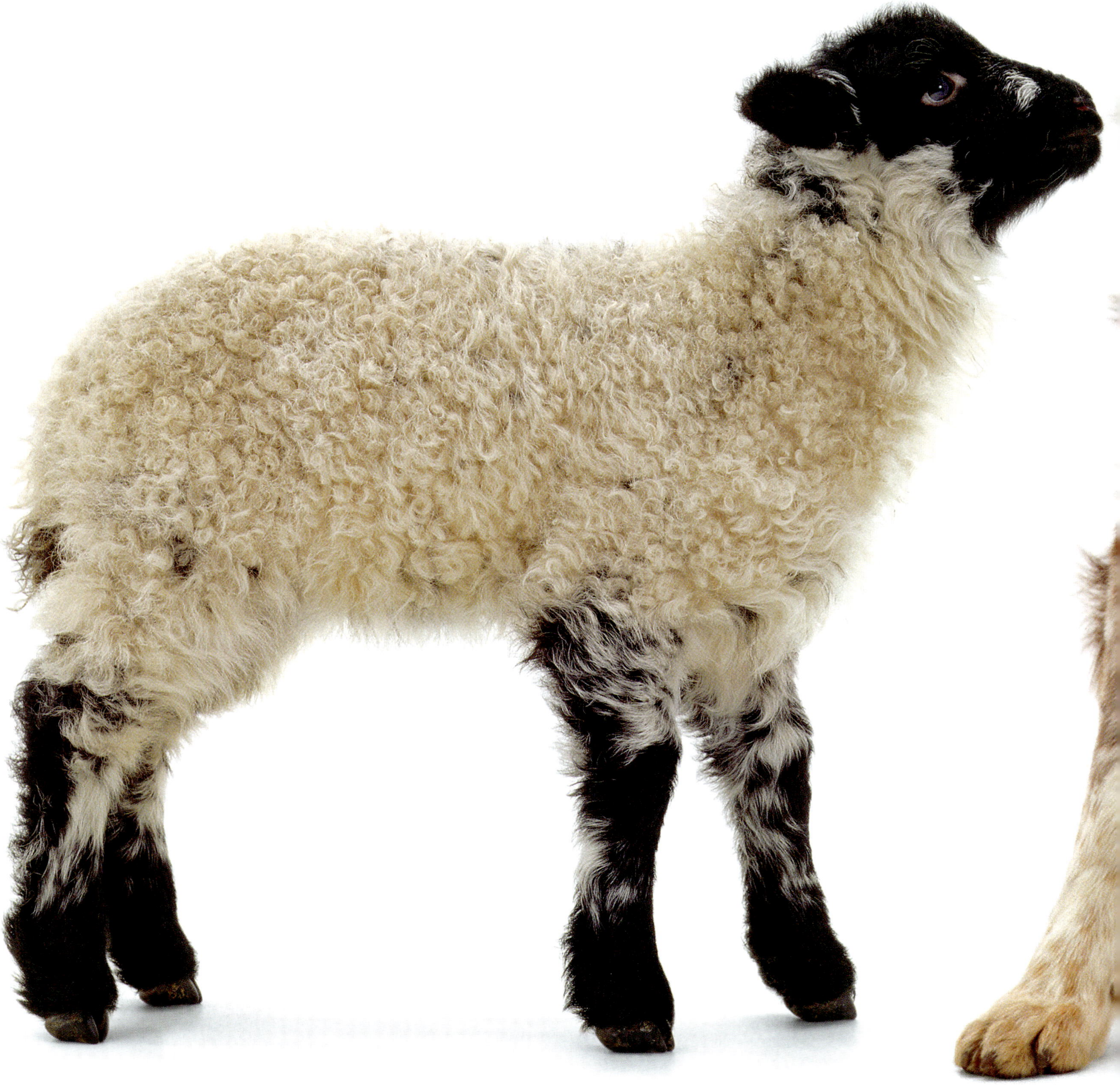

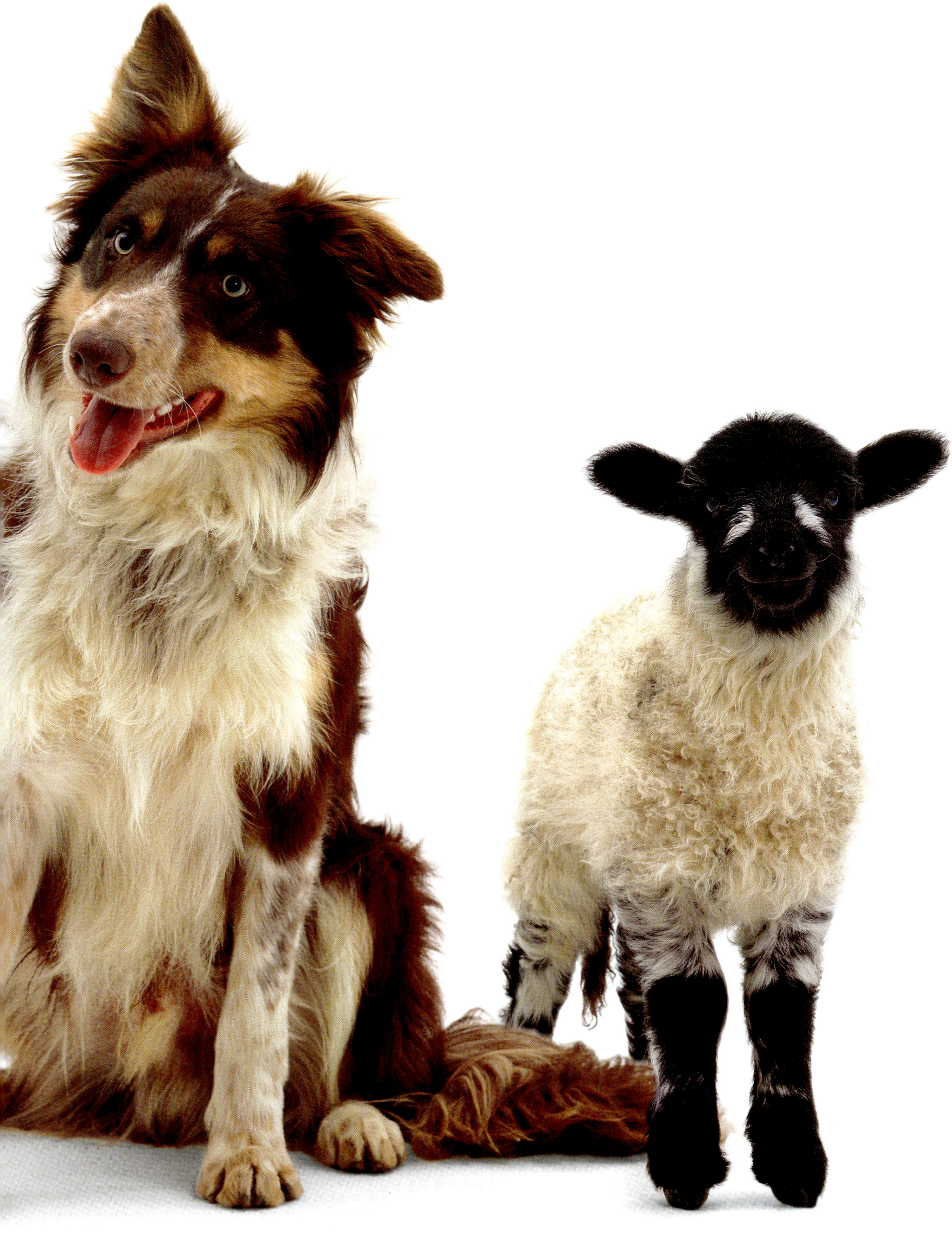

LOOKING FOR A COW-BOY

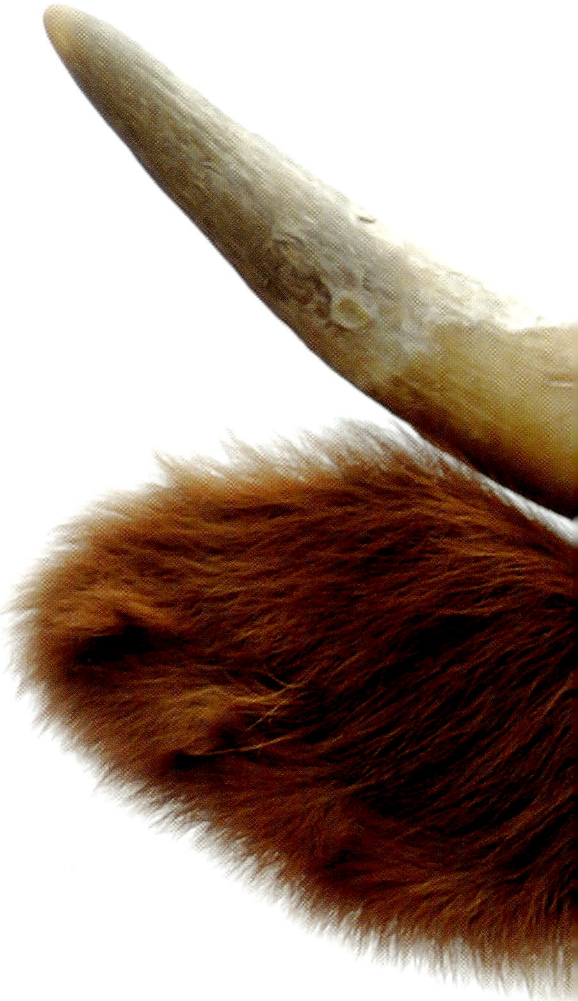

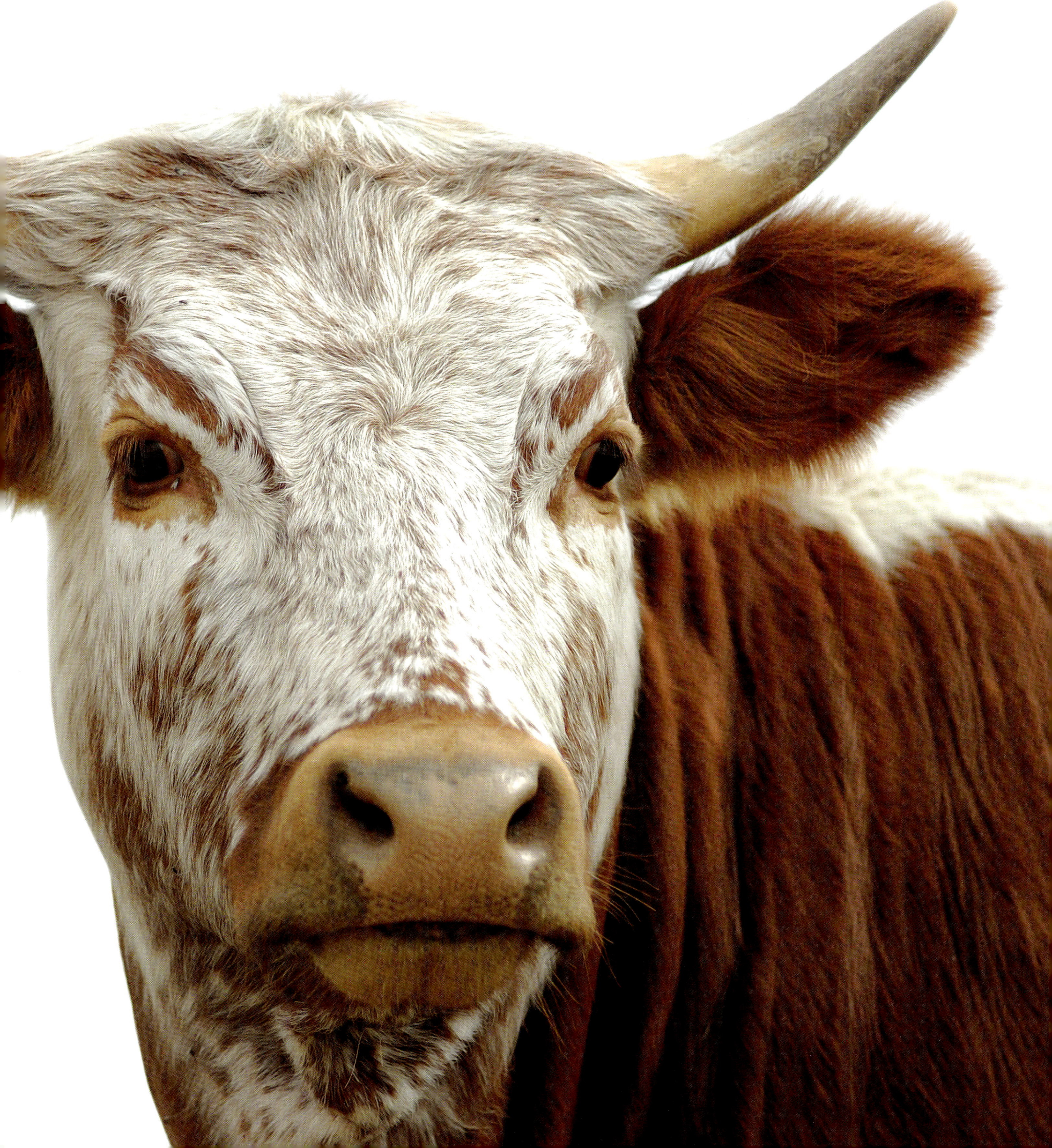

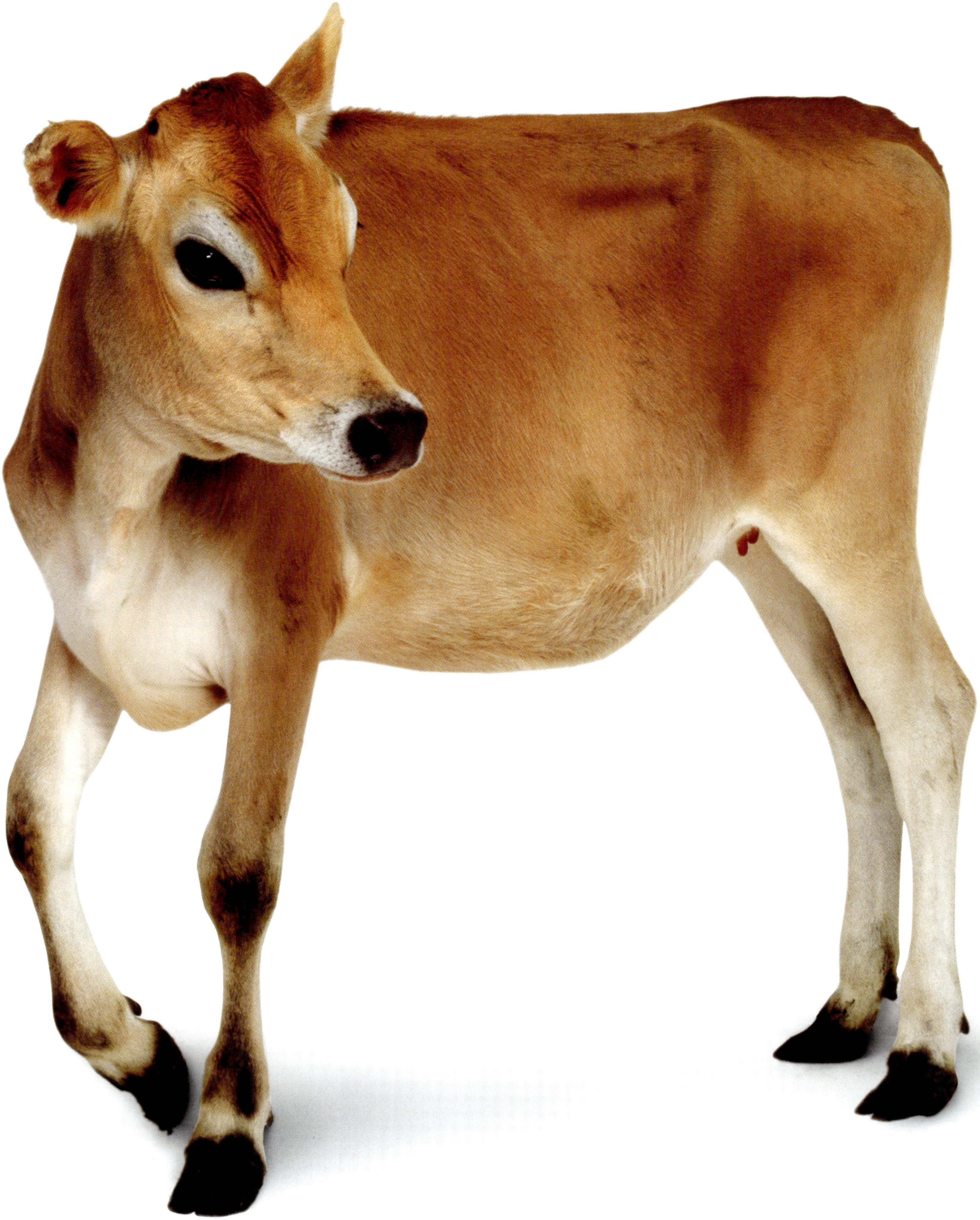

THE PIED PIPER

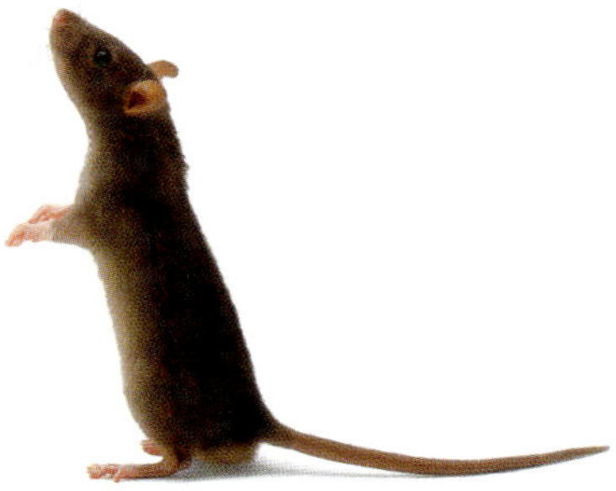

“I never met a cow I didn’t like.”

(Jennie Garth)

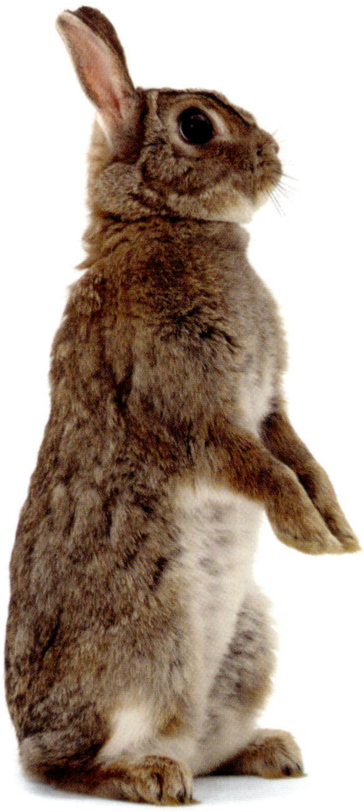

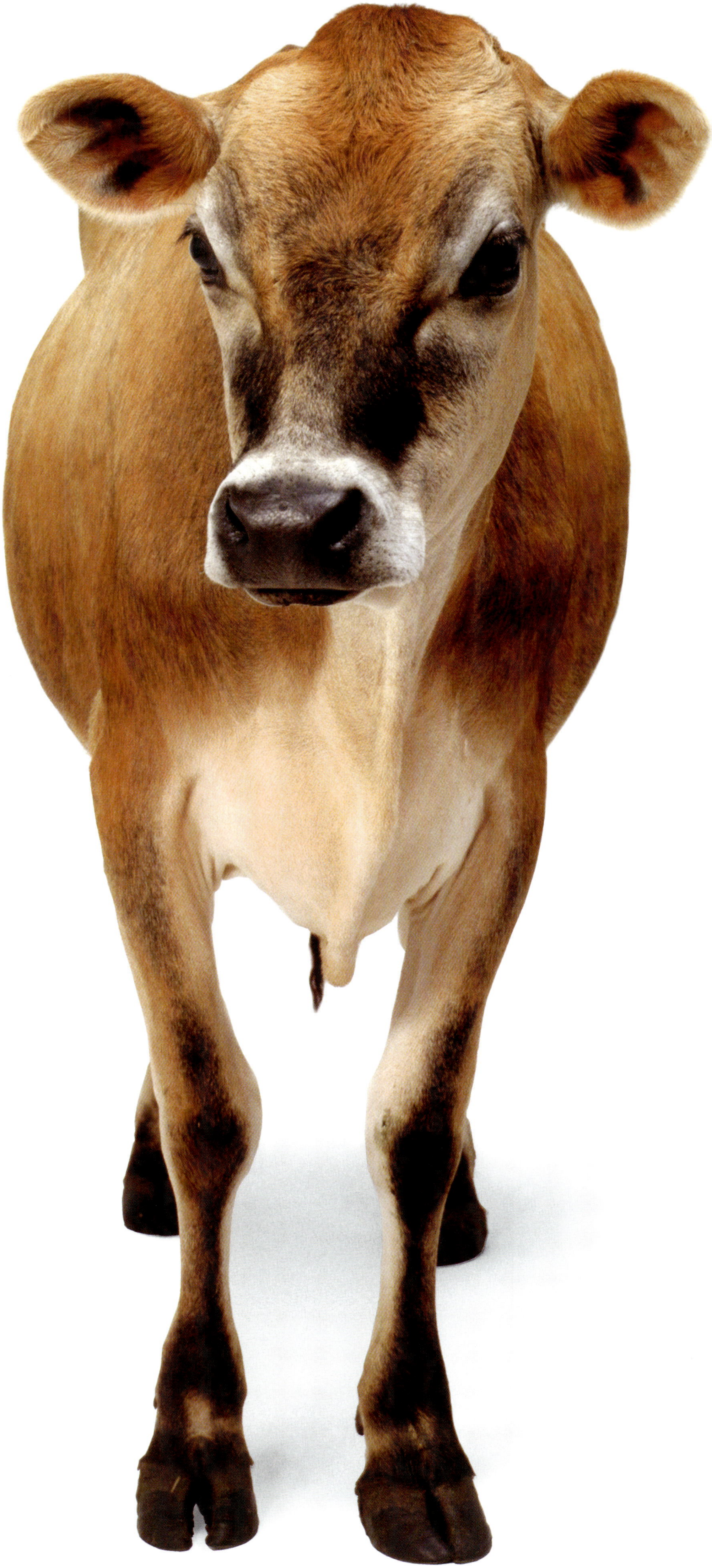

"The new art takes its elements from nature. Existence, nature and material are a perfect unity."

(Lucio Fontana)

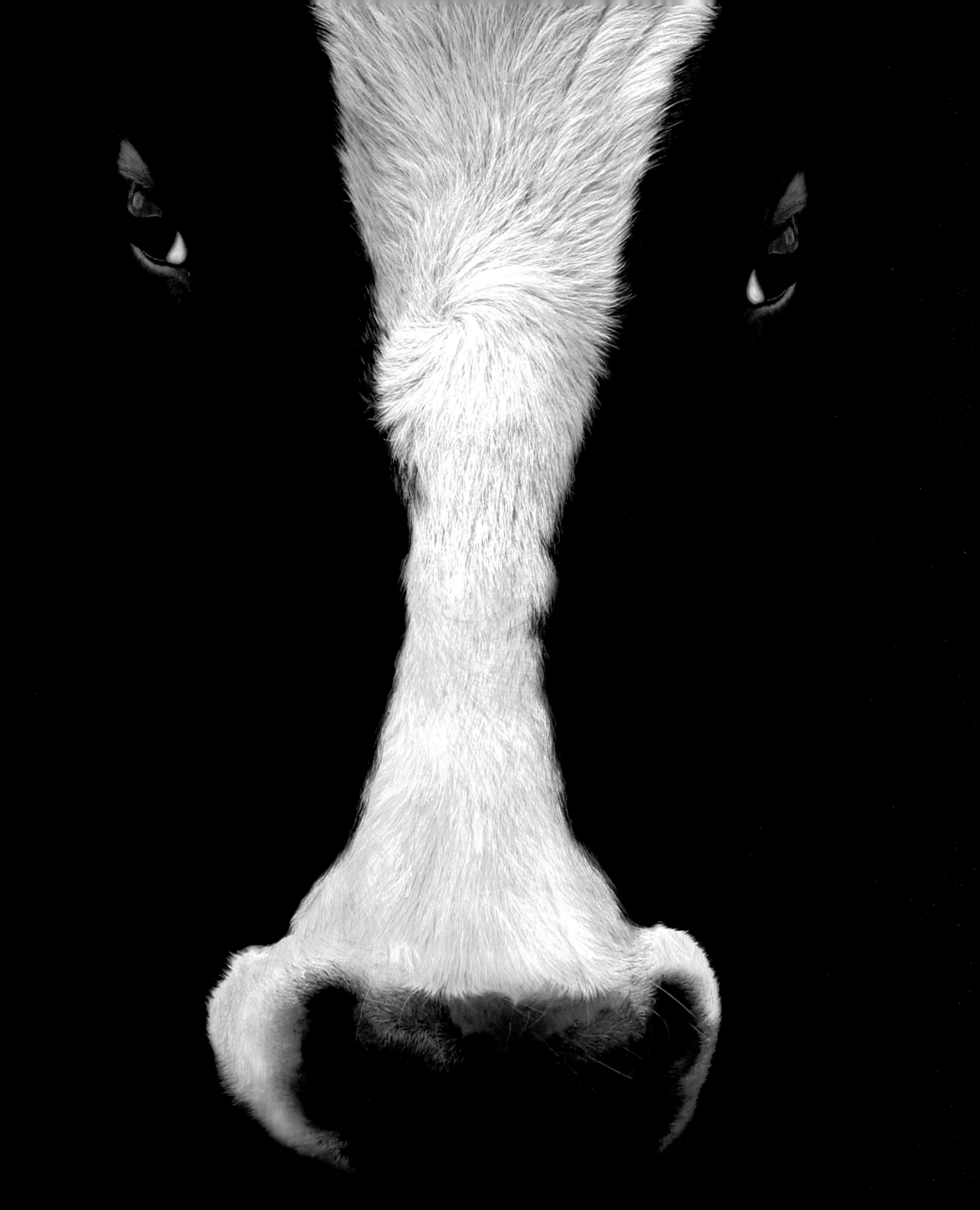

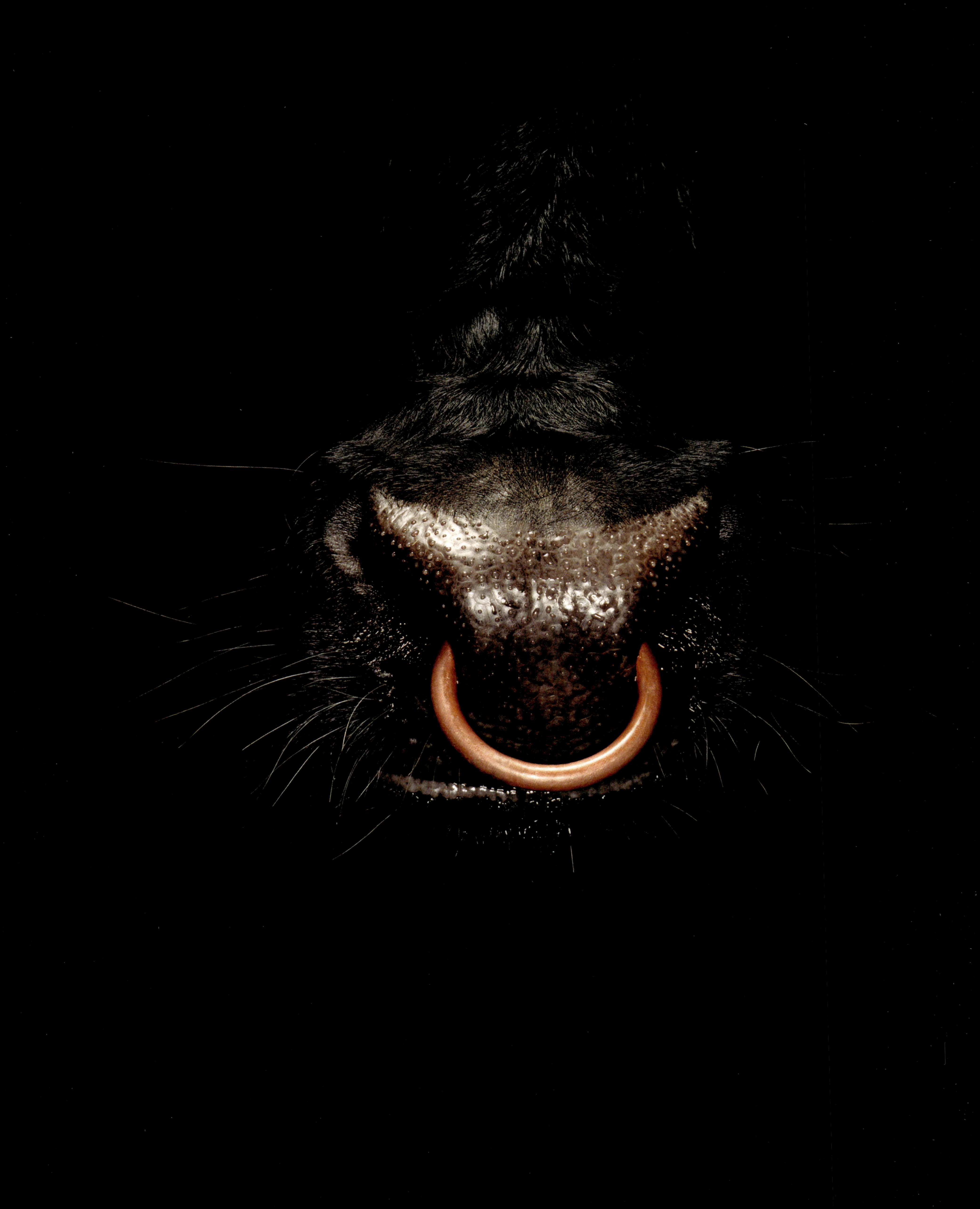

IN THE DARKNESS OF THE LABYRINTH

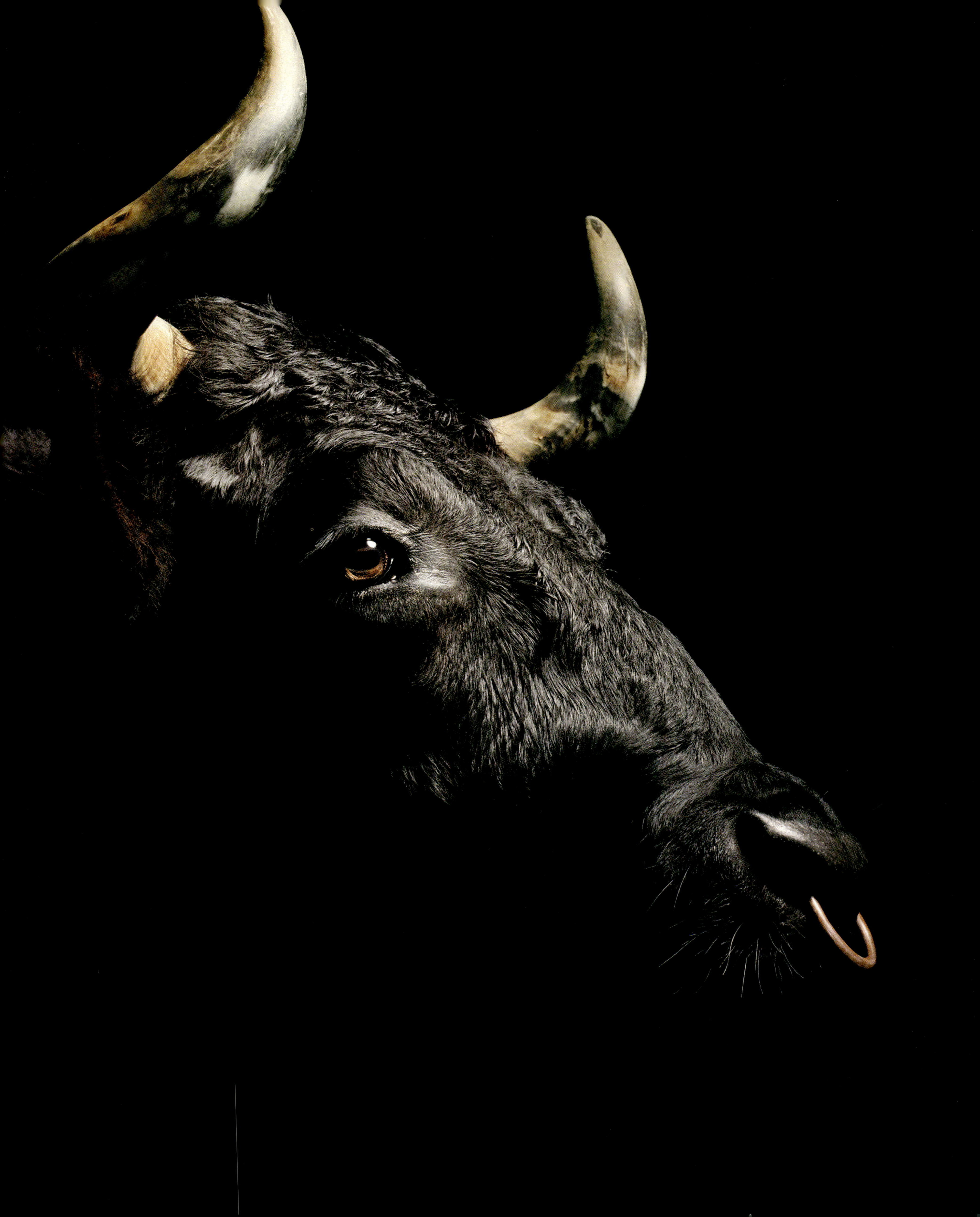

“Don’t take the bull by the horns, take him by the tail;
then you can let go when you want to.”

(Josh Billings)

STABLE FRIENDS

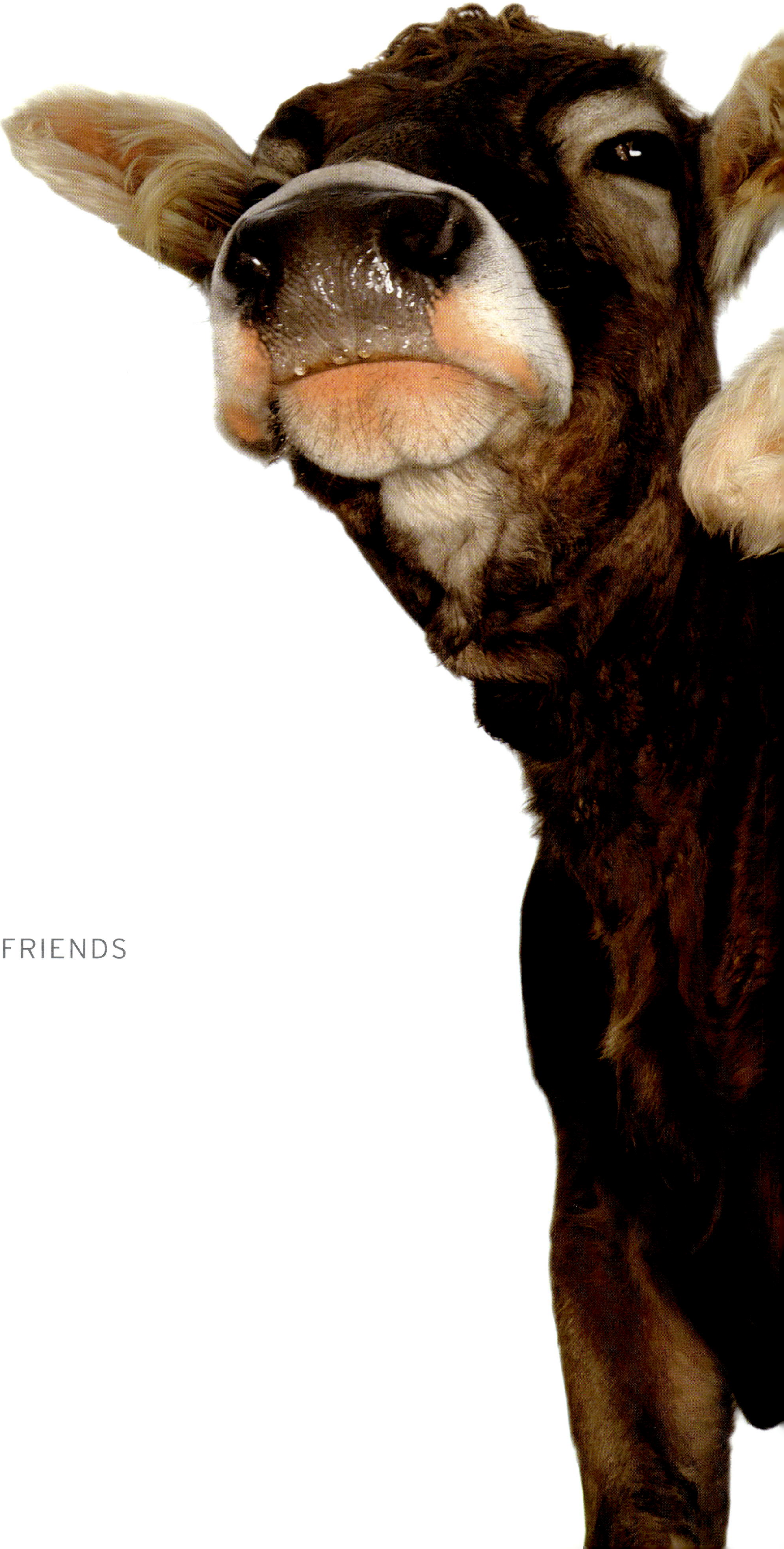

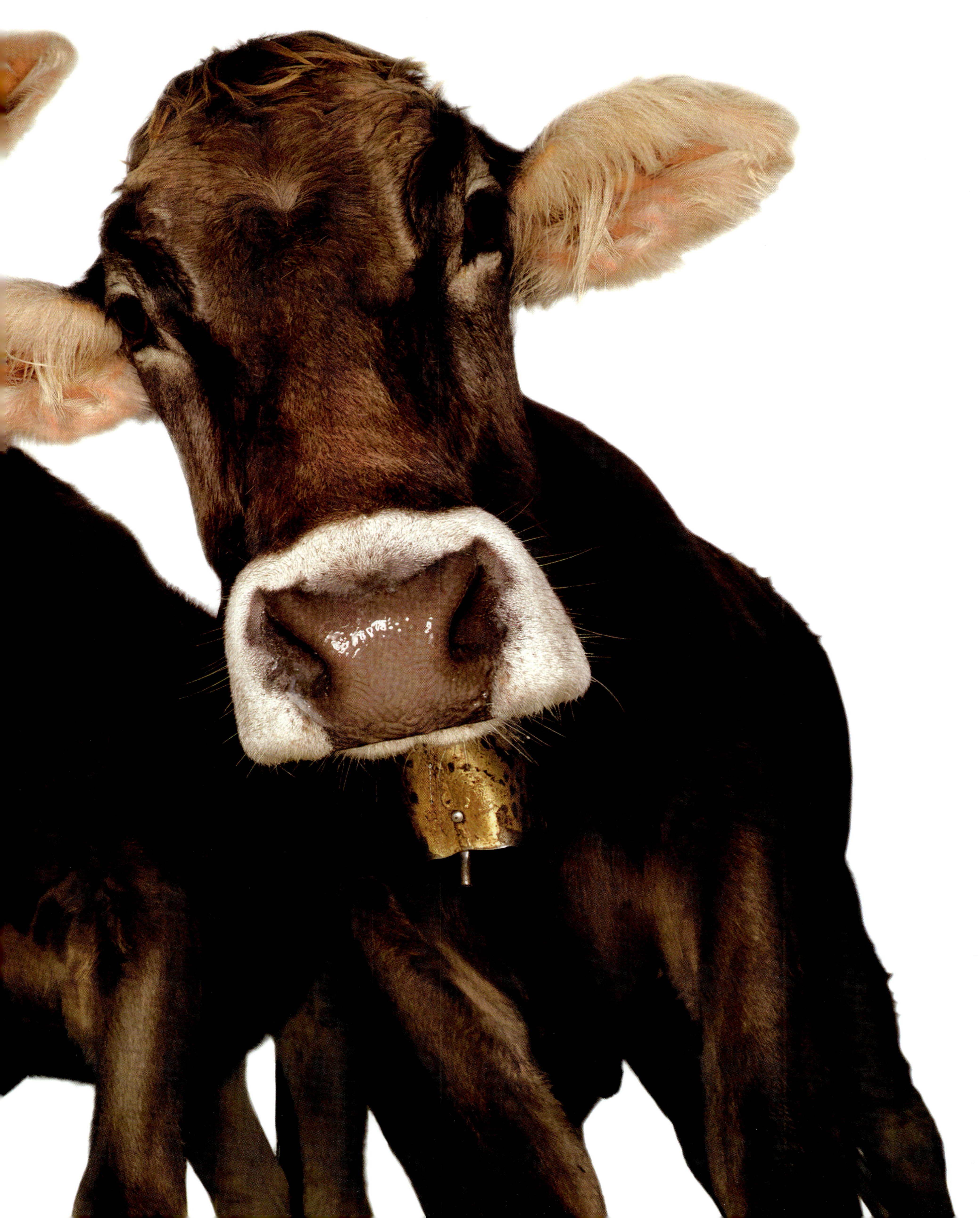

“COW PARADE”

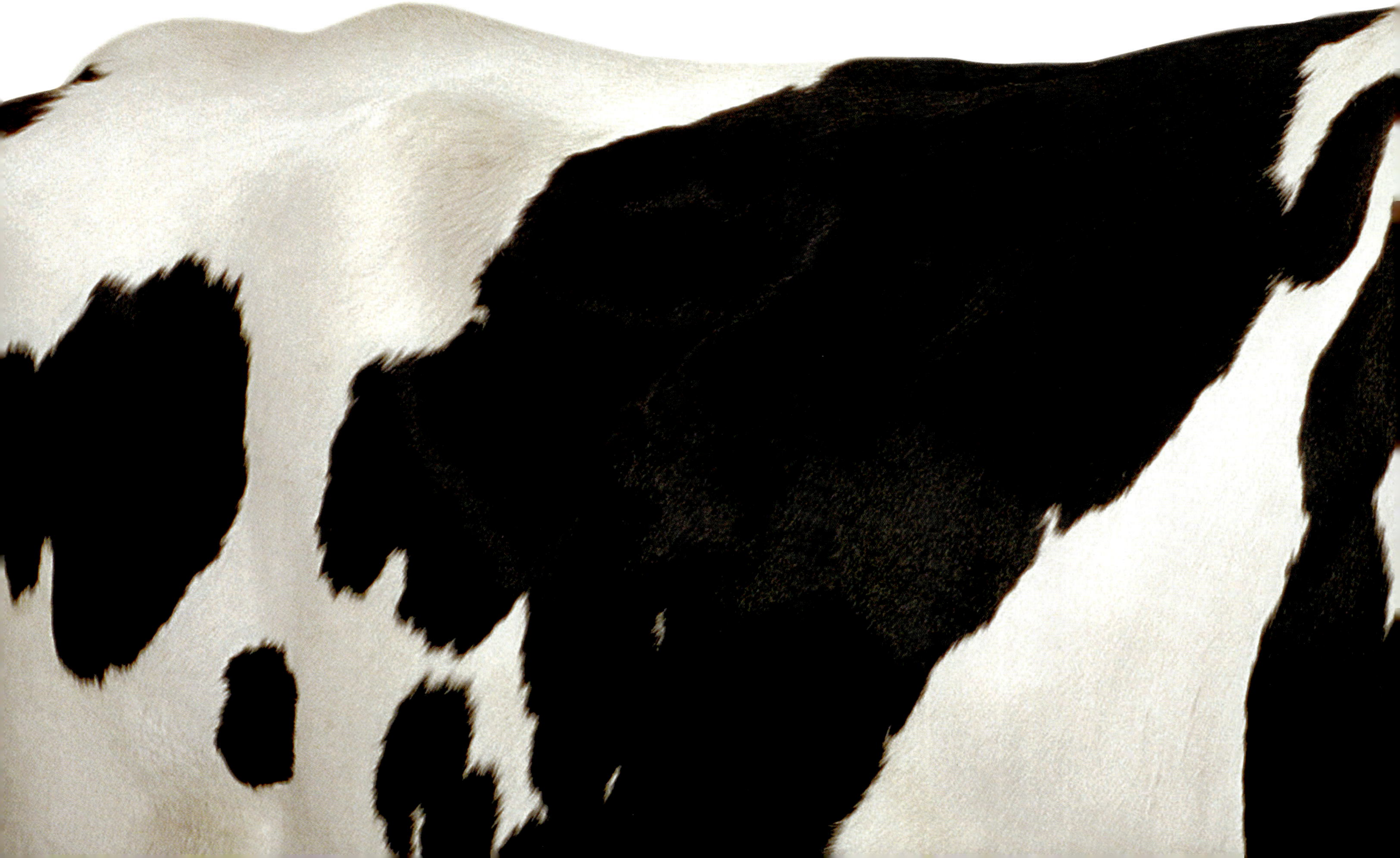

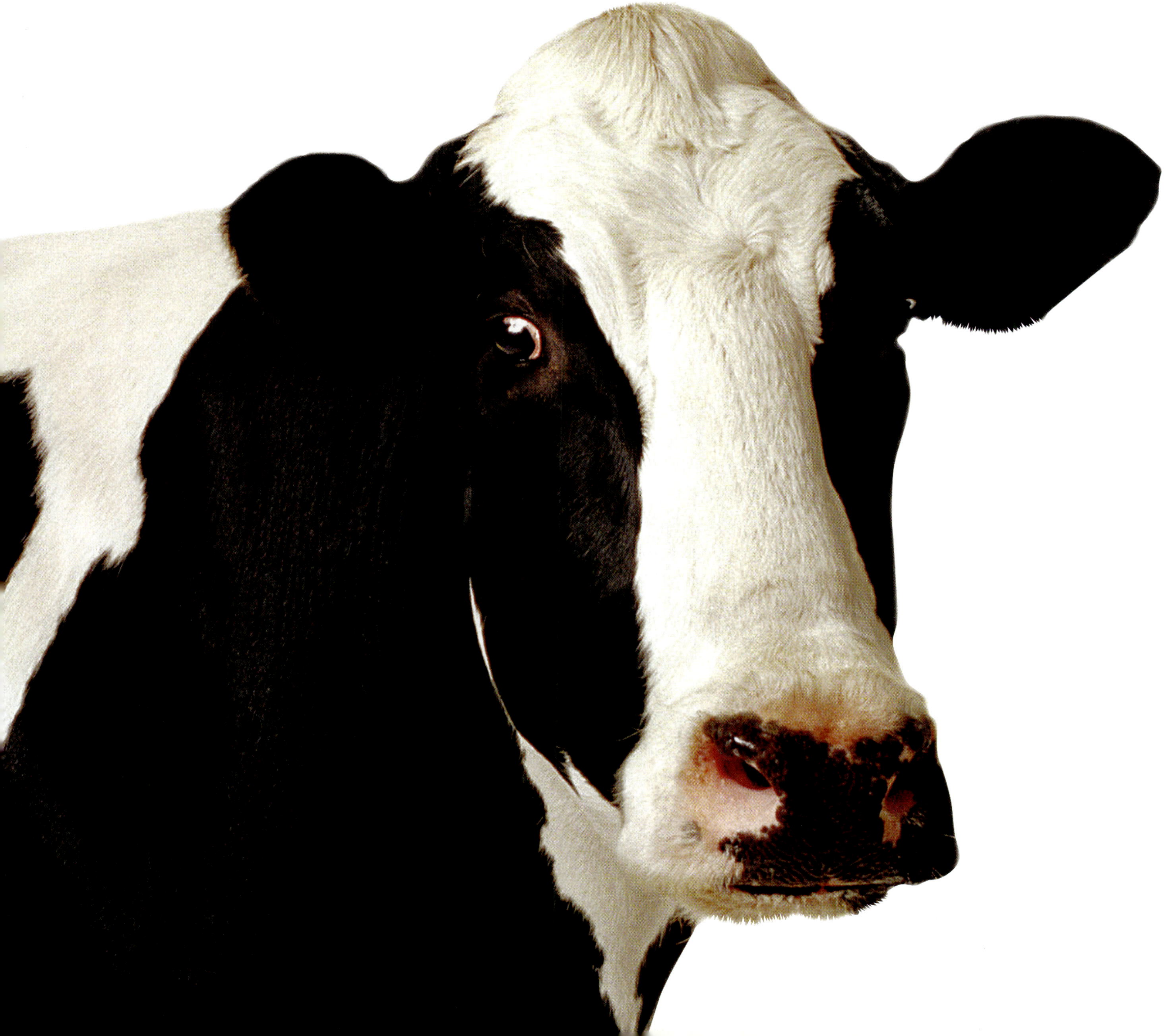

YOUTHFUL IRREVERENCE

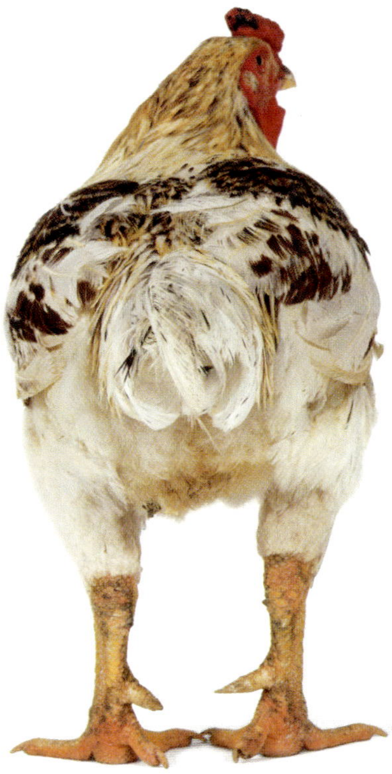

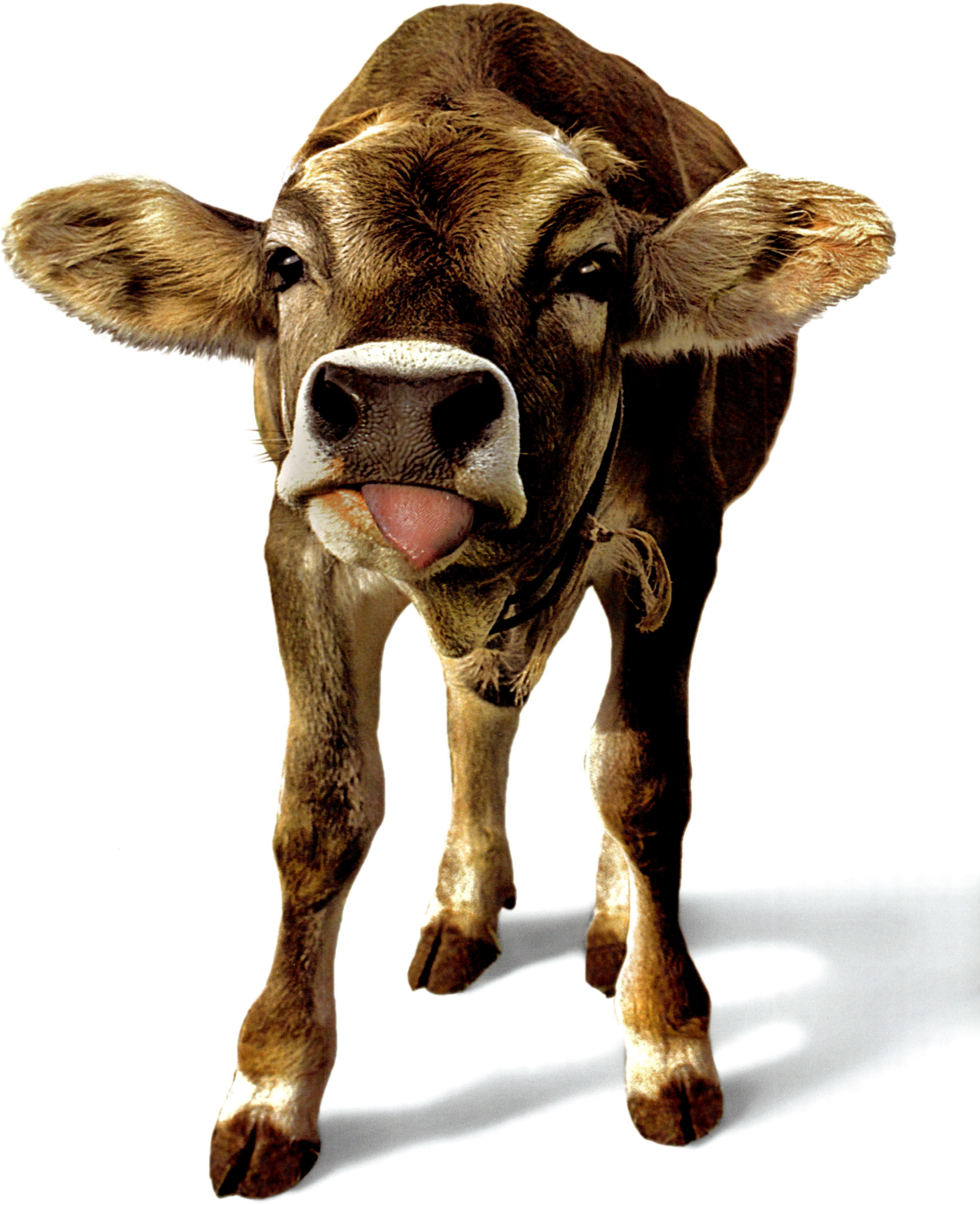

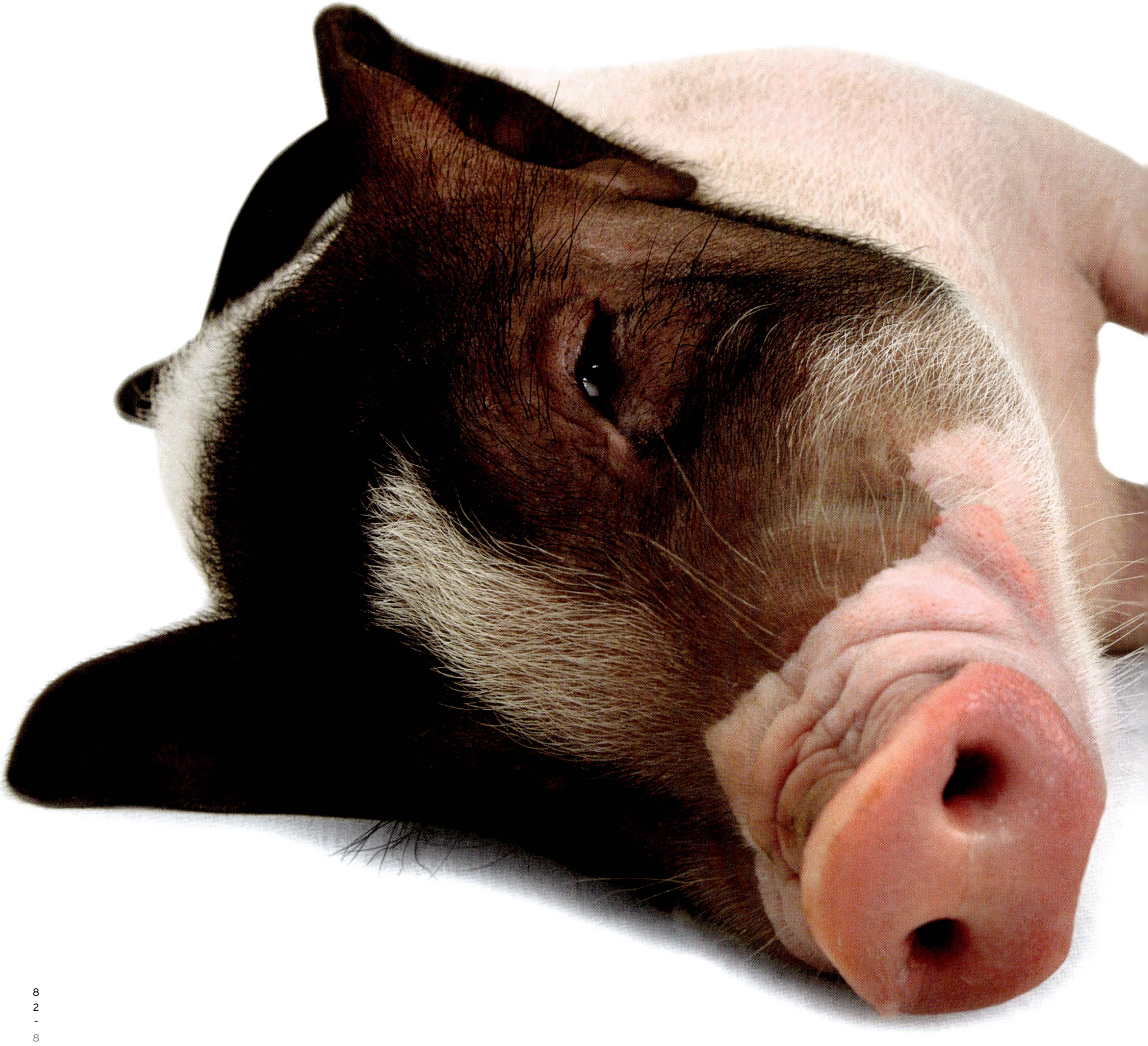

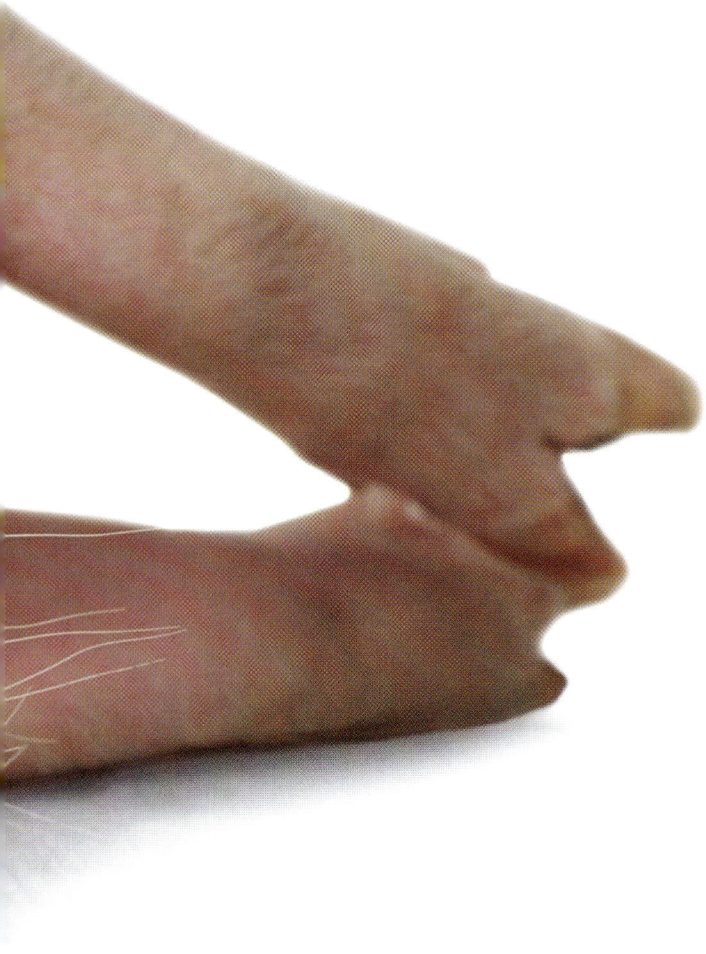

"If you want your dreams to come true,
don't sleep."

(Yiddish proverb)

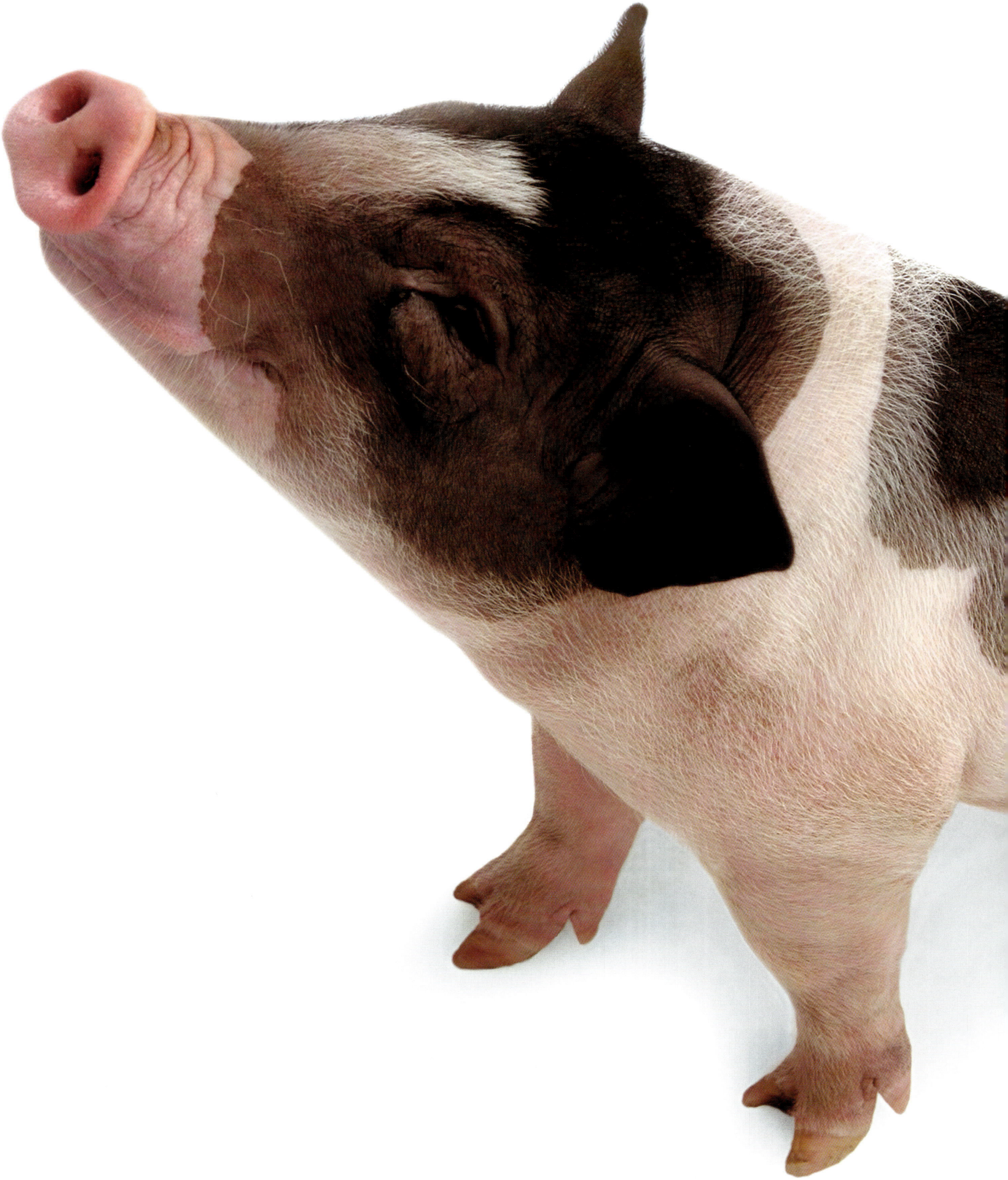

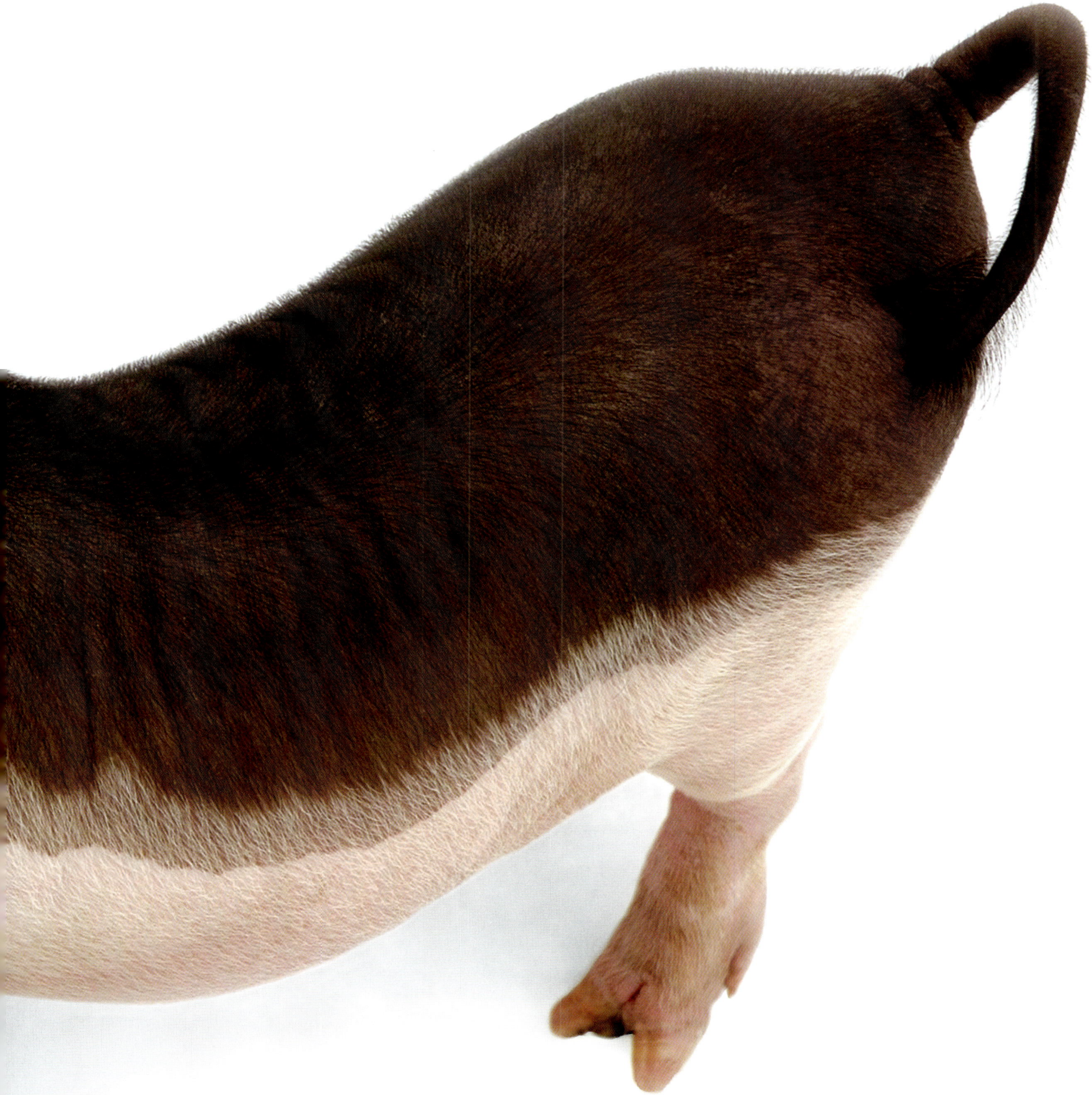

“One has a nose. The nose scents and it chooses.
An artist is simply a kind of pig snouting truffles.”

(Igor Stravinsky)

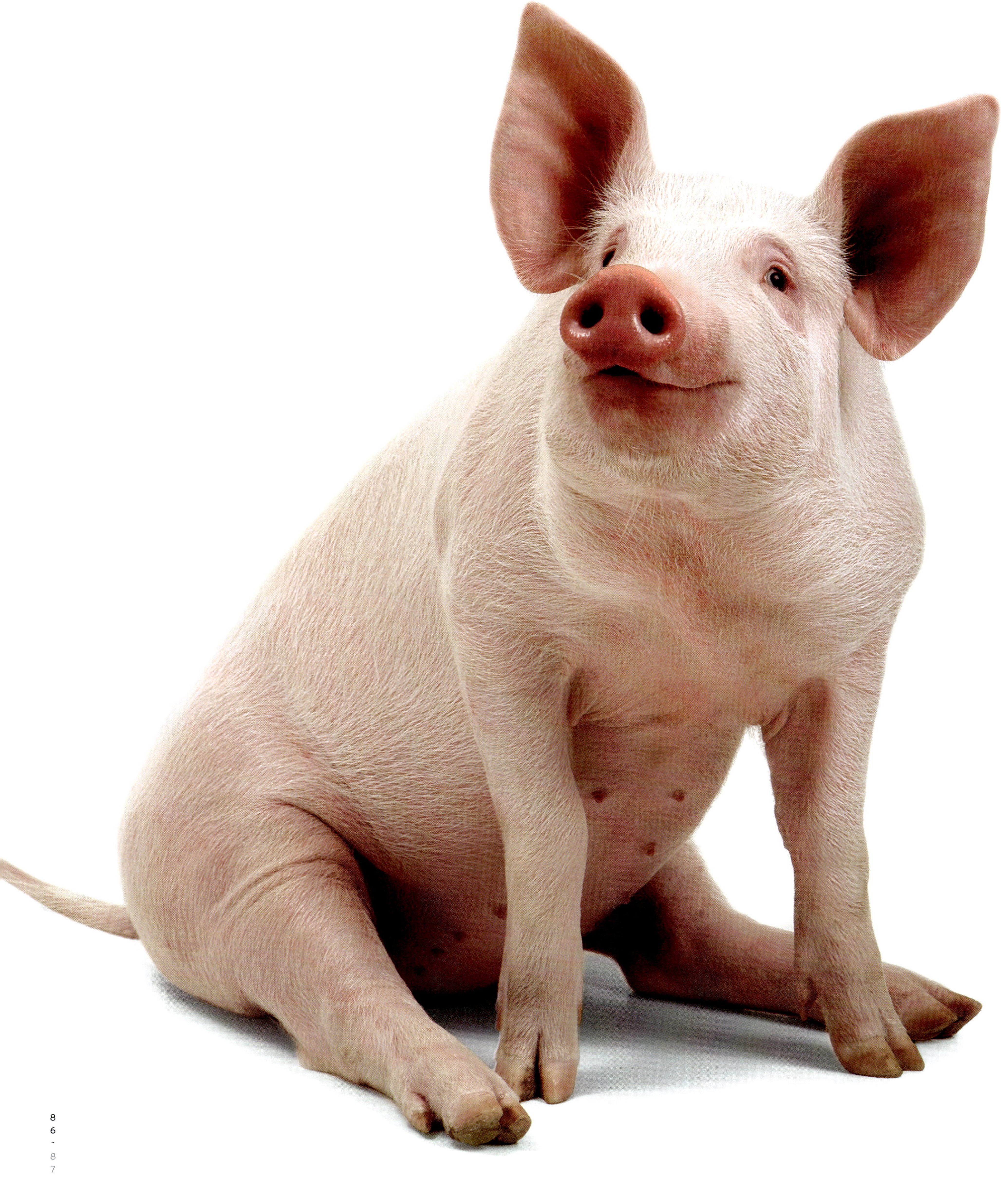

“If it looks like a pig,
sounds like a pig,
acts like pig,
don't be mistaken:
it is a pig!”

(Anonymous)

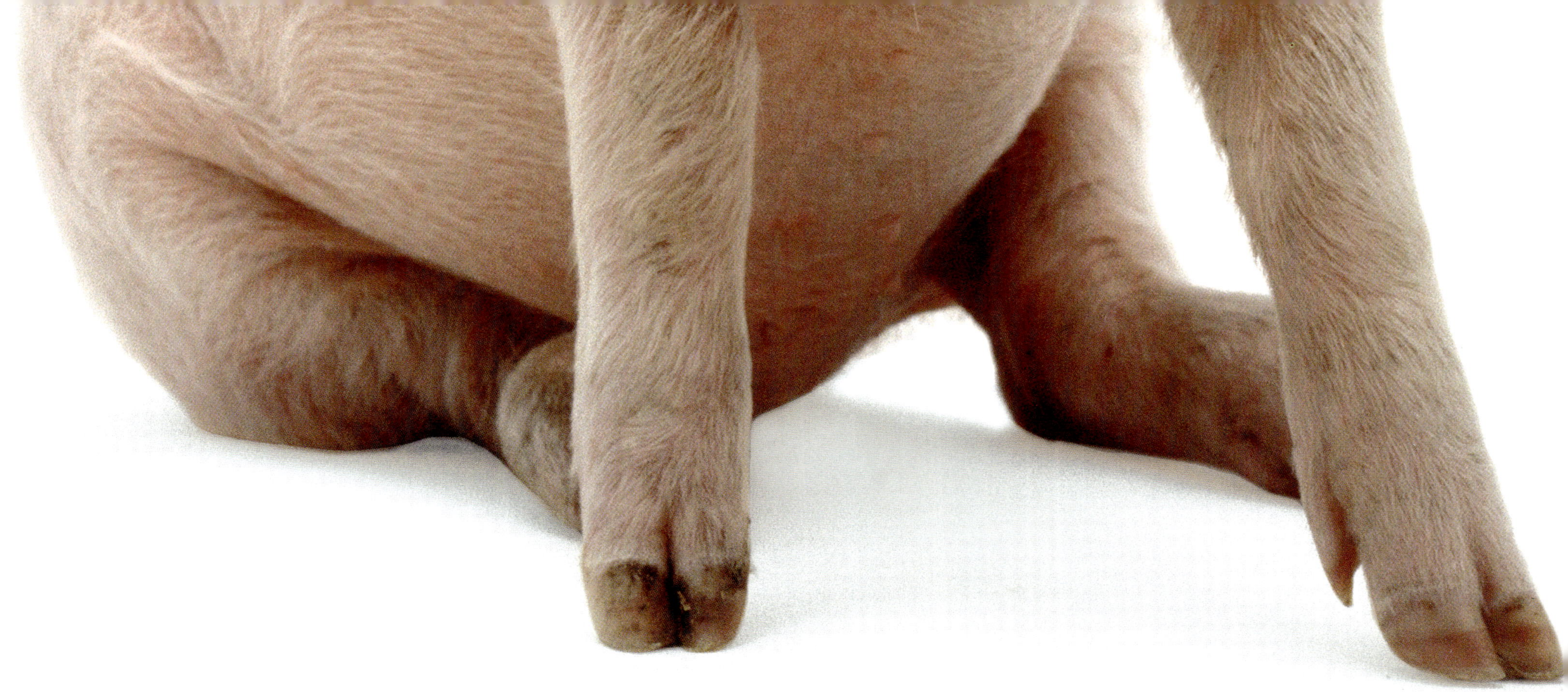

“Pigs are very beautiful animals. . . .
There is no point of view from which a really corpulent pig
is not full of sumptuous and satisfying curves.”

(G.K. Chesterton)

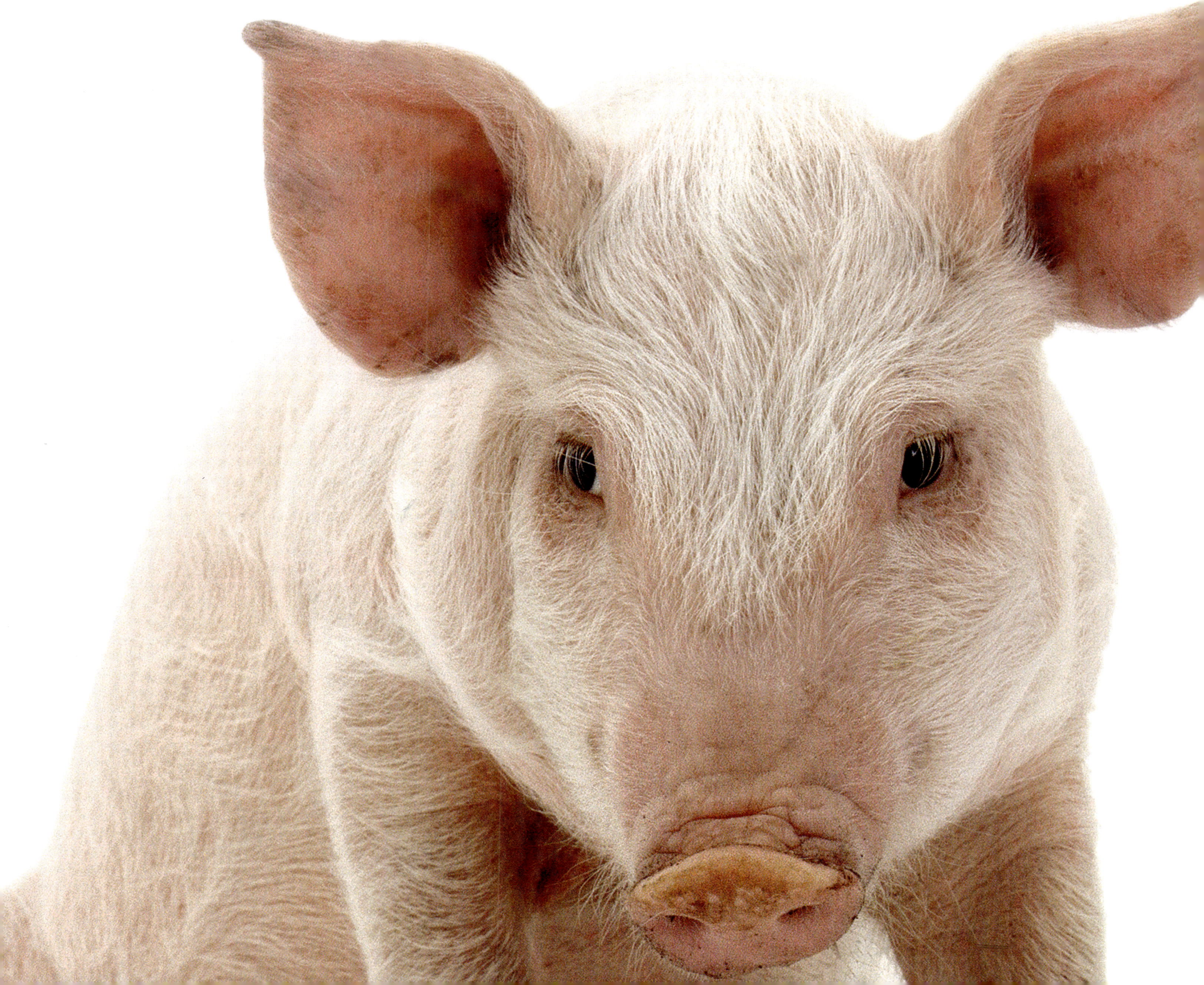

PIG TAMER

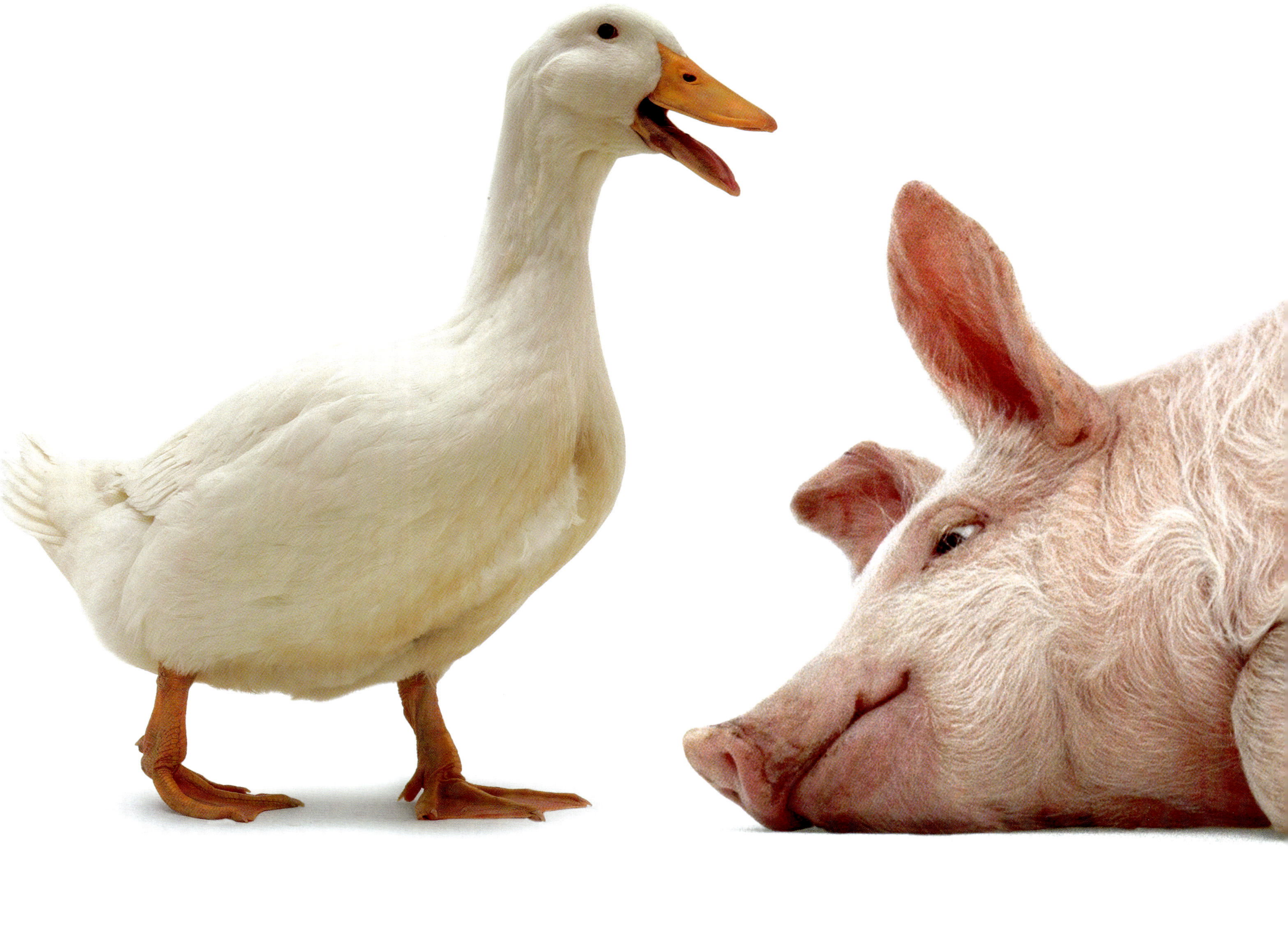

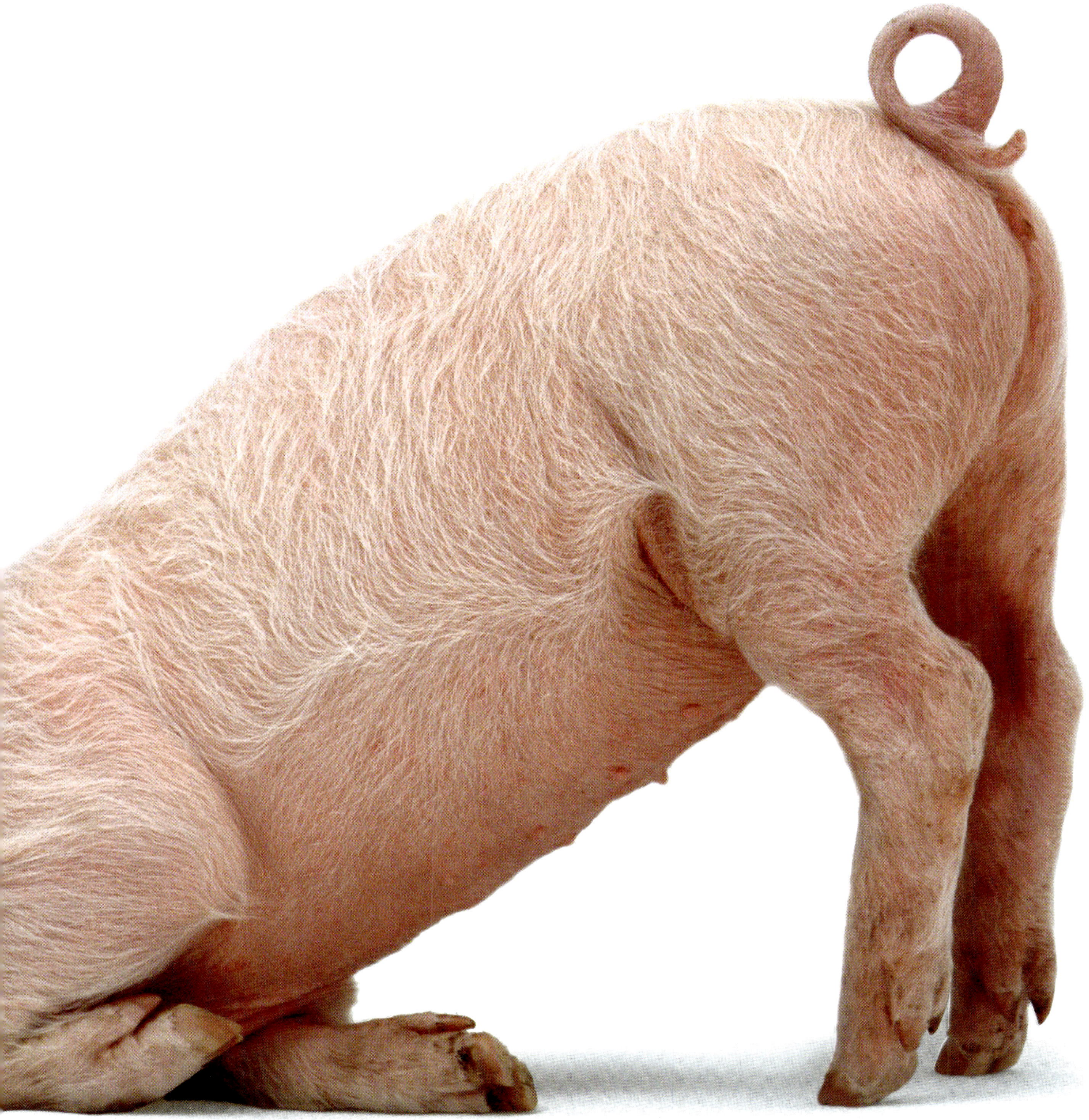

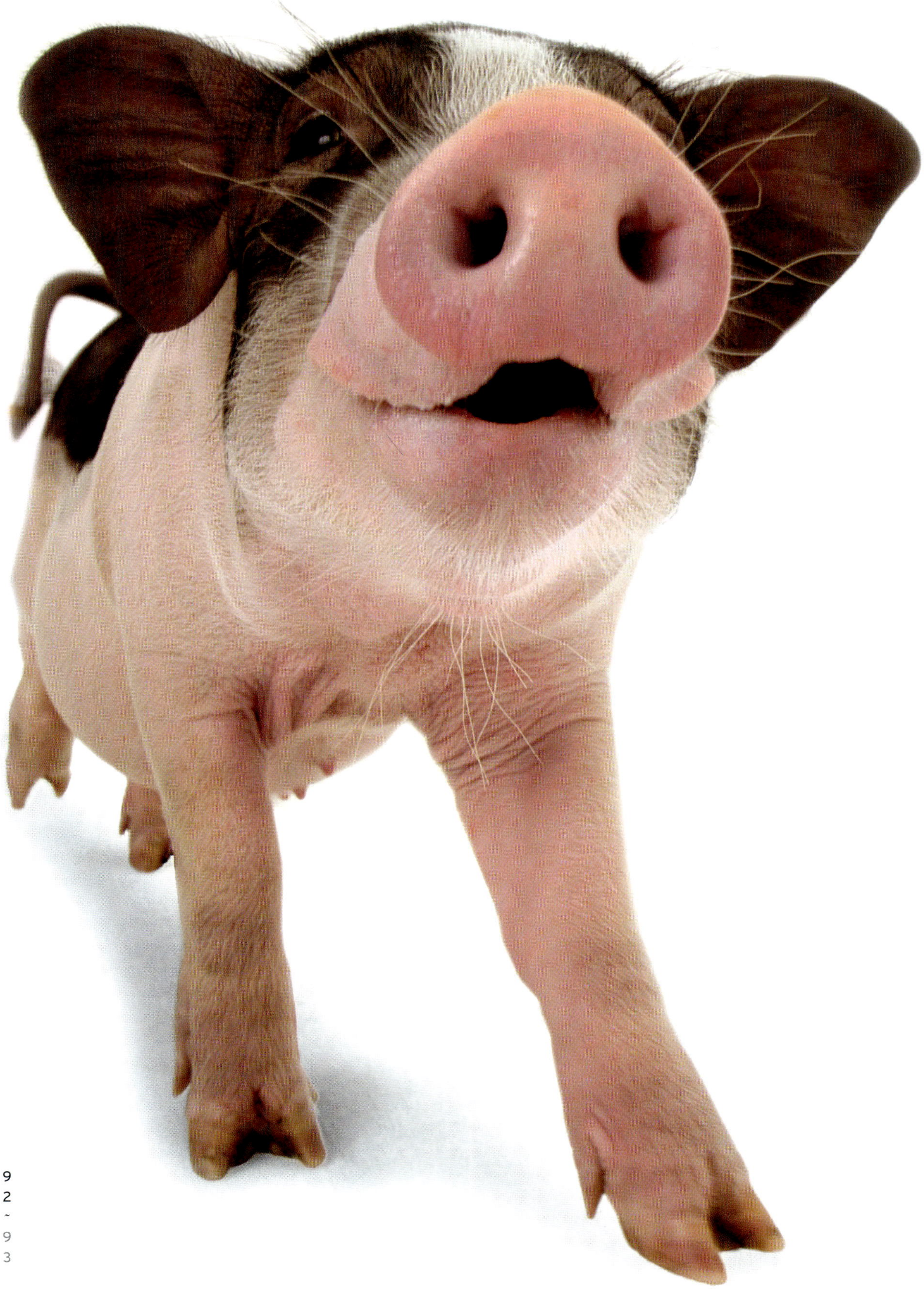

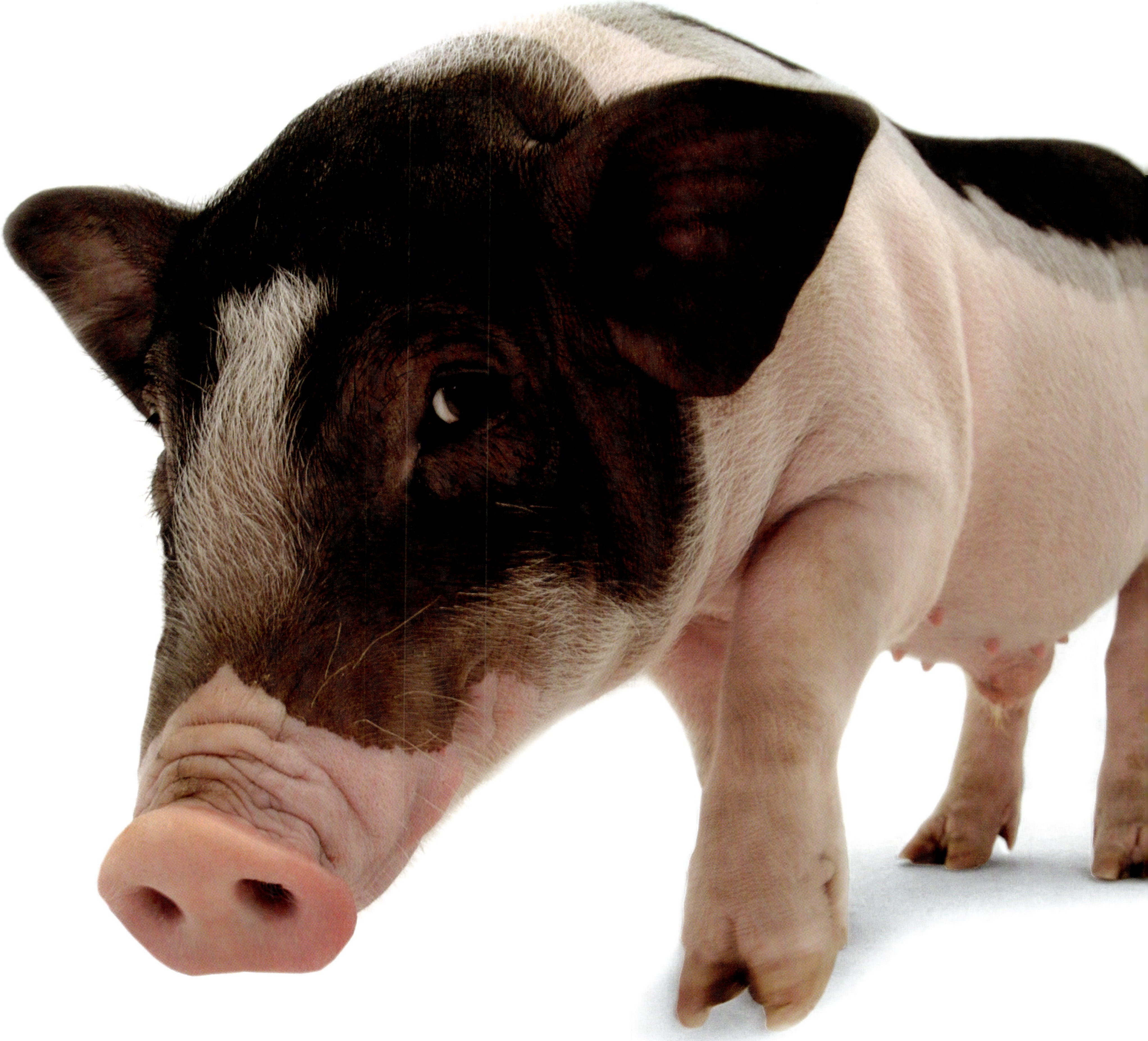

“Never try to teach a pig to sing;
it wastes your time and it annoys the pig.”
(Robert Heinlein)

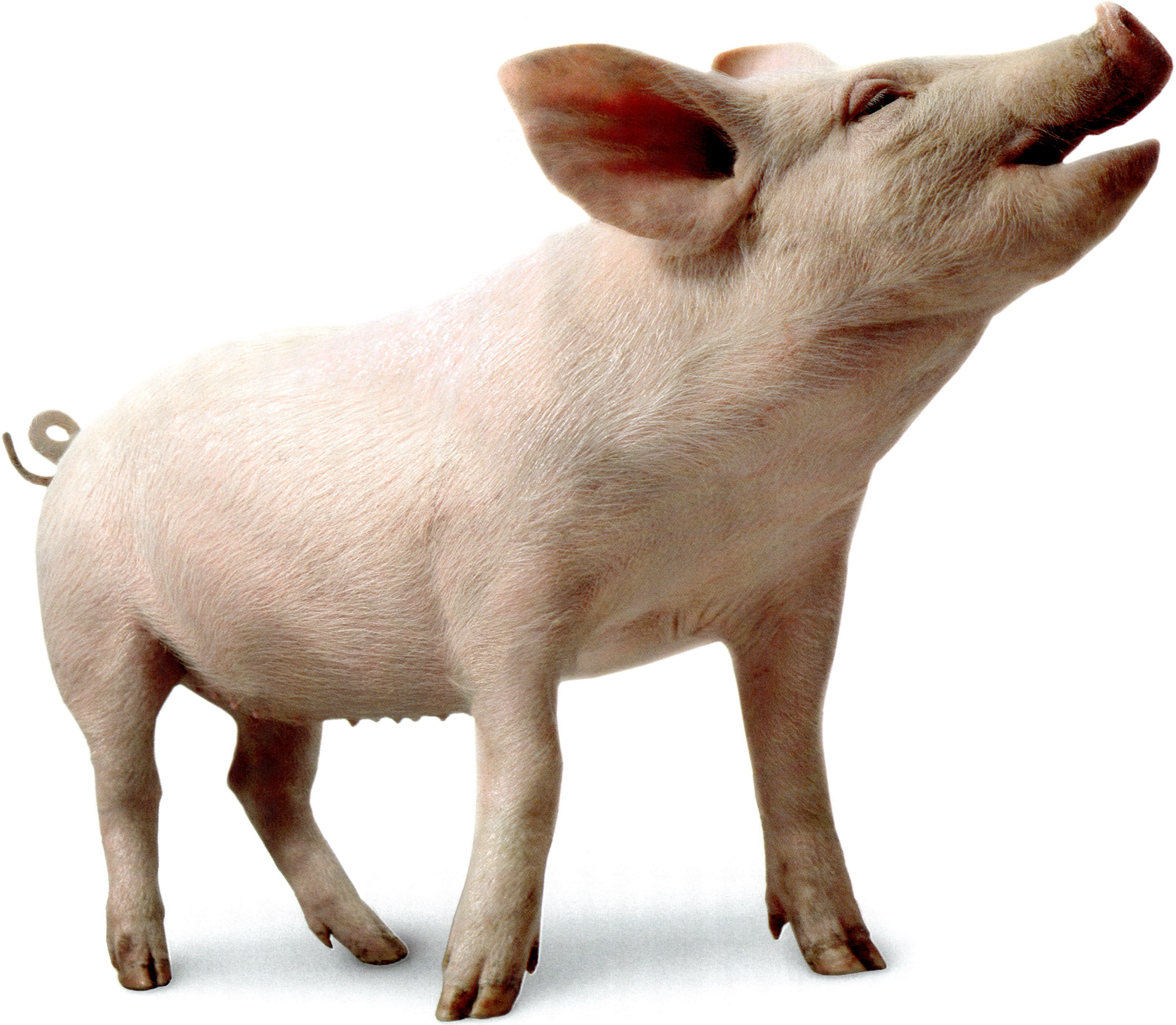

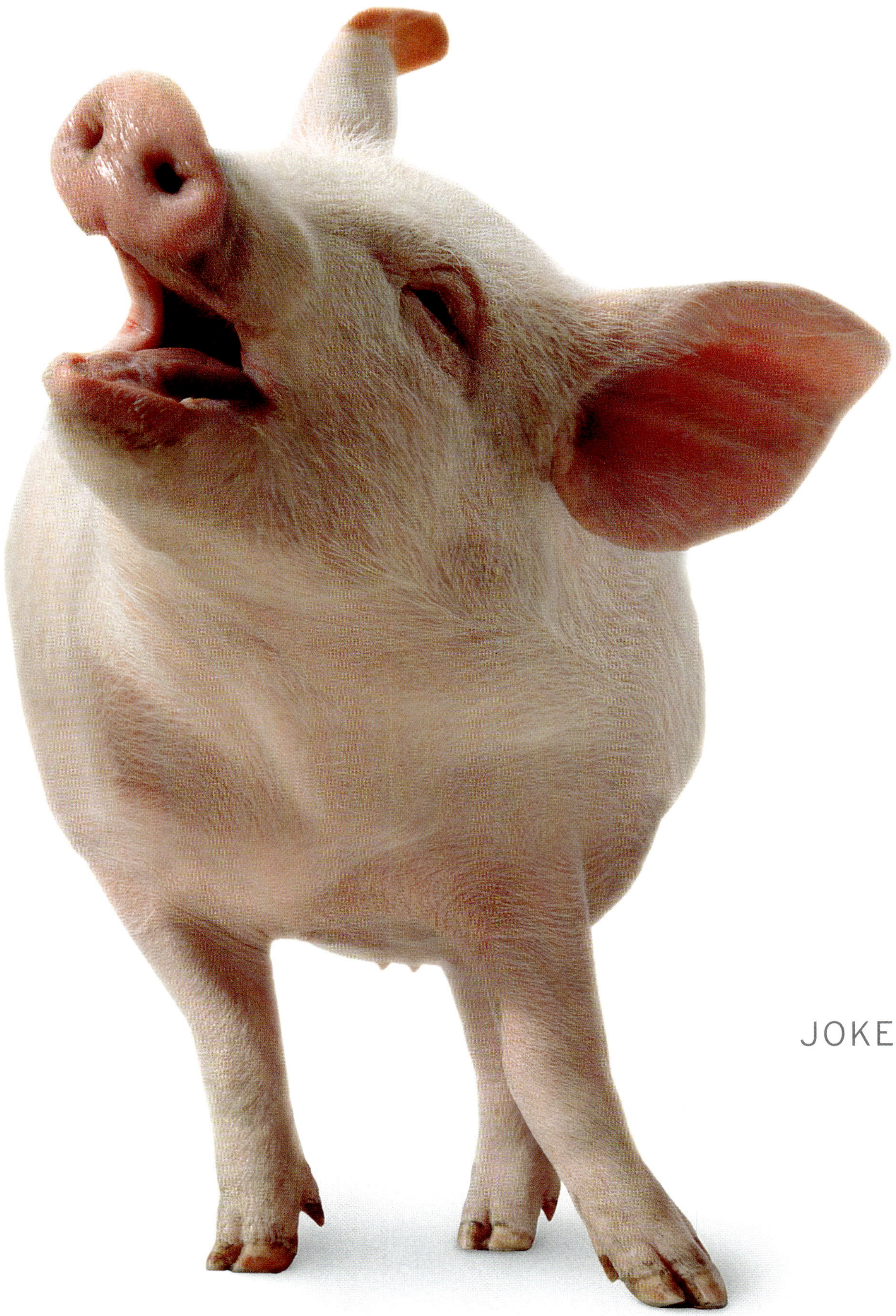

JOKES AMONG FRIENDS

“If I’m free, it’s because I’m always running.” (Jimi Hendrix)

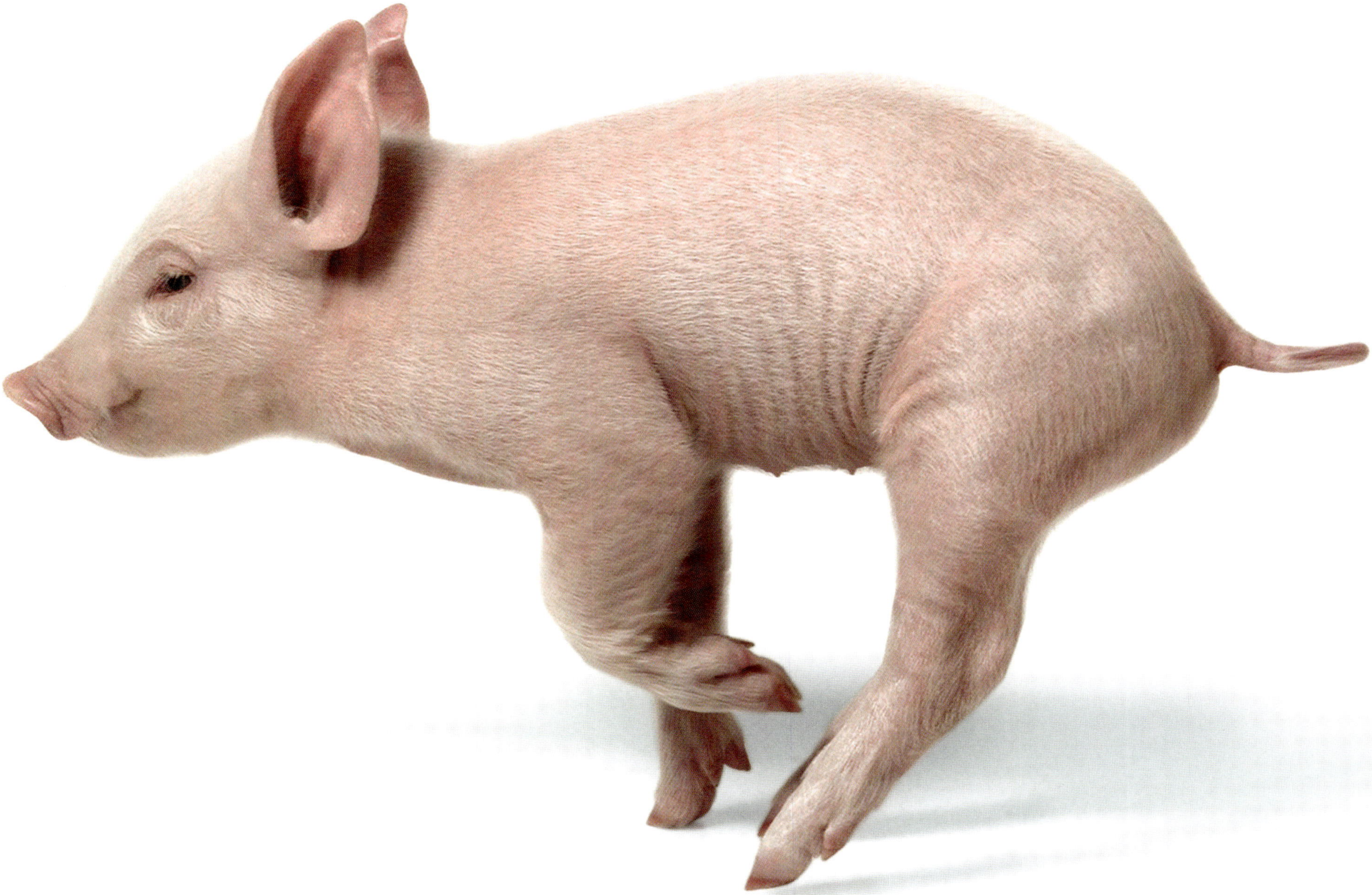

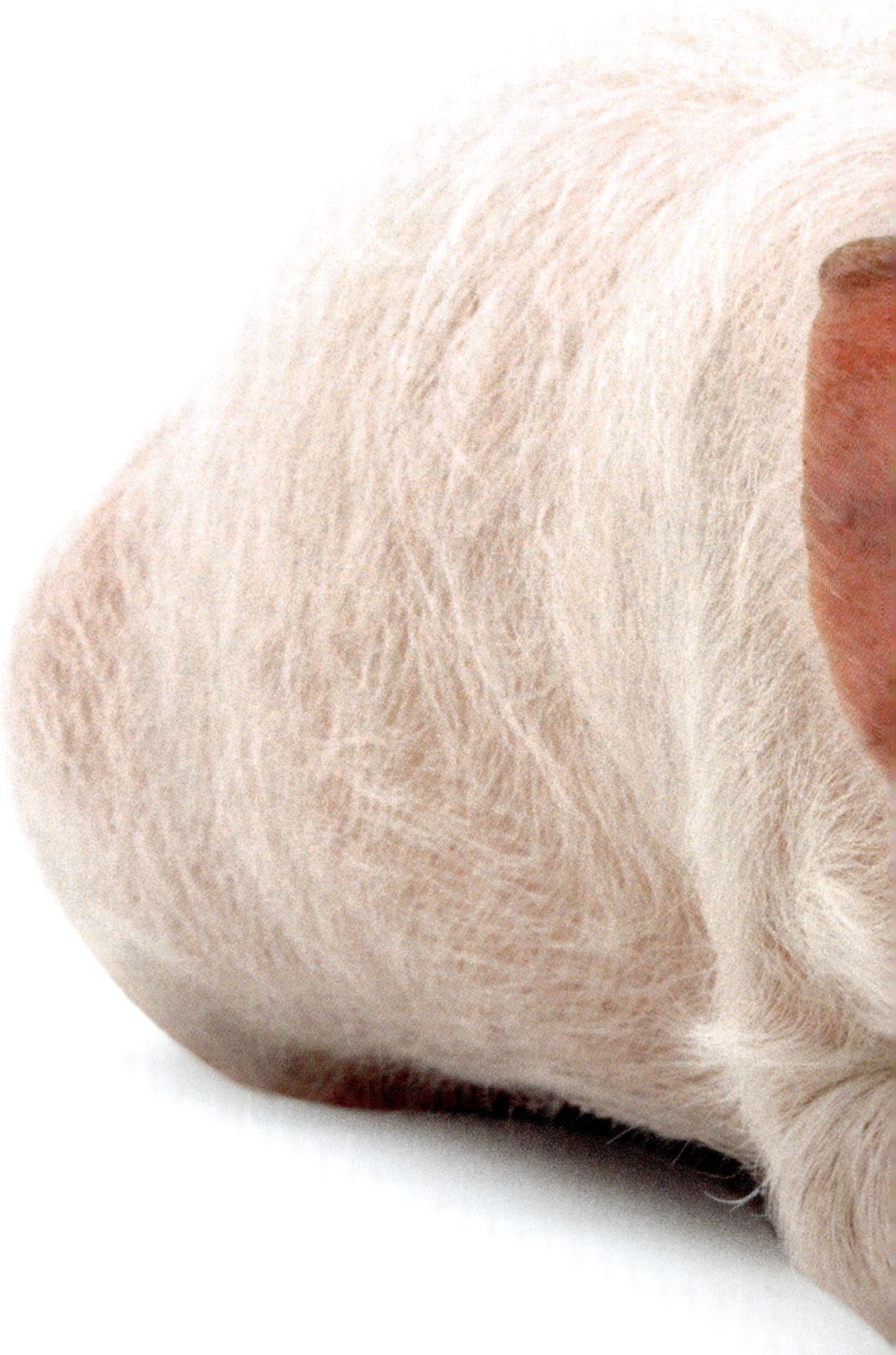

"In this world there is always danger
for those who are afraid of it."

(George Bernard Shaw)

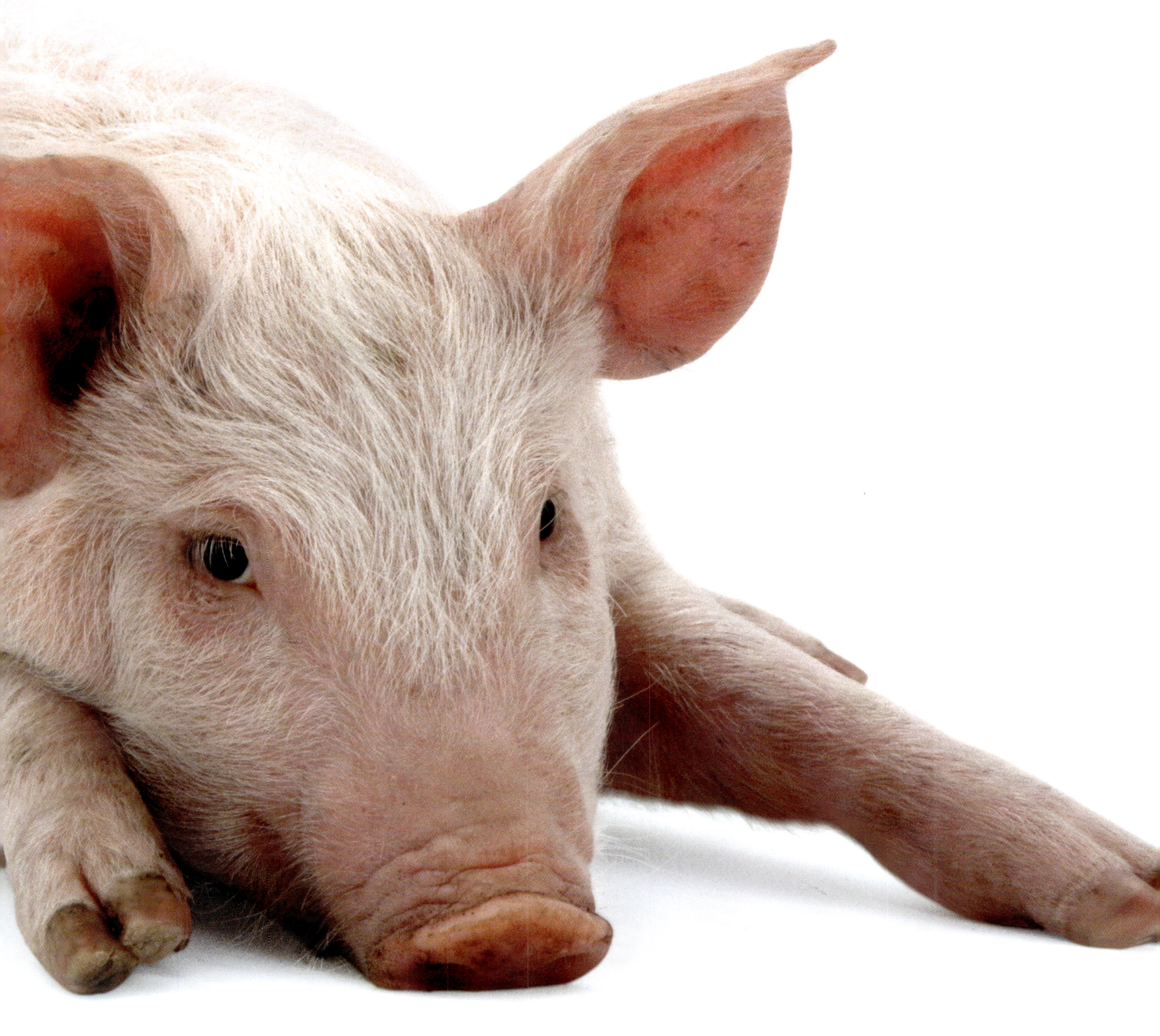

INCOMPATIBLE DIFFERENCES

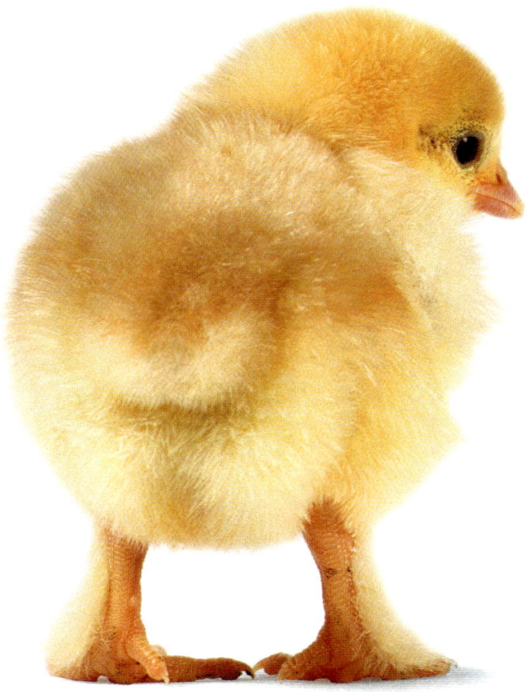

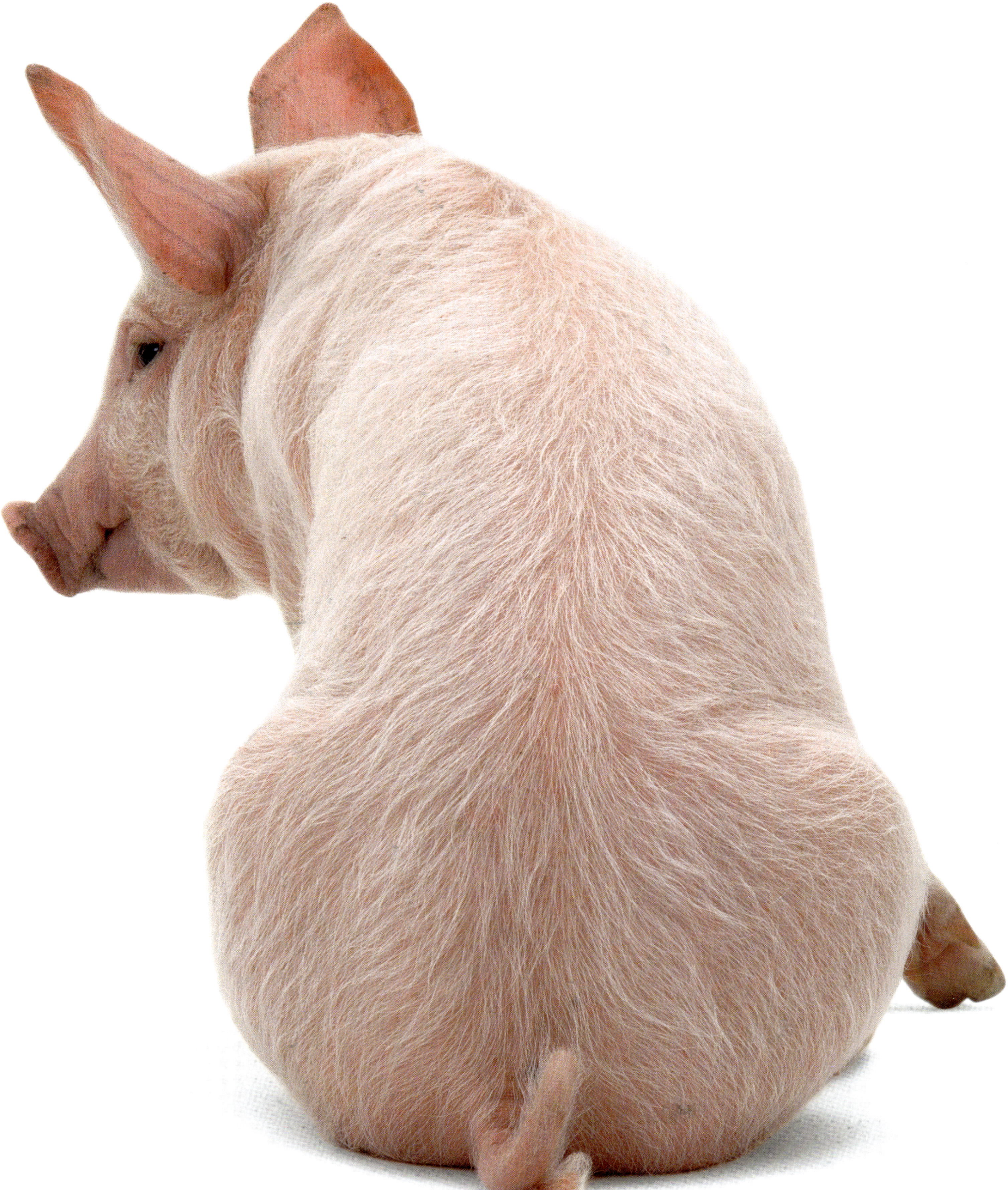

FINAL FANTASY

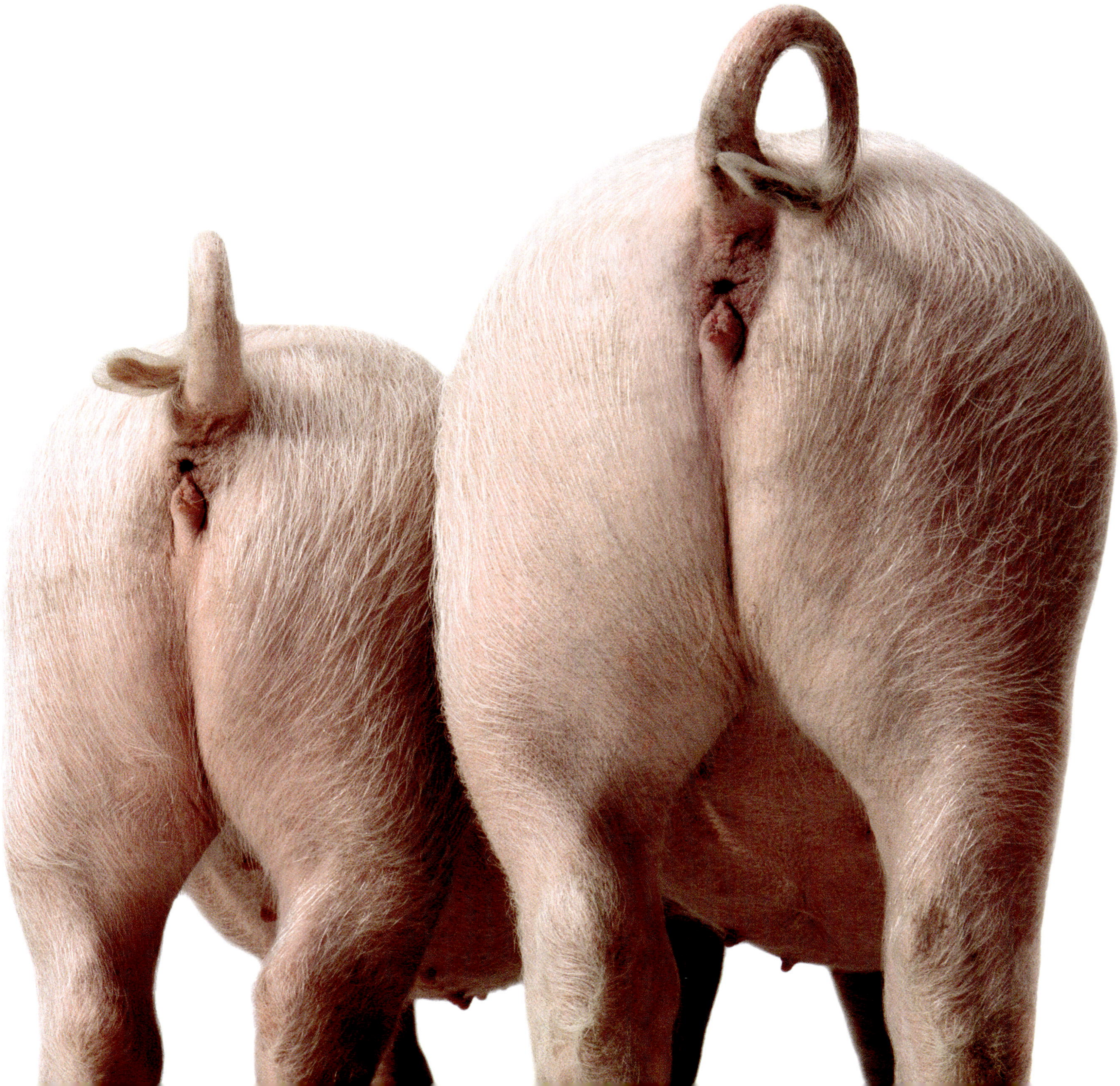

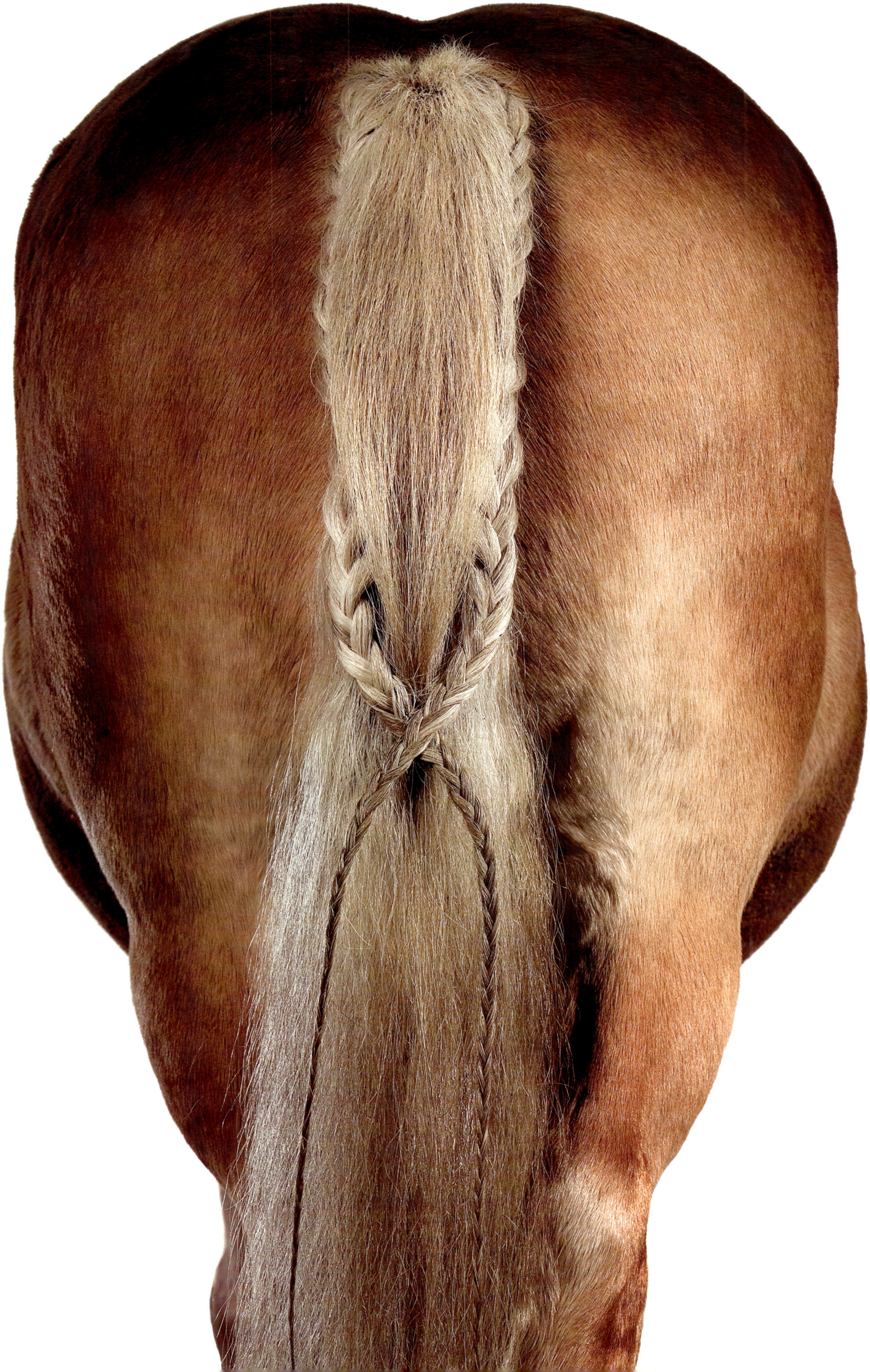

PUNK SOUL

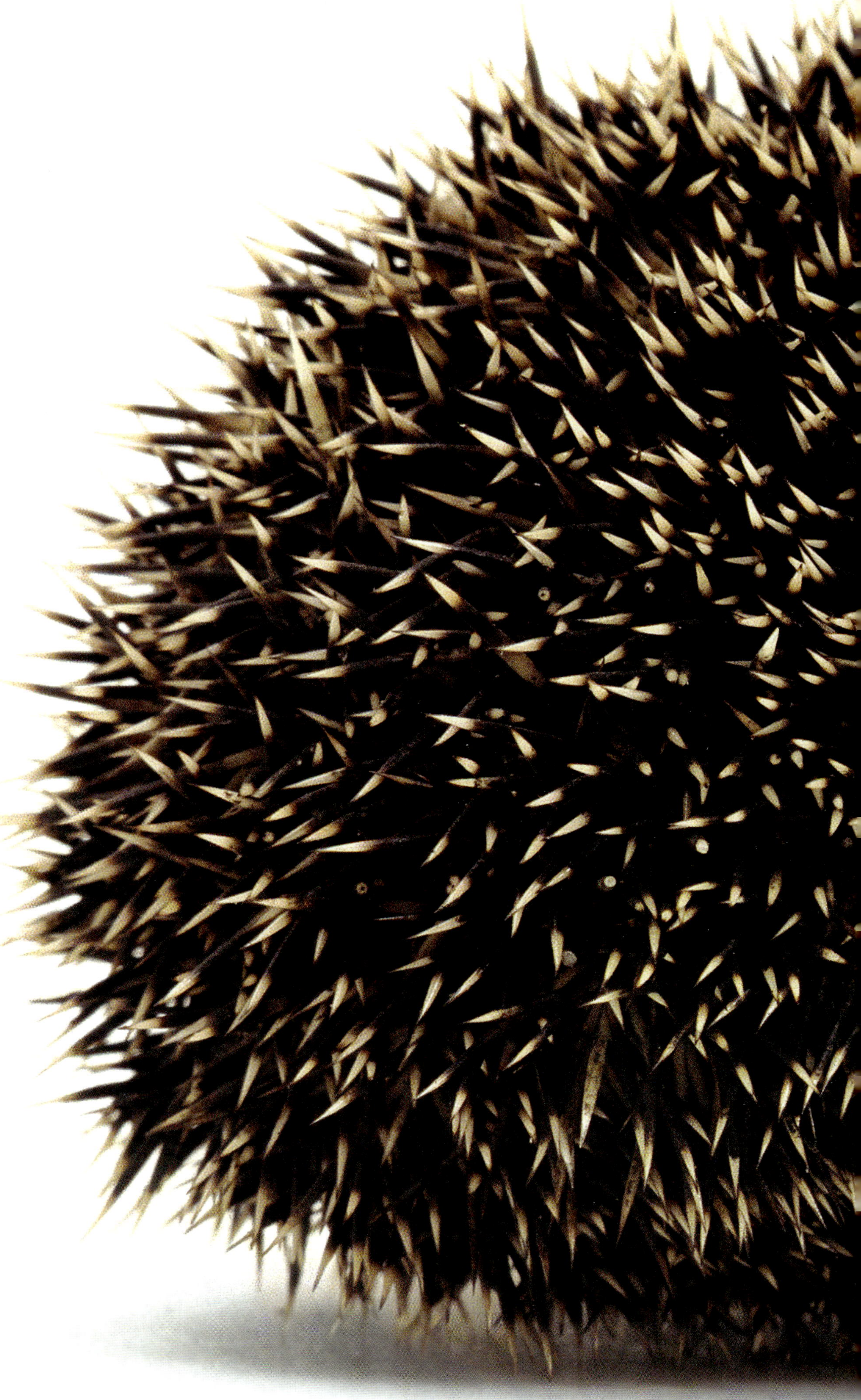

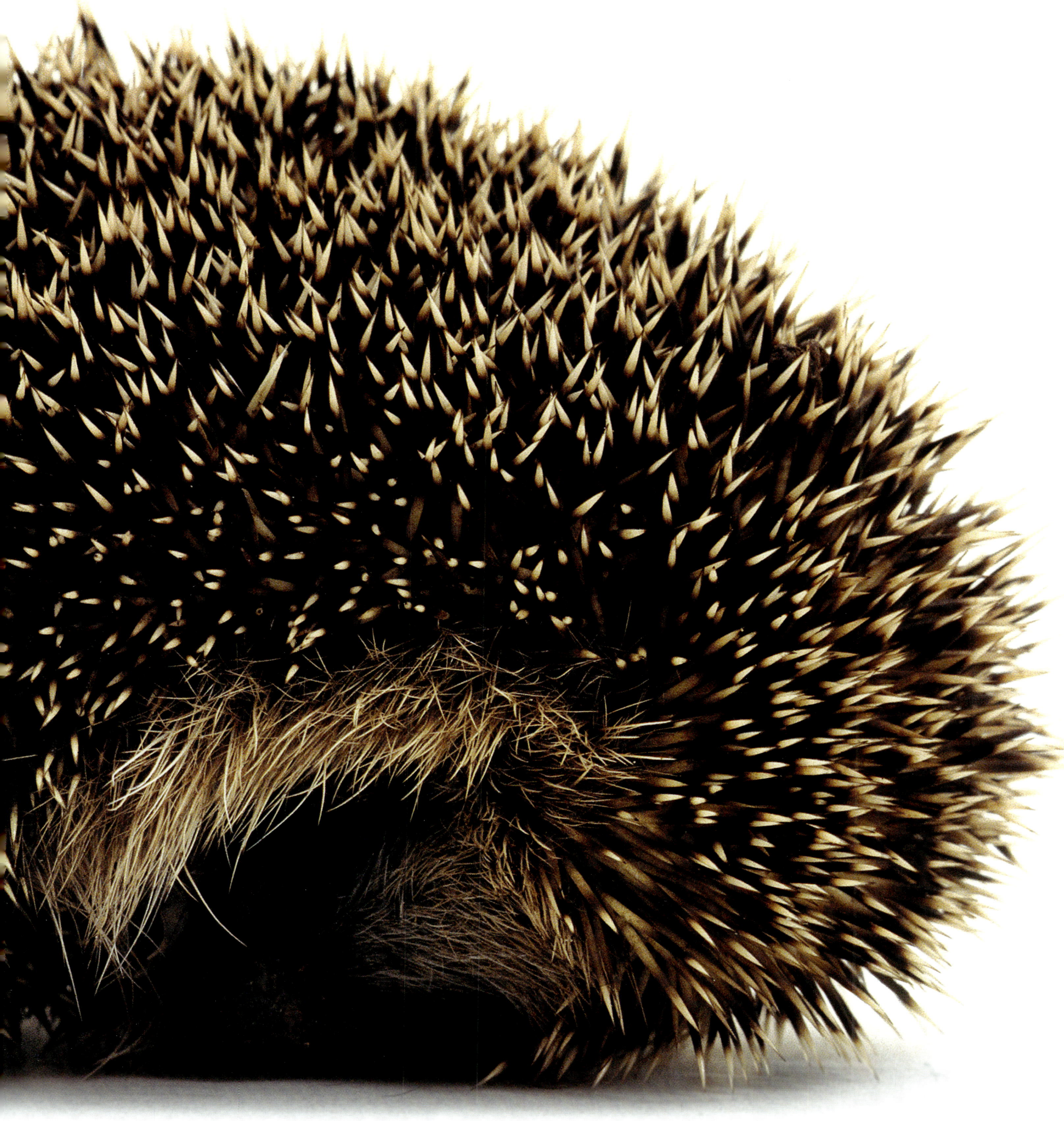

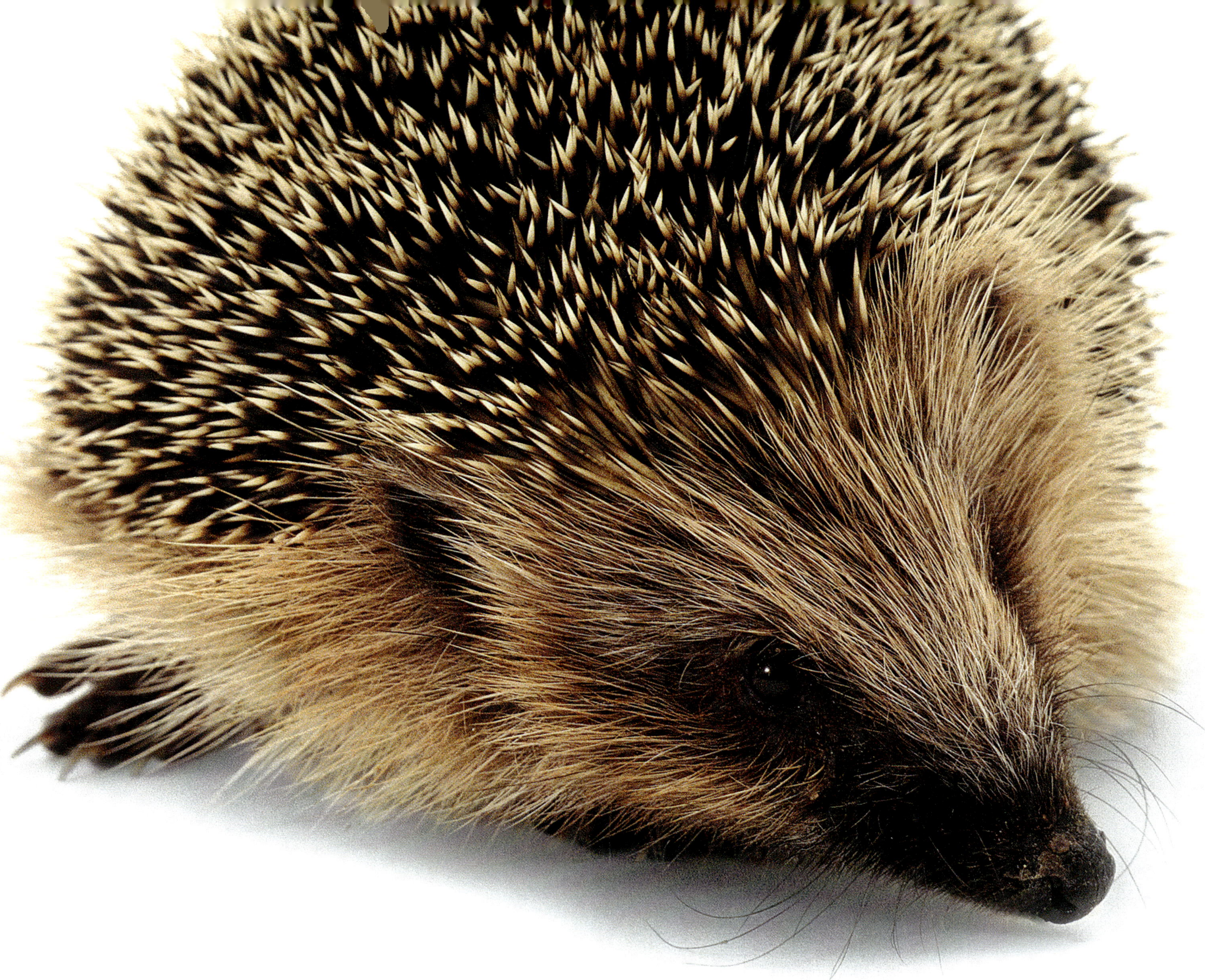

CIRCUMSPECT MOVEMENTS

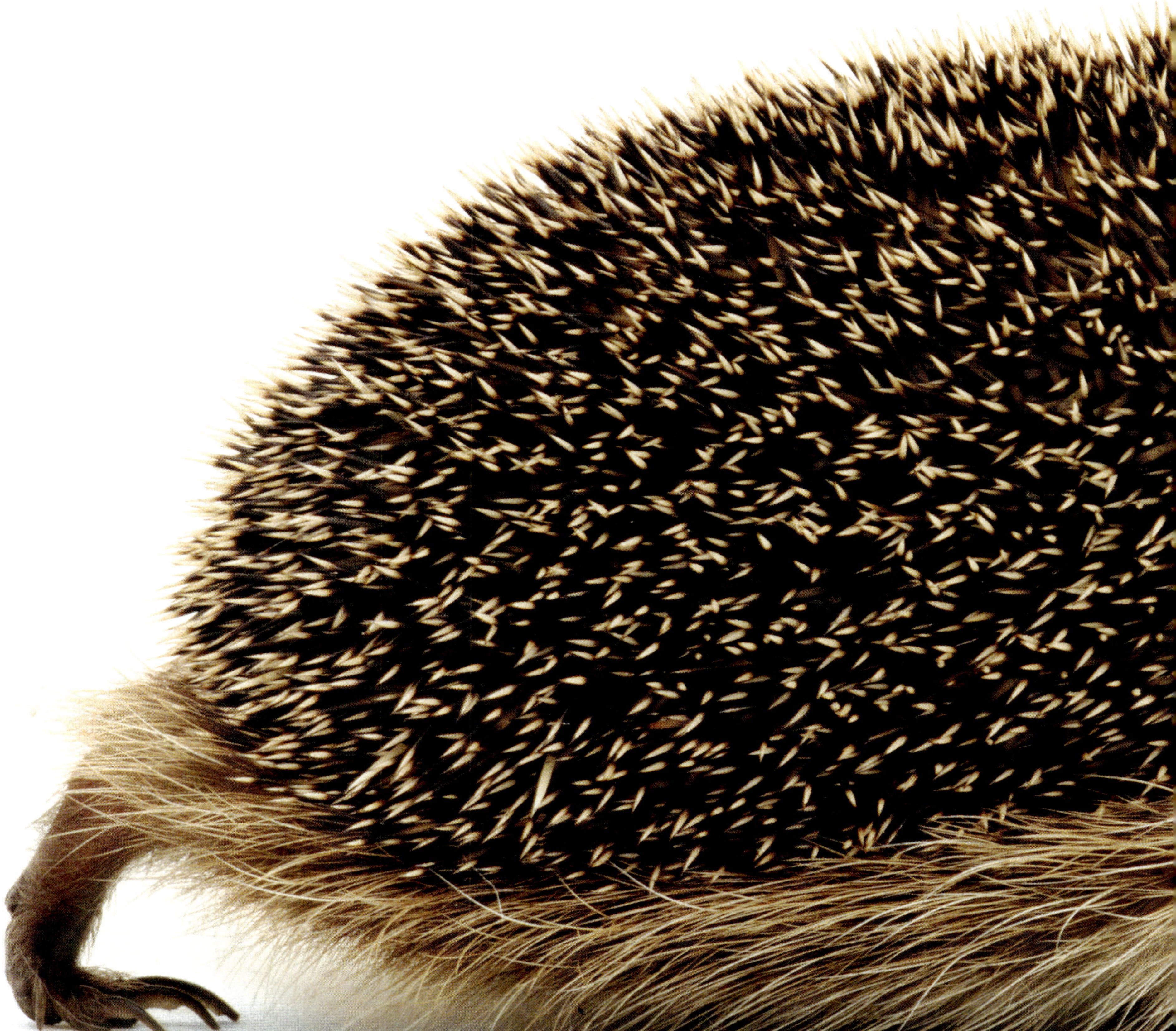

“Don’t be too timid and squeamish about your actions.
All life is an experiment.”

(Ralph Waldo Emerson)

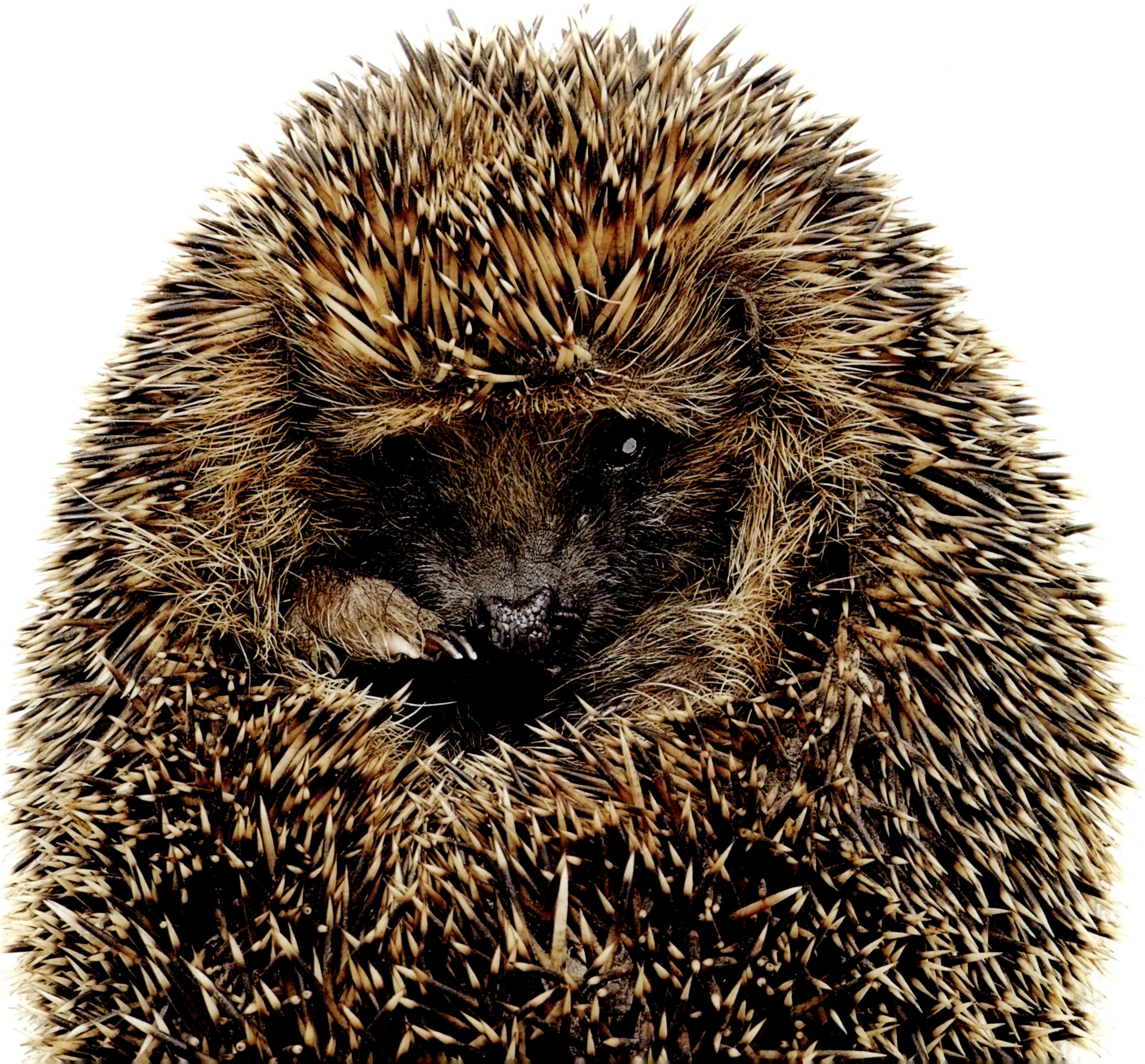

IMPENETRABLE PERSONALITY

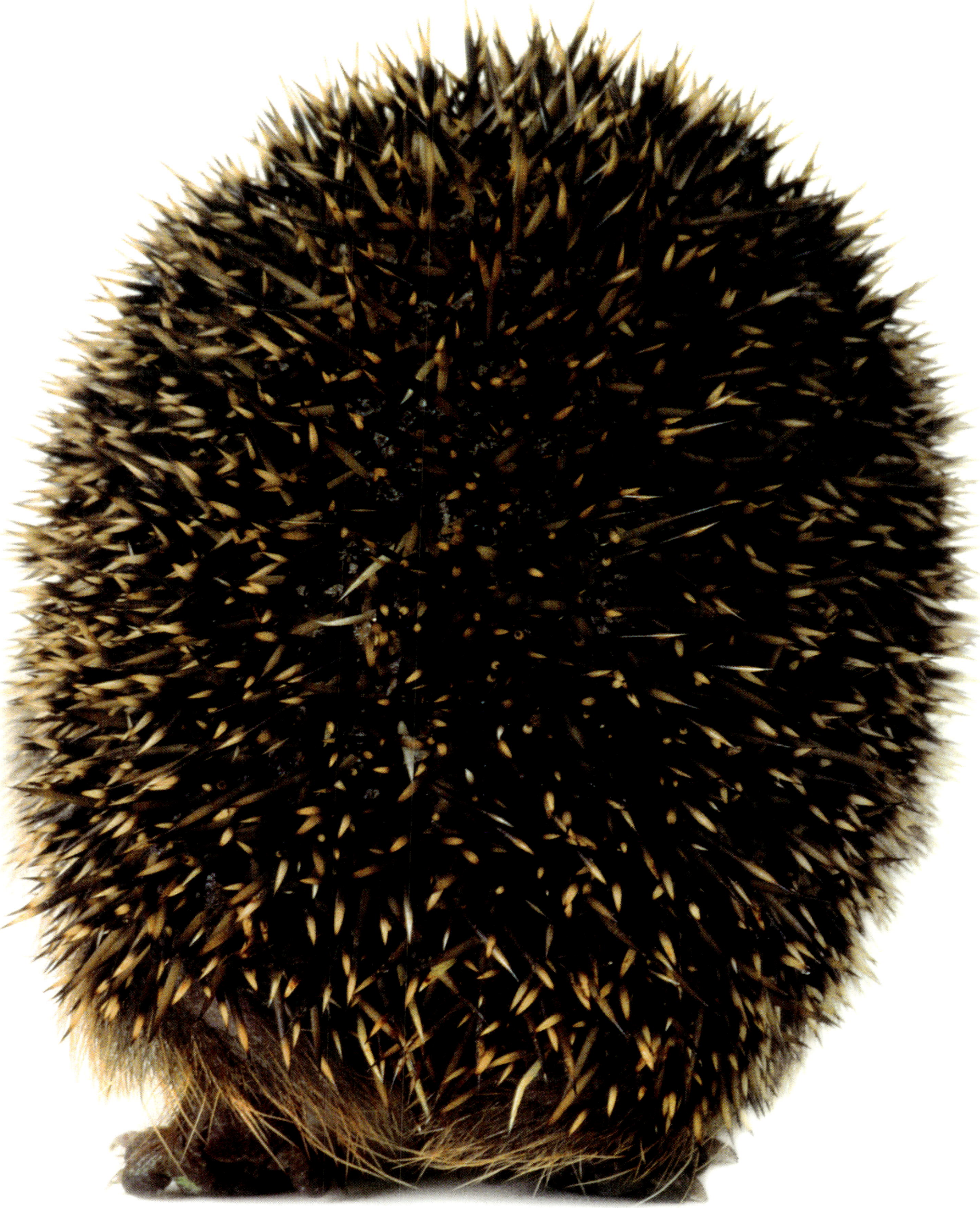

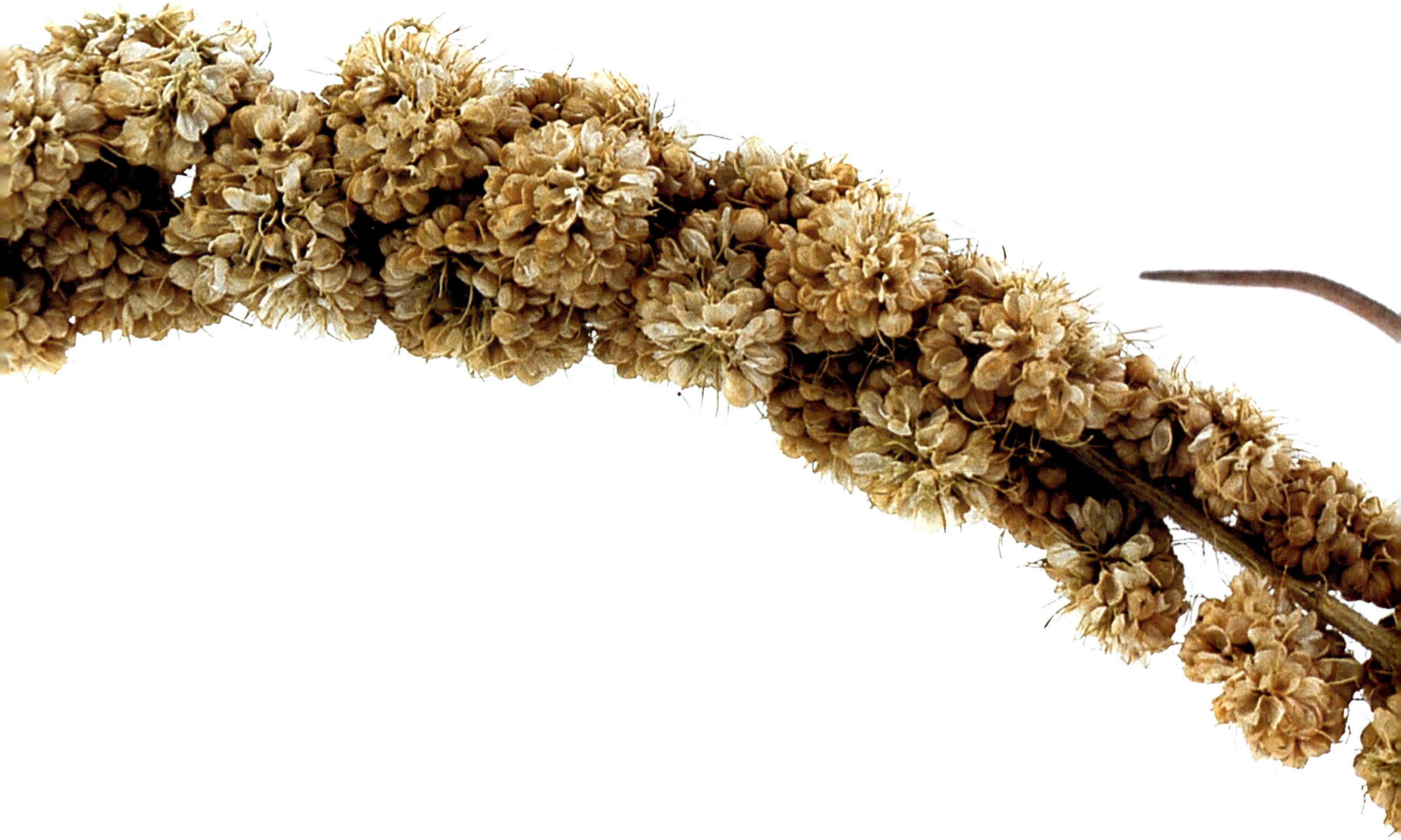

"… because life is a shiver that flies away, it's all a balance over madness… " (Vasco Rossi)

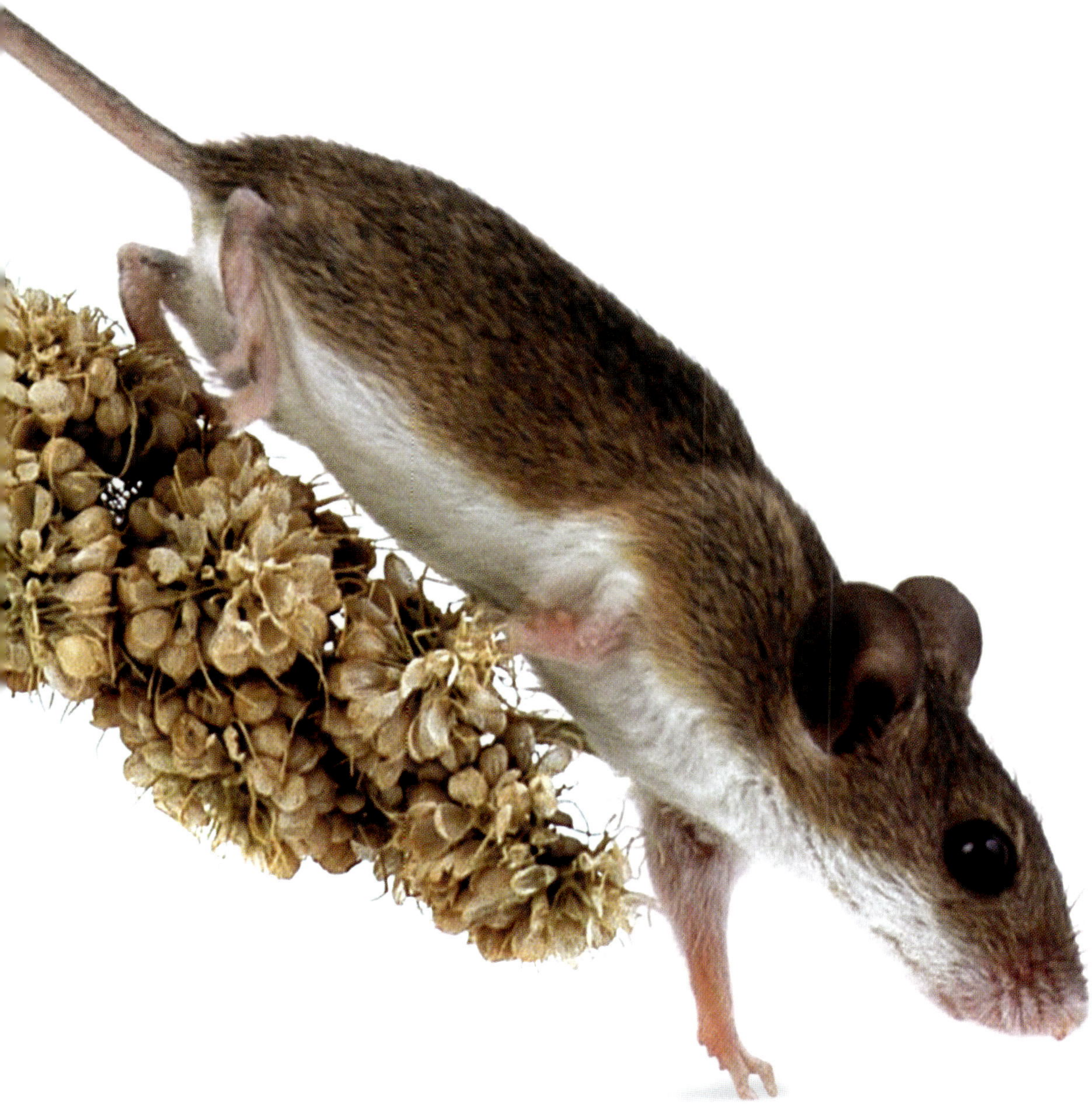

“It’s not what they say about you, it’s what they whisper.”

(Errol Flynn)

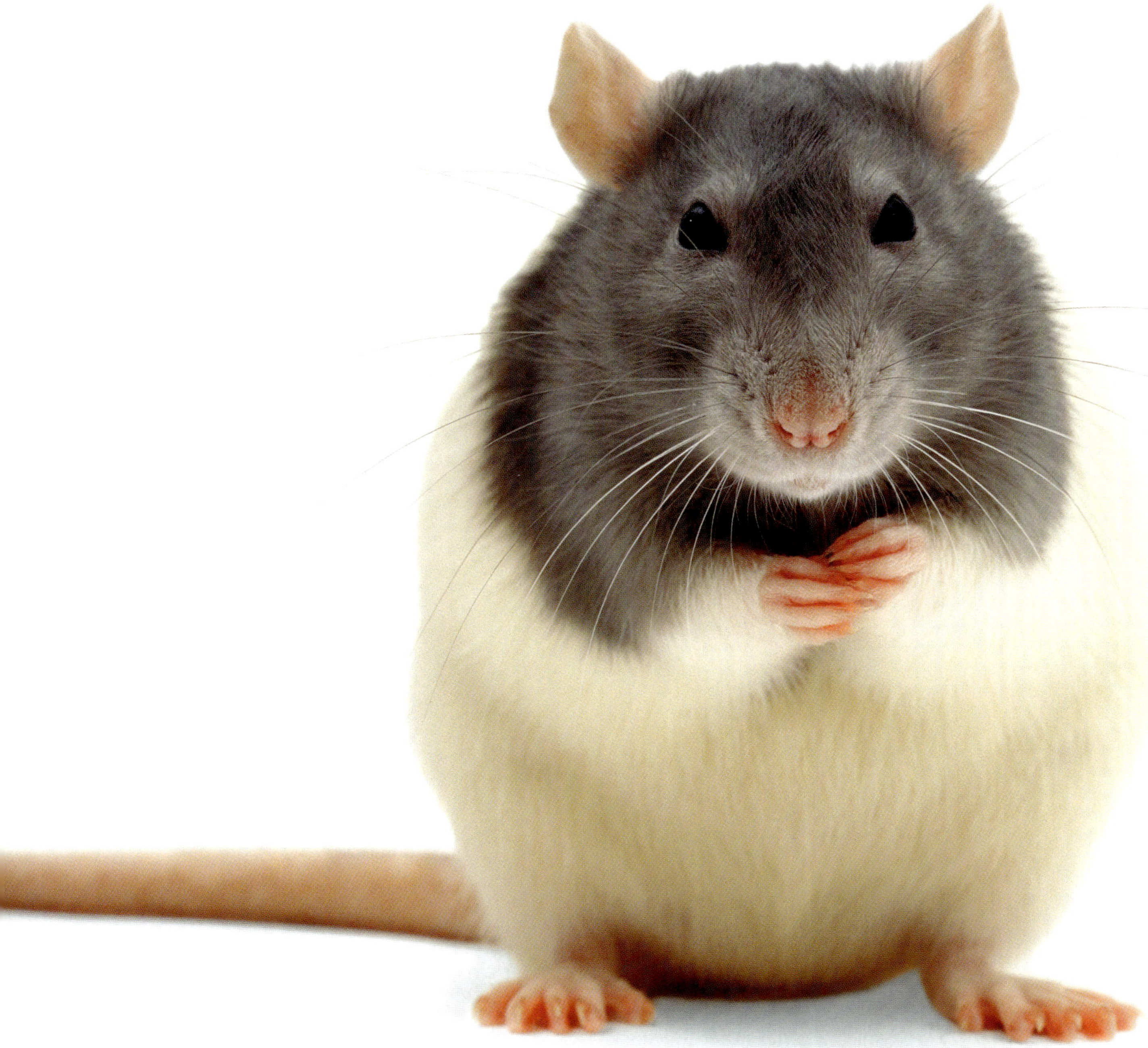

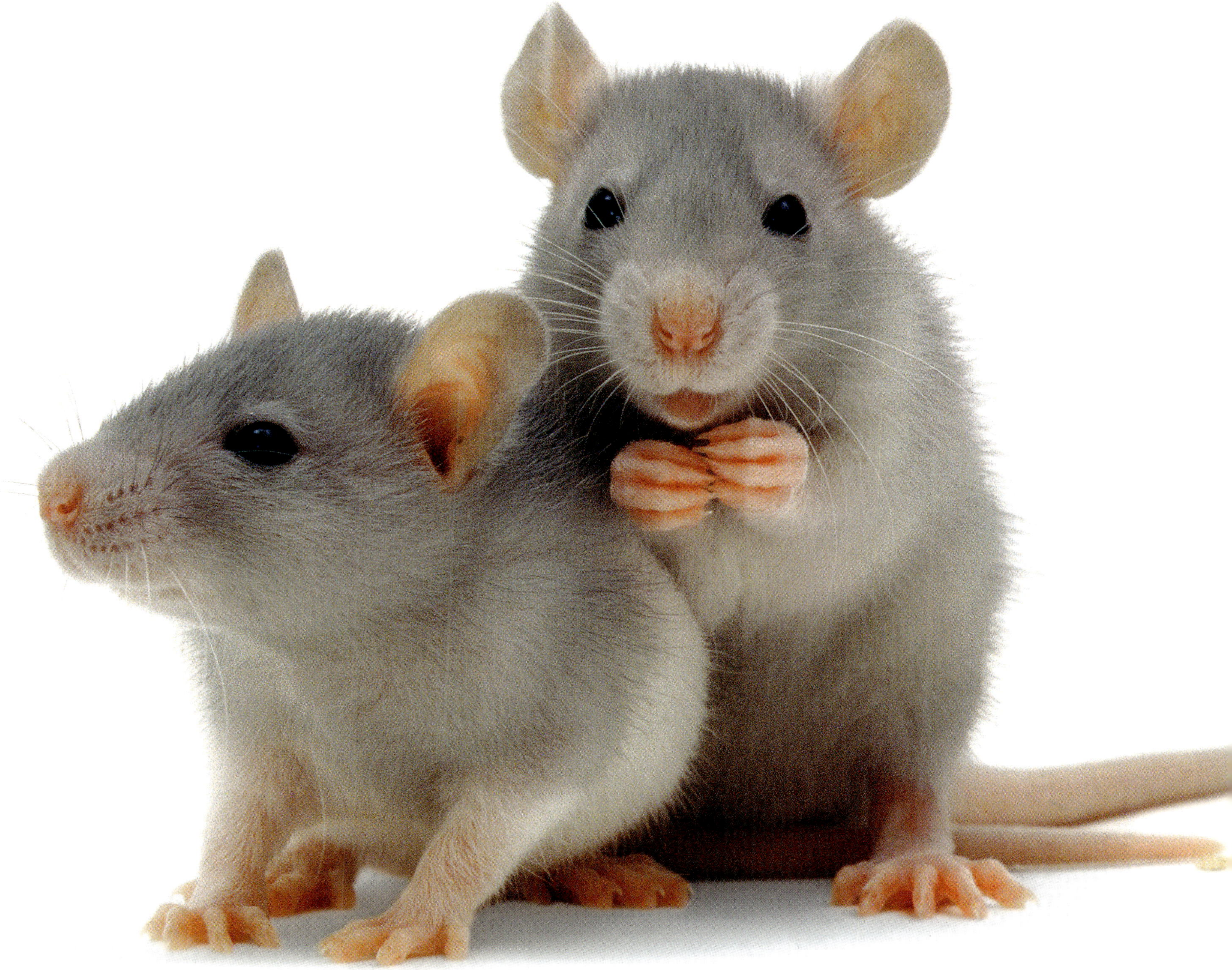

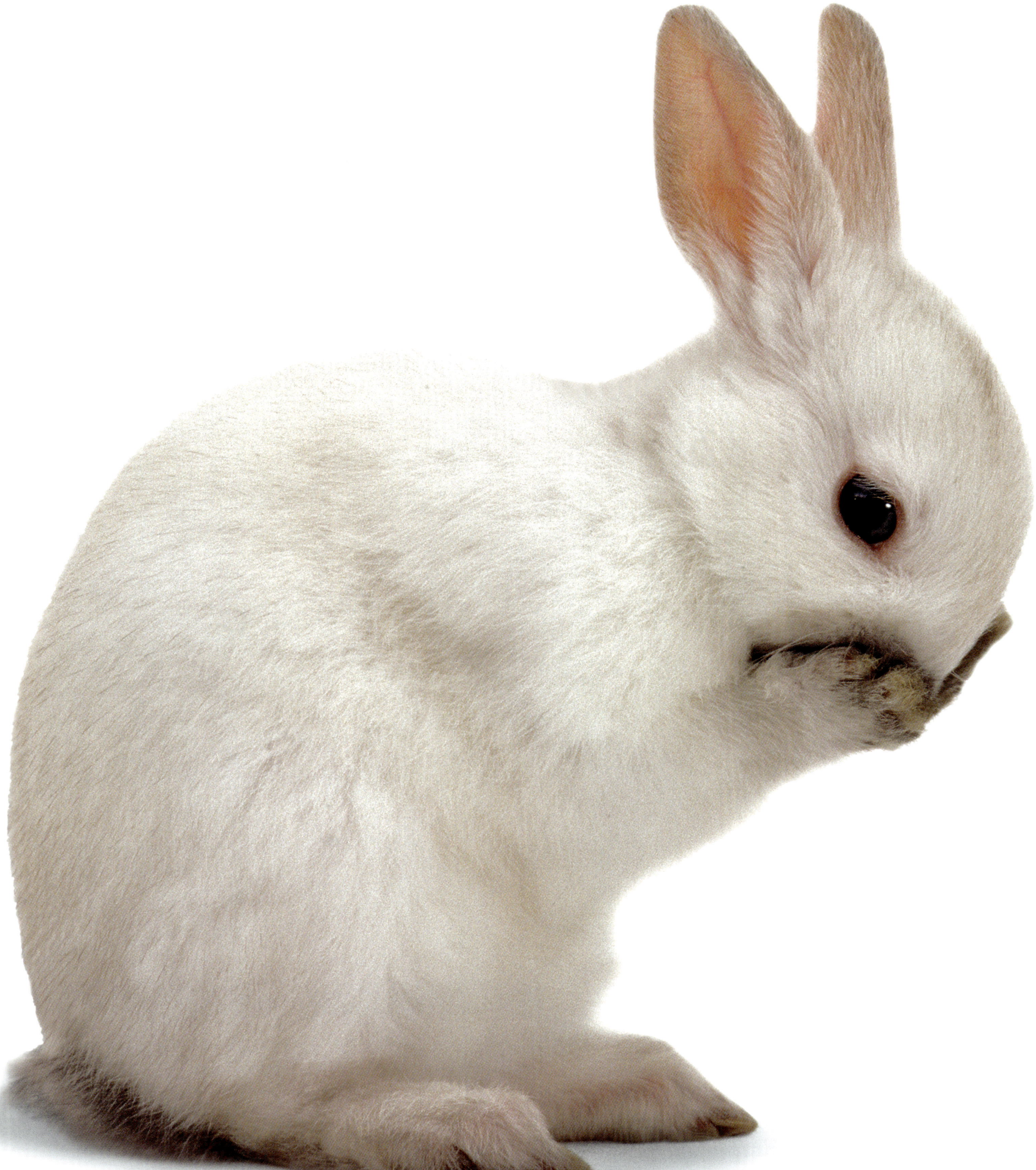

“Fear doesn't exist anywhere except in the mind.”
(Dale Carnegie)

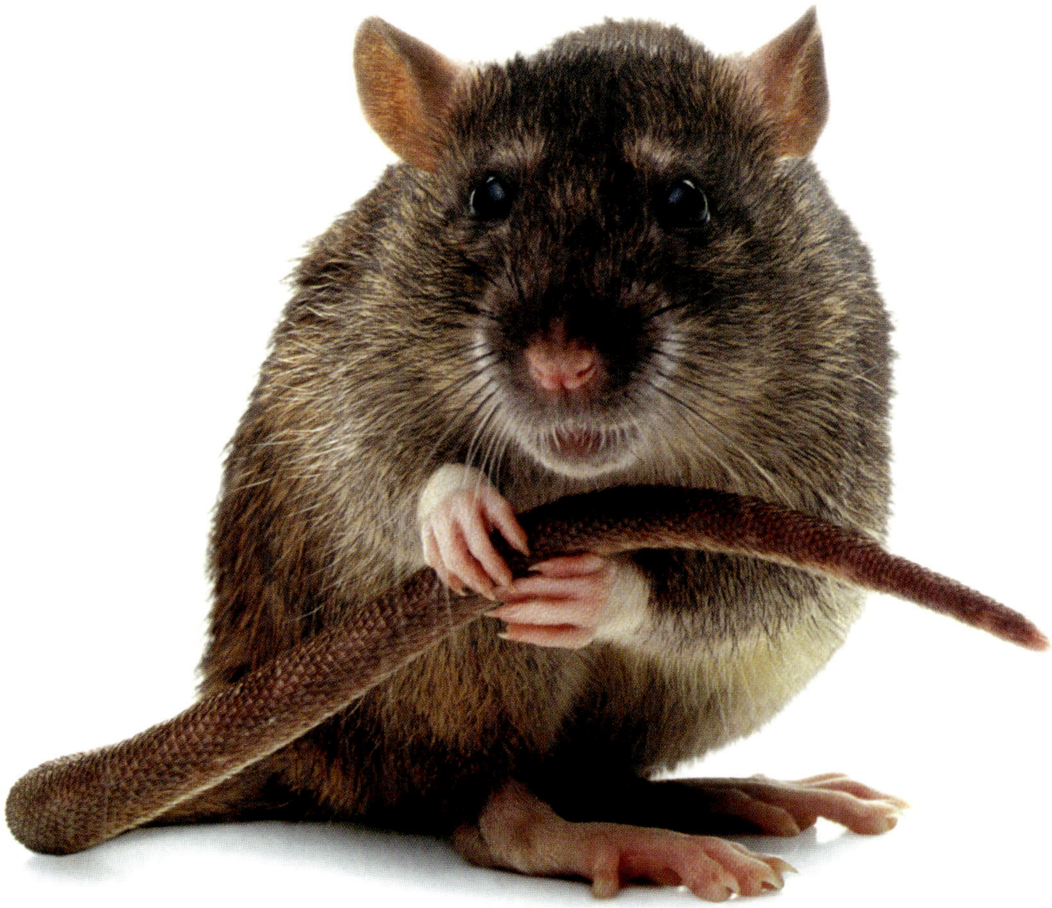

THORNY RELATIONSHIPS

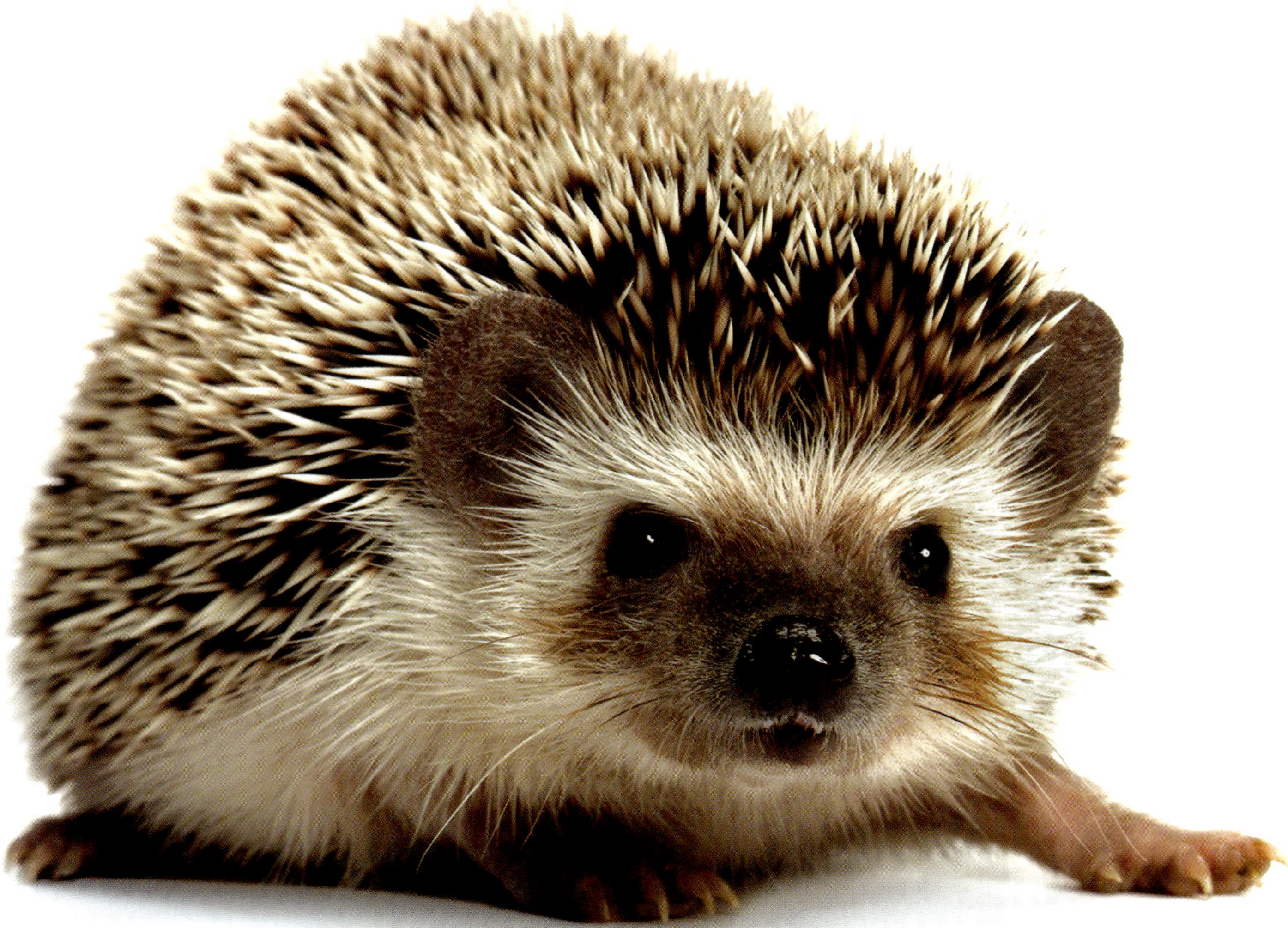

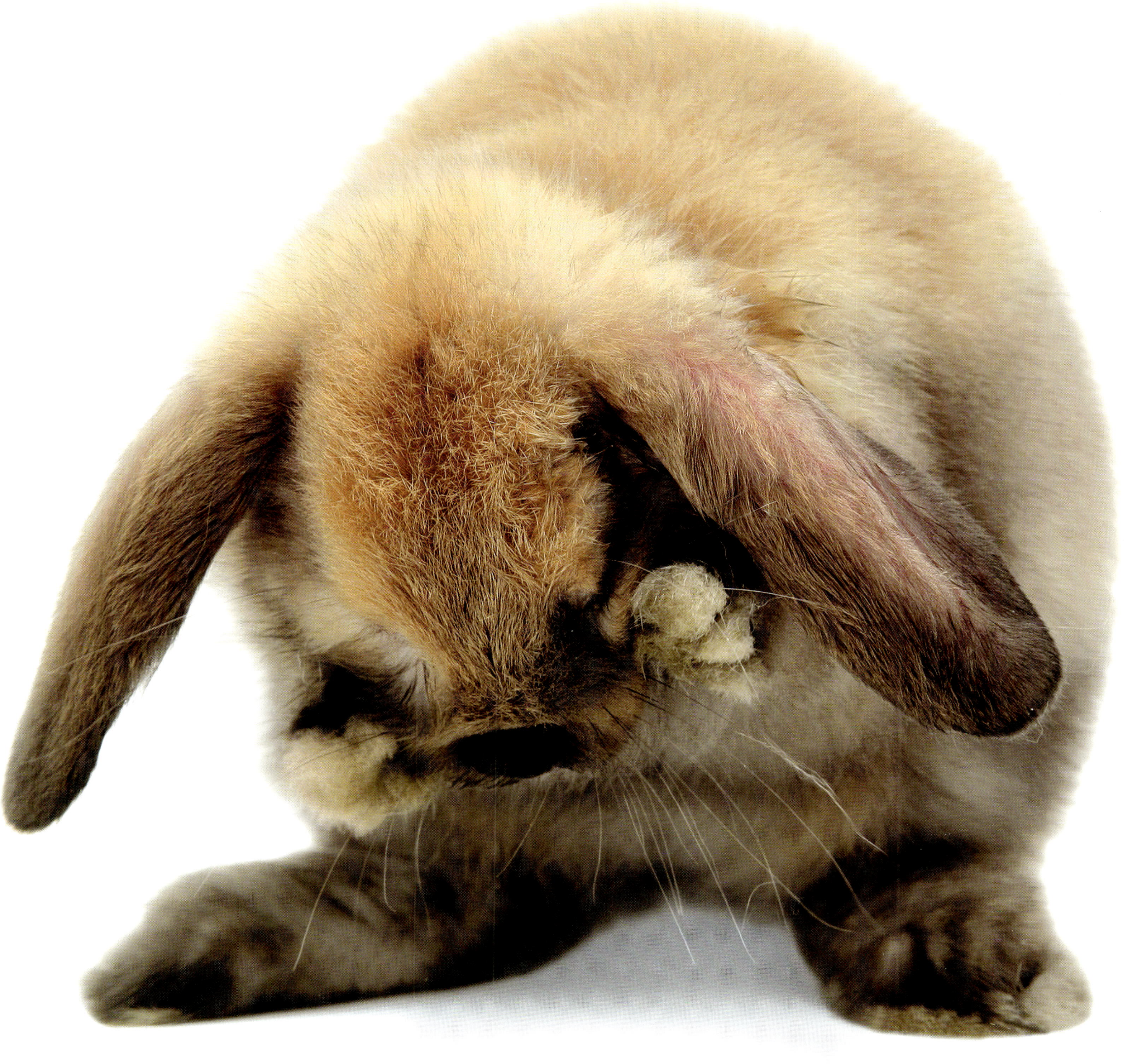

CLOSE ENCOUNTERS

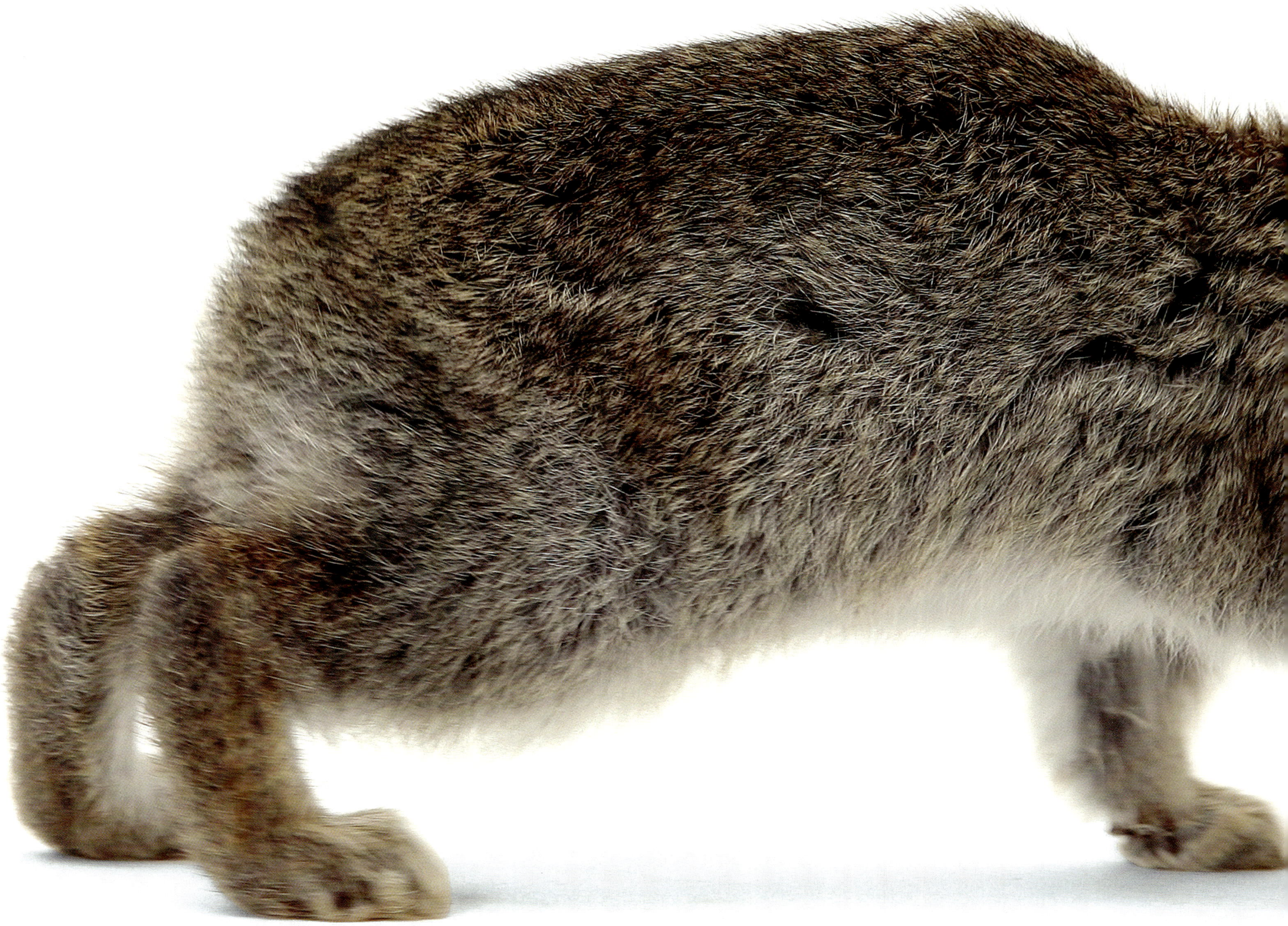

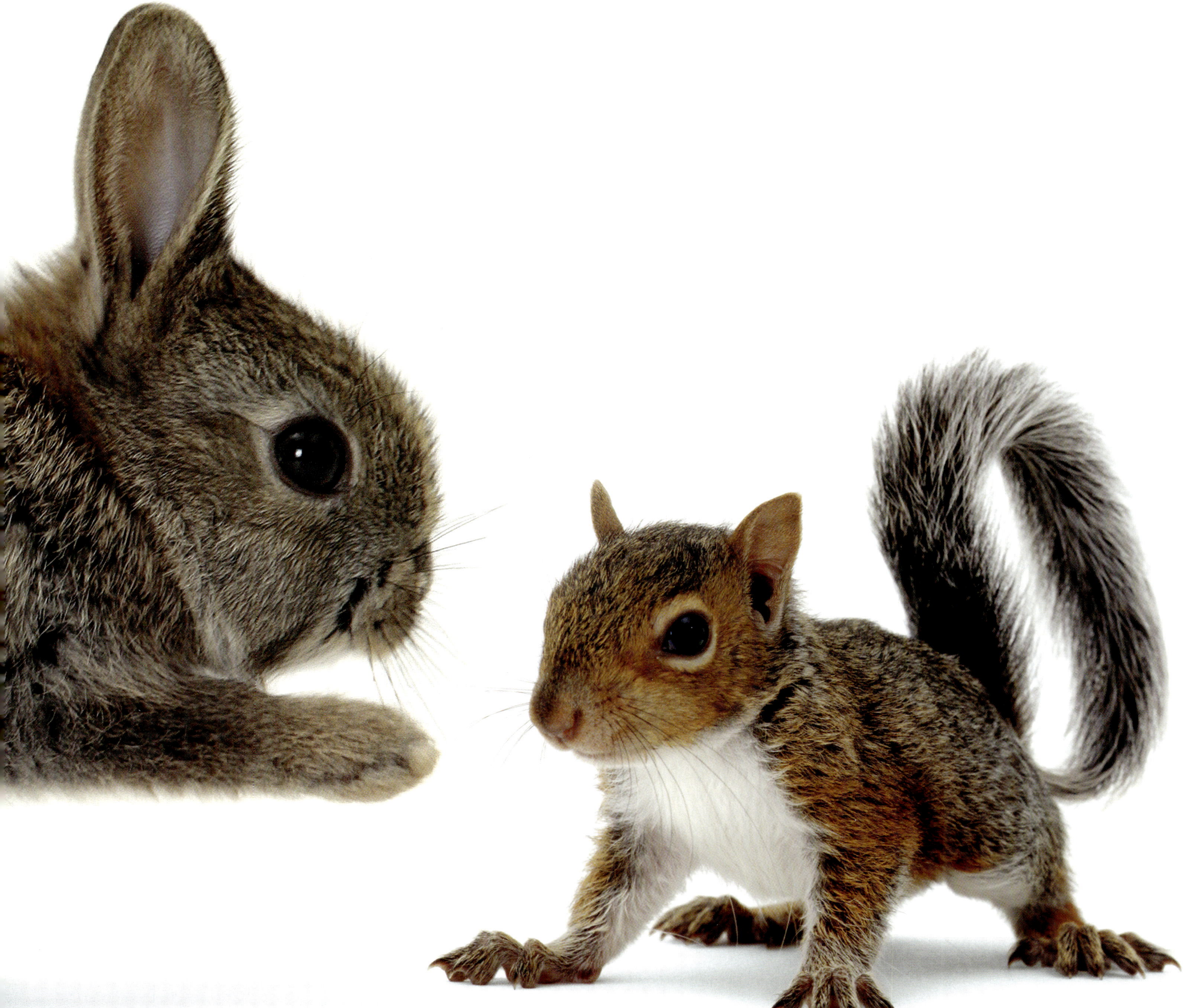

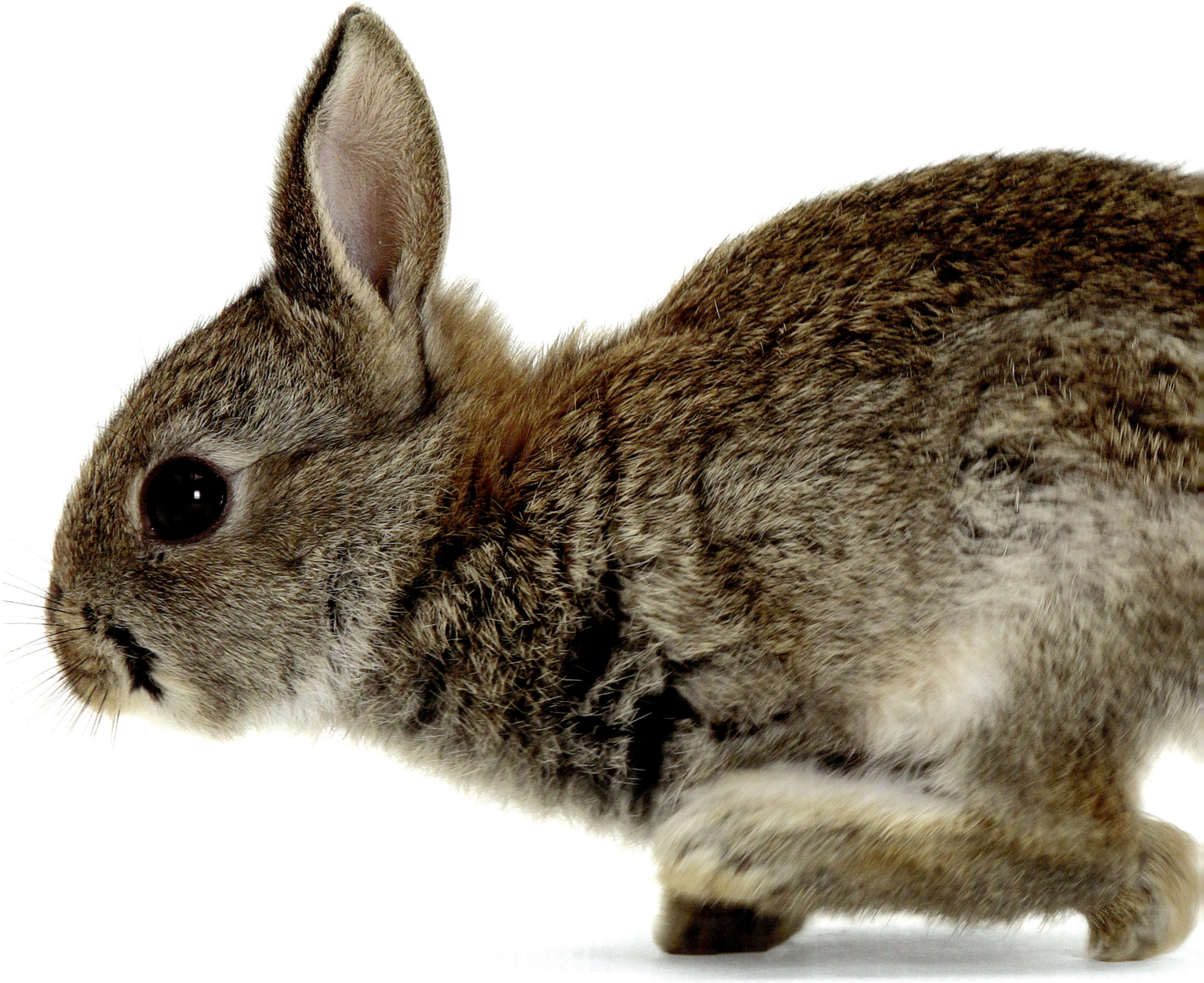

“The best way to escape from your problem is to solve it.” (Robert Anthony)

“No man sees far, most see no farther than their noses.” (Thomas Carlyle)

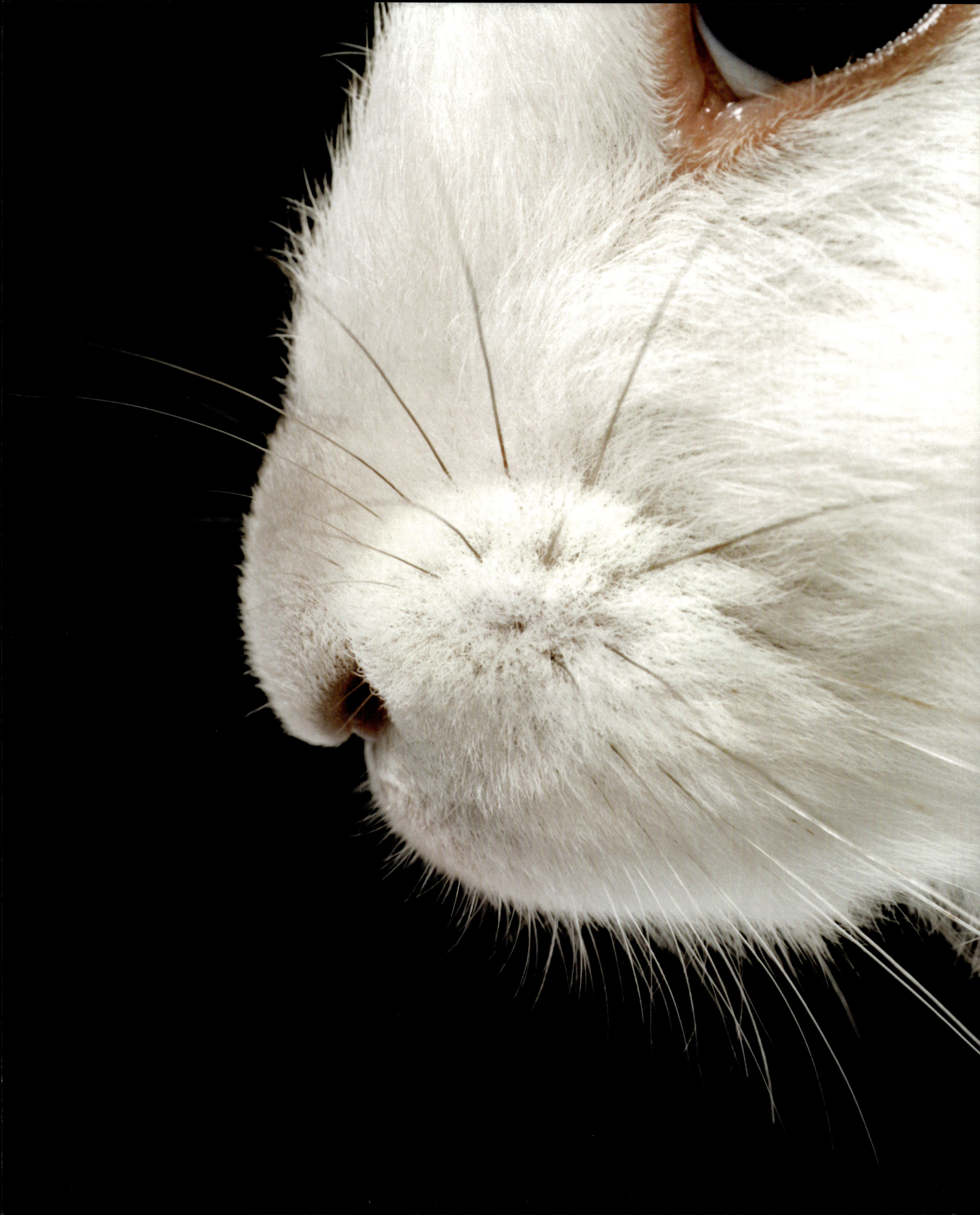

THE RABBIT . . .

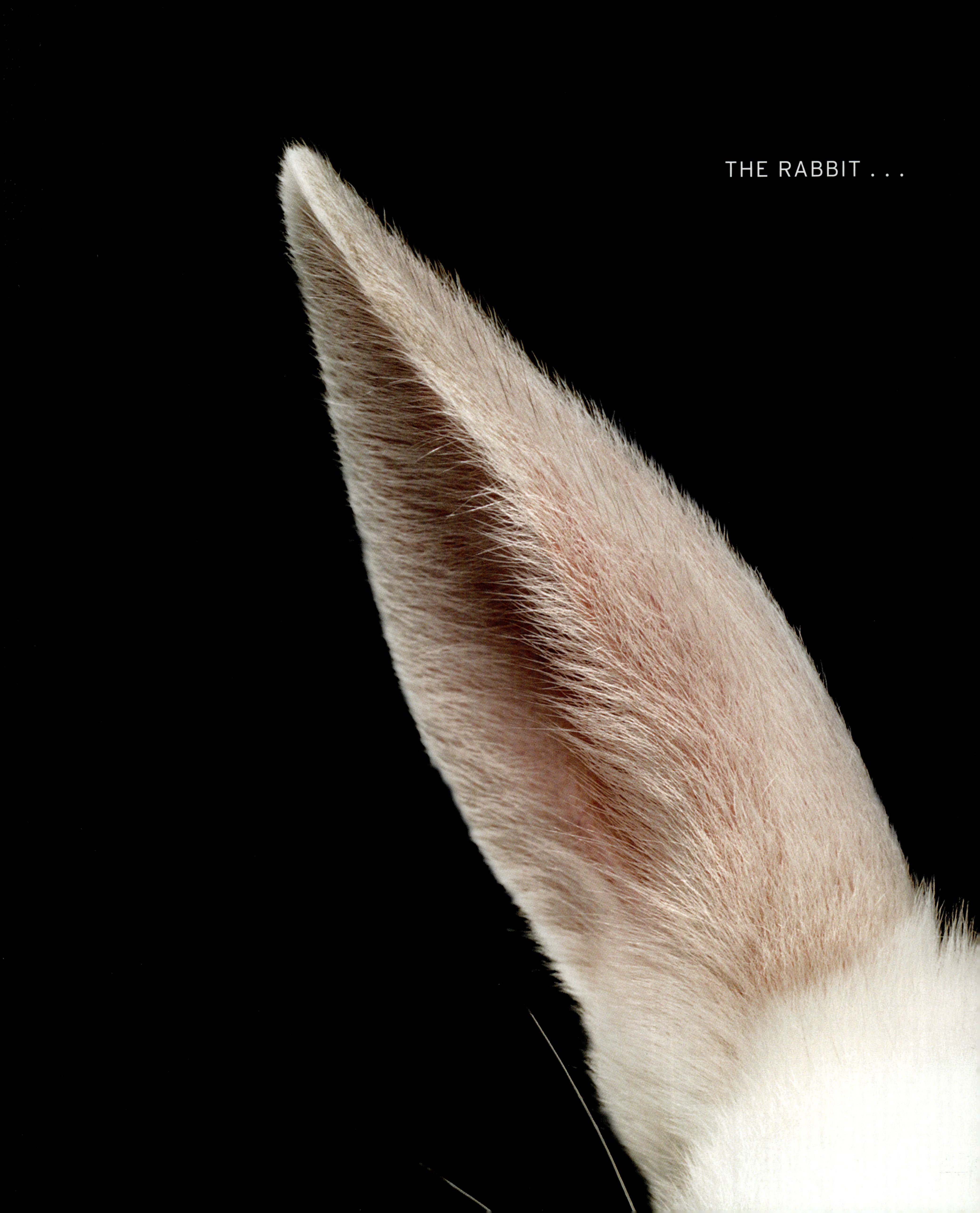

INSIDE THE CYLINDER HAT

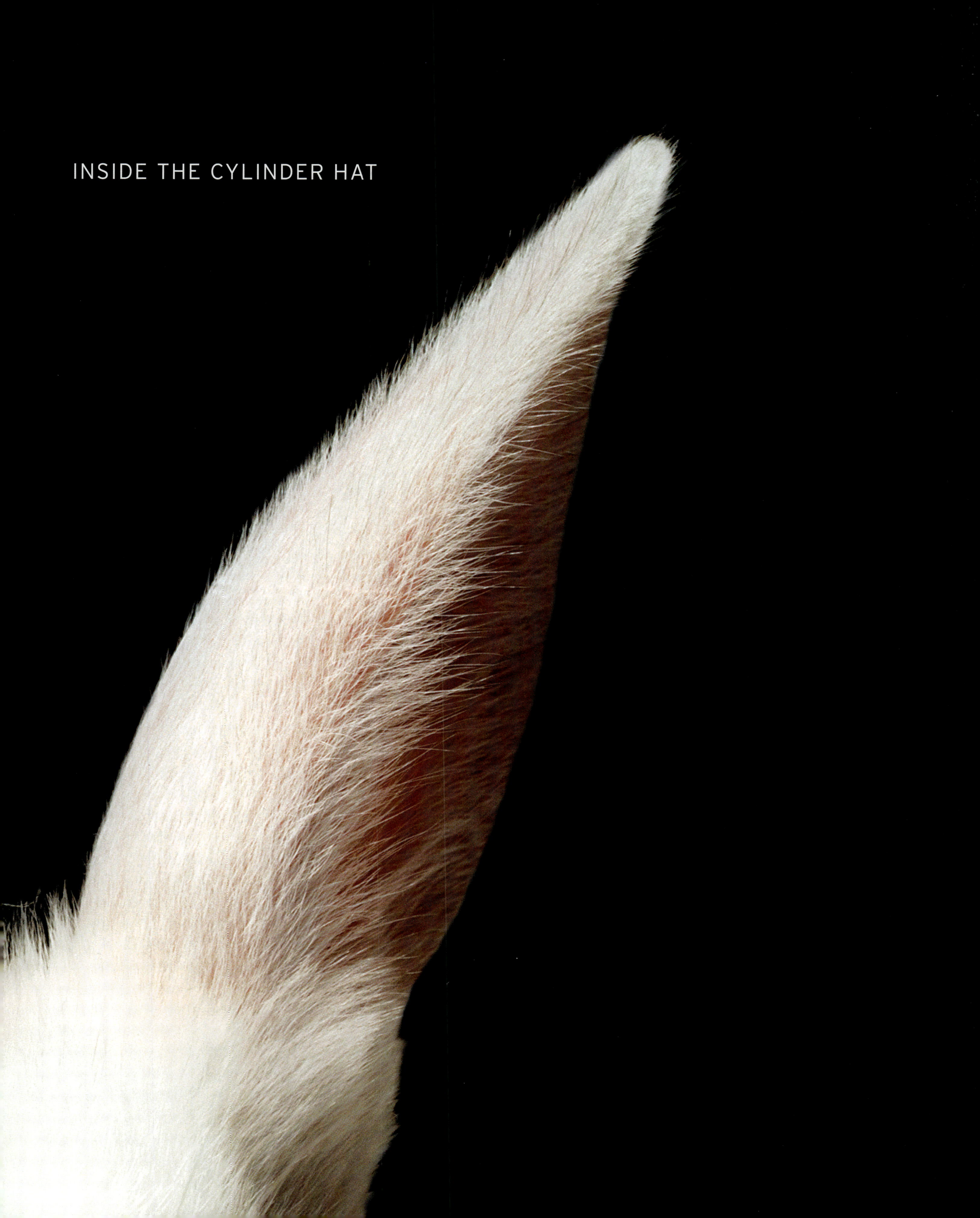

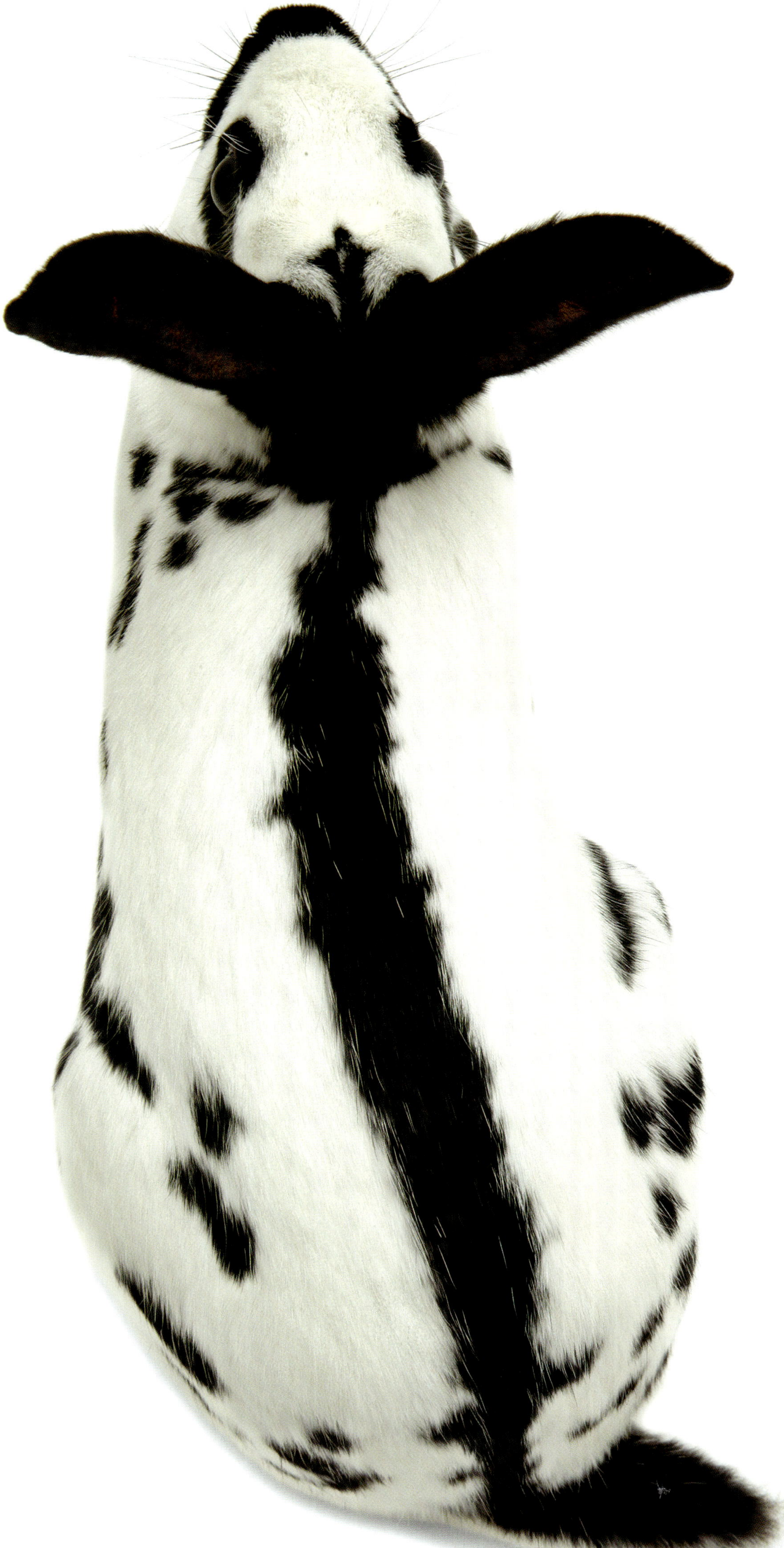

ONE HUNDRED AND ONE . . . RABBITS

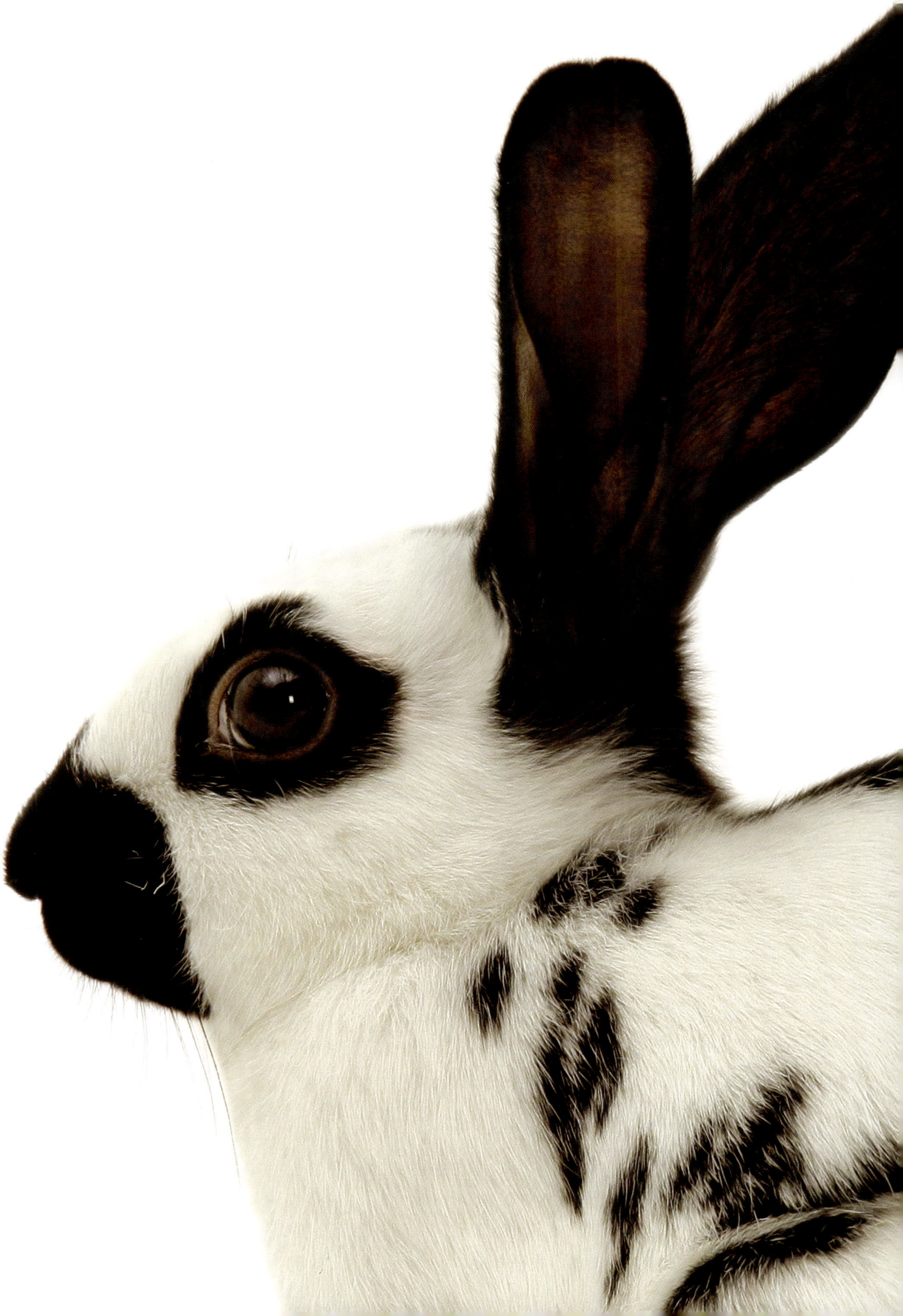

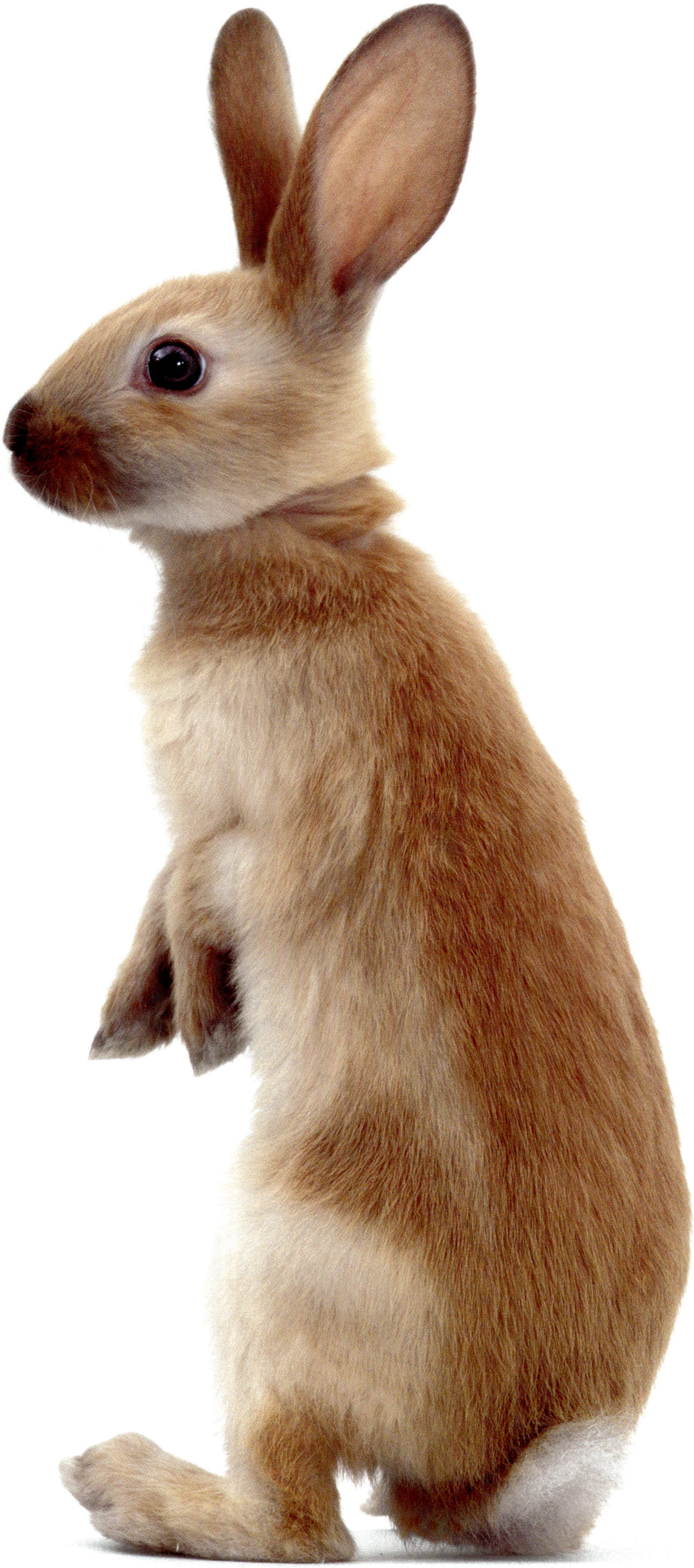

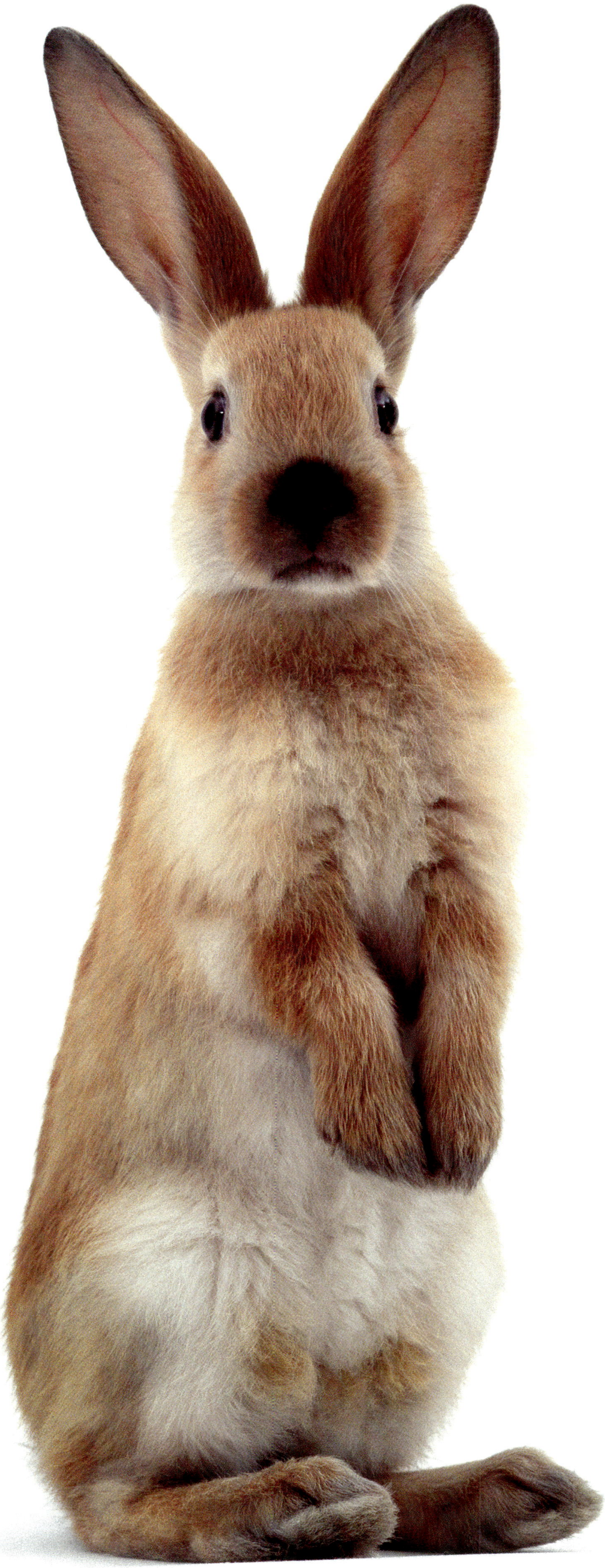

WAITING FOR GODOT

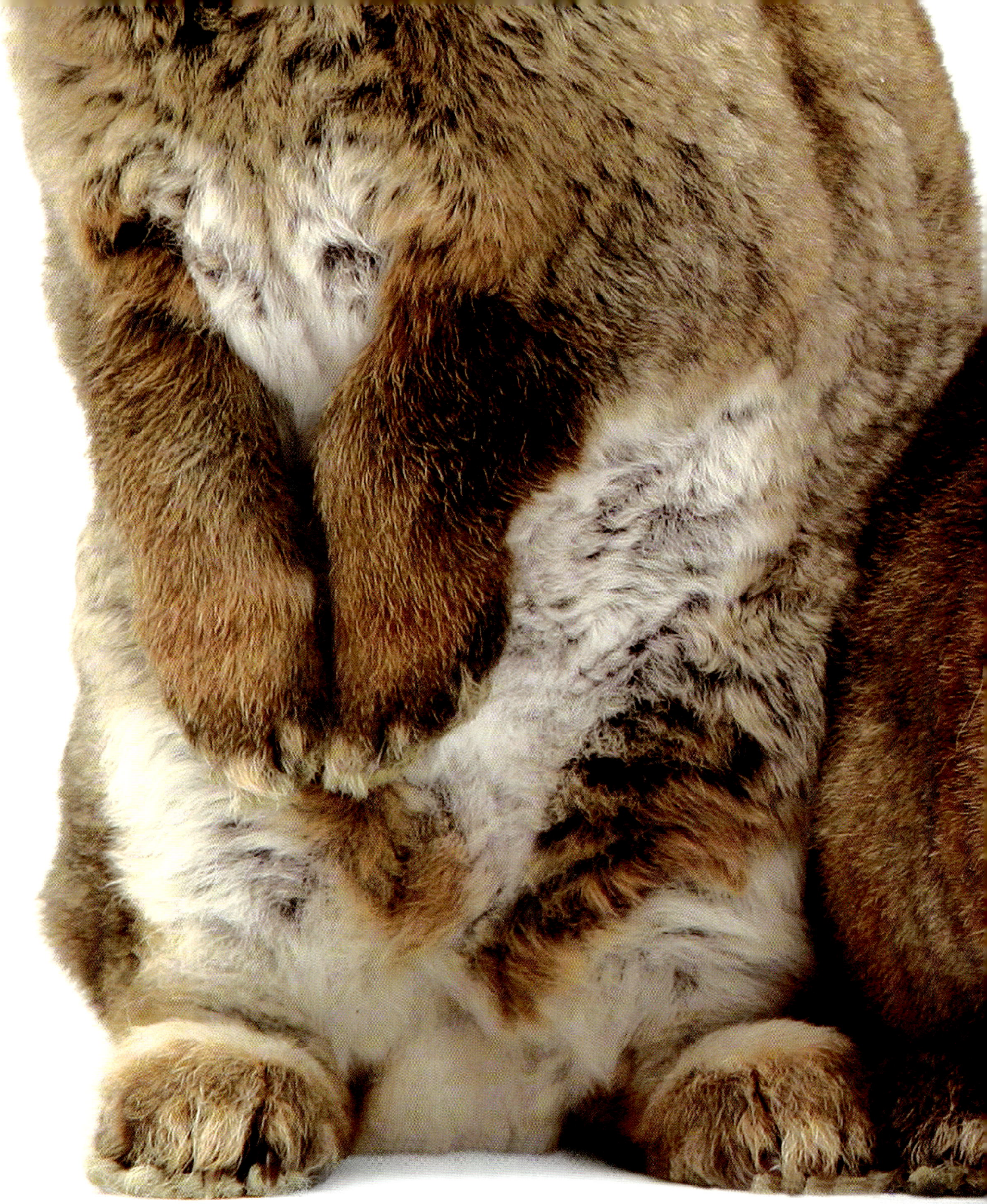

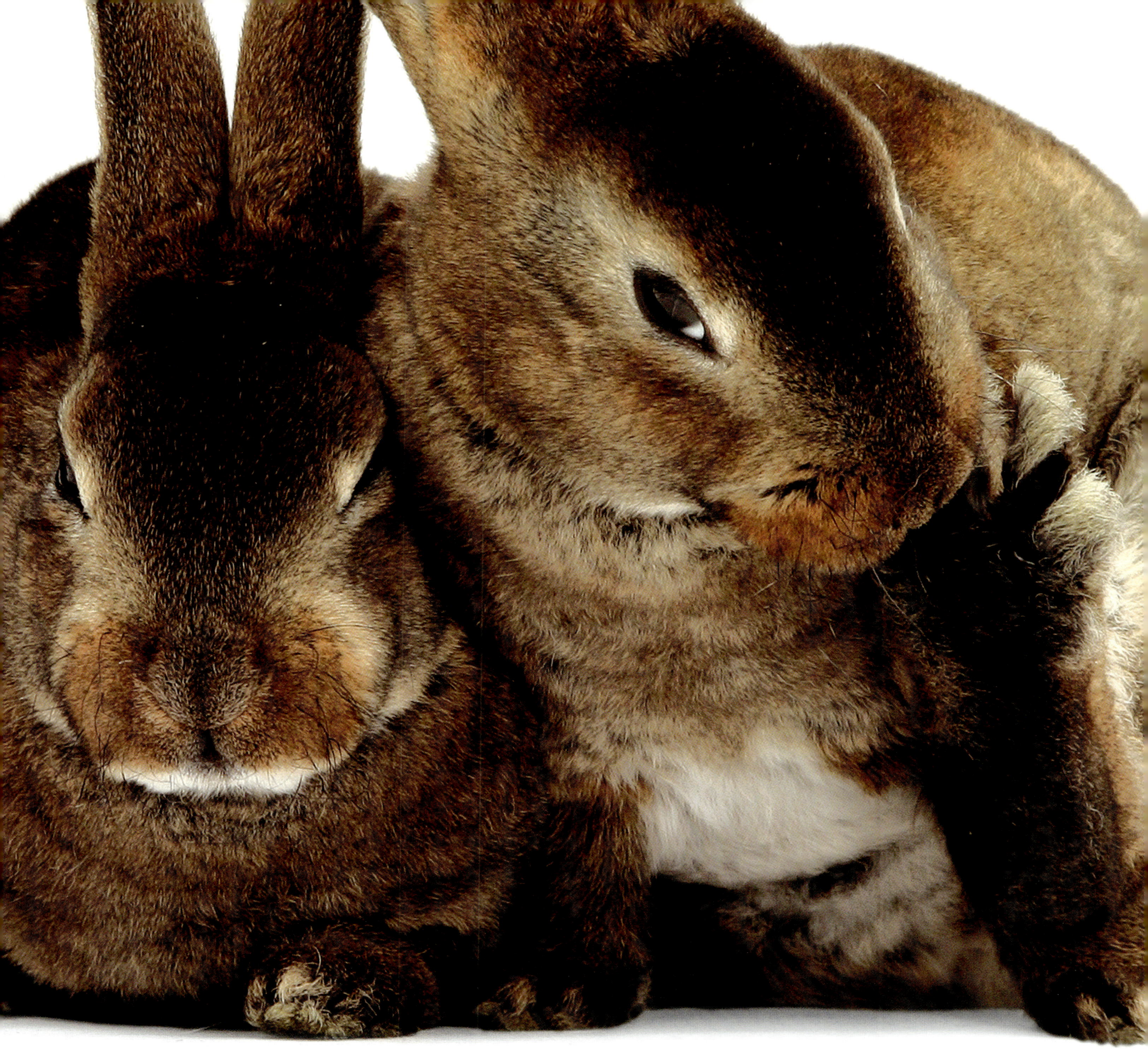

The Good, the Bad and the Ugly
(Sergio Leone)

A LORD WEARING A WIG

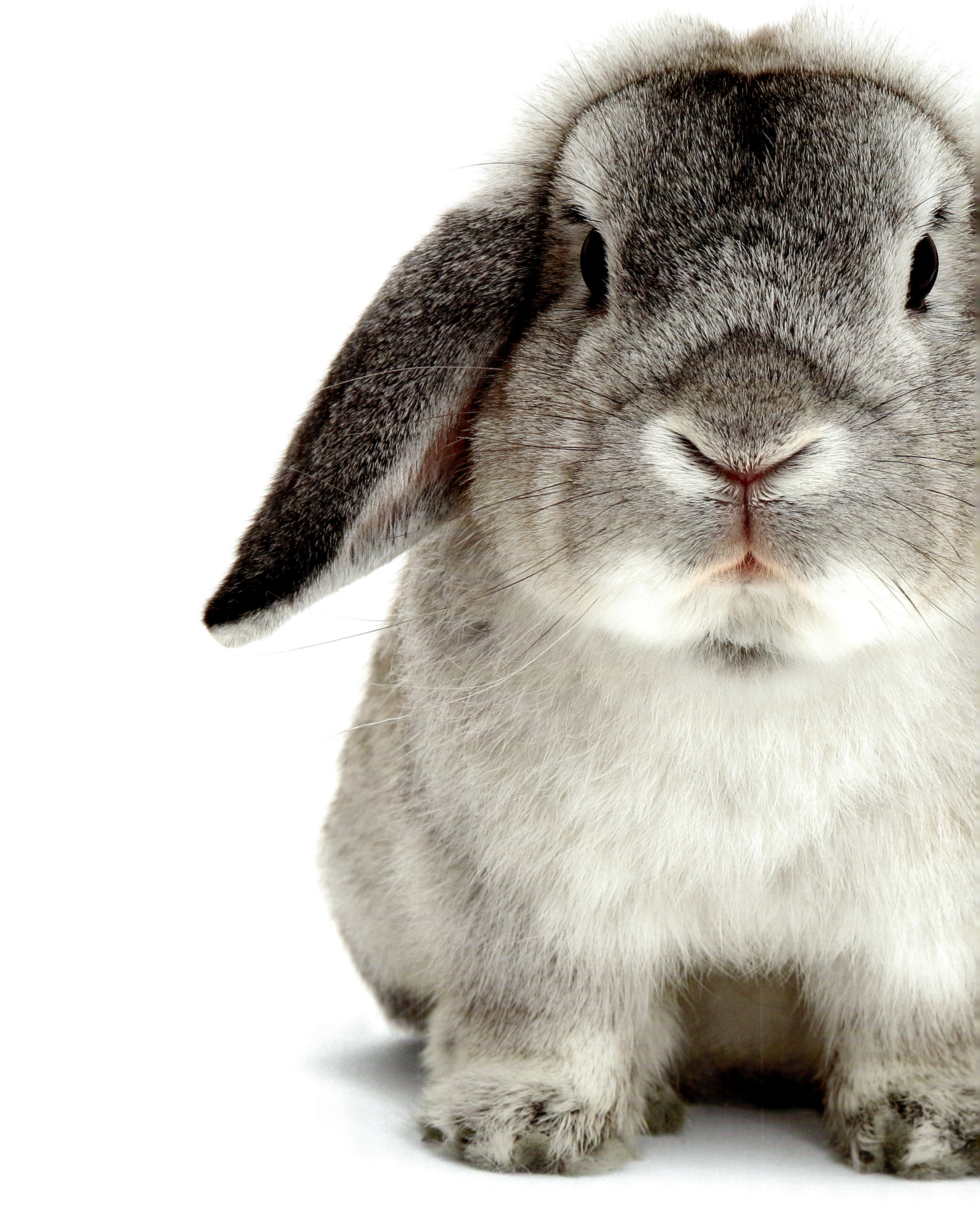

A RABBIT AS A PILLOW

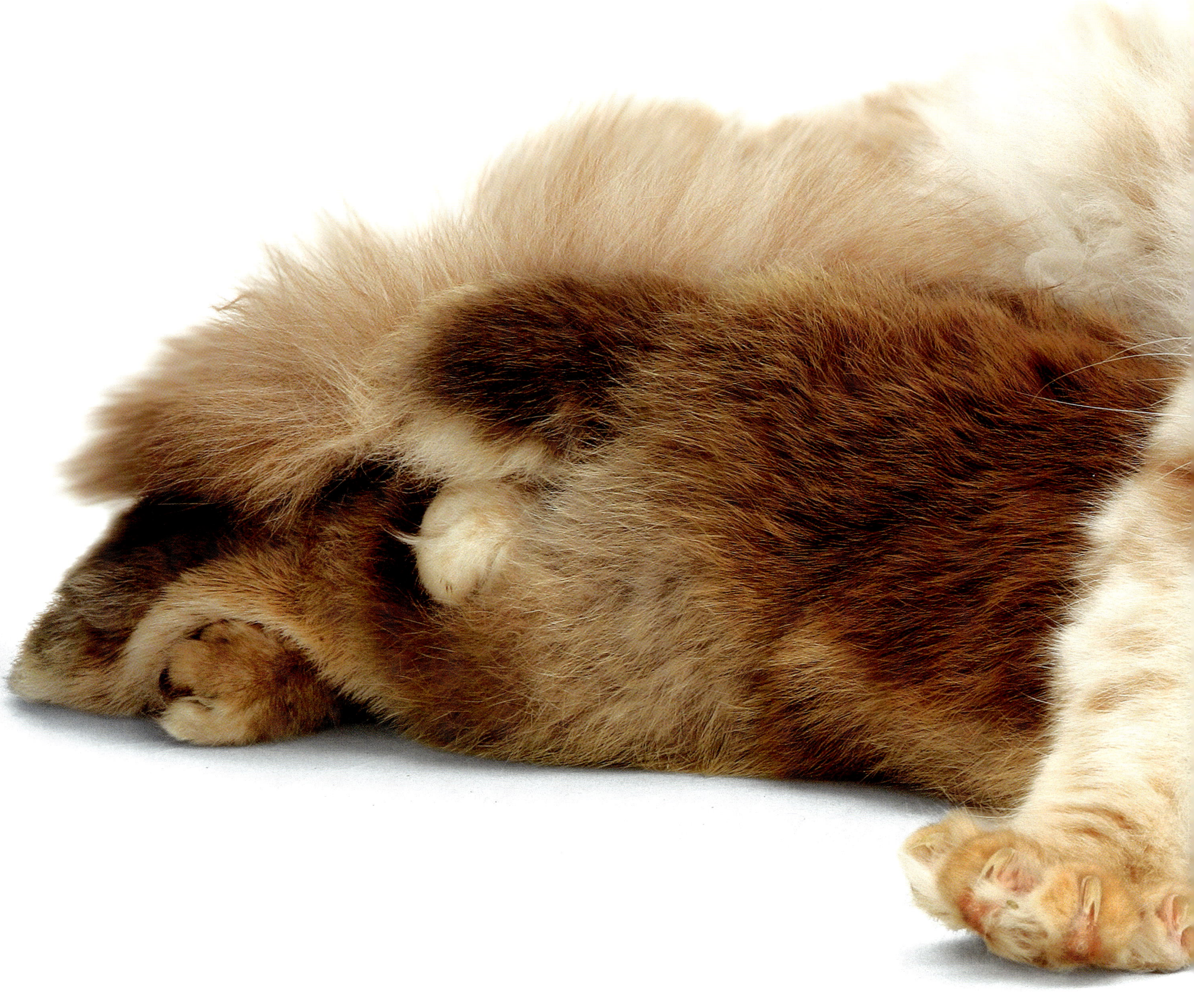

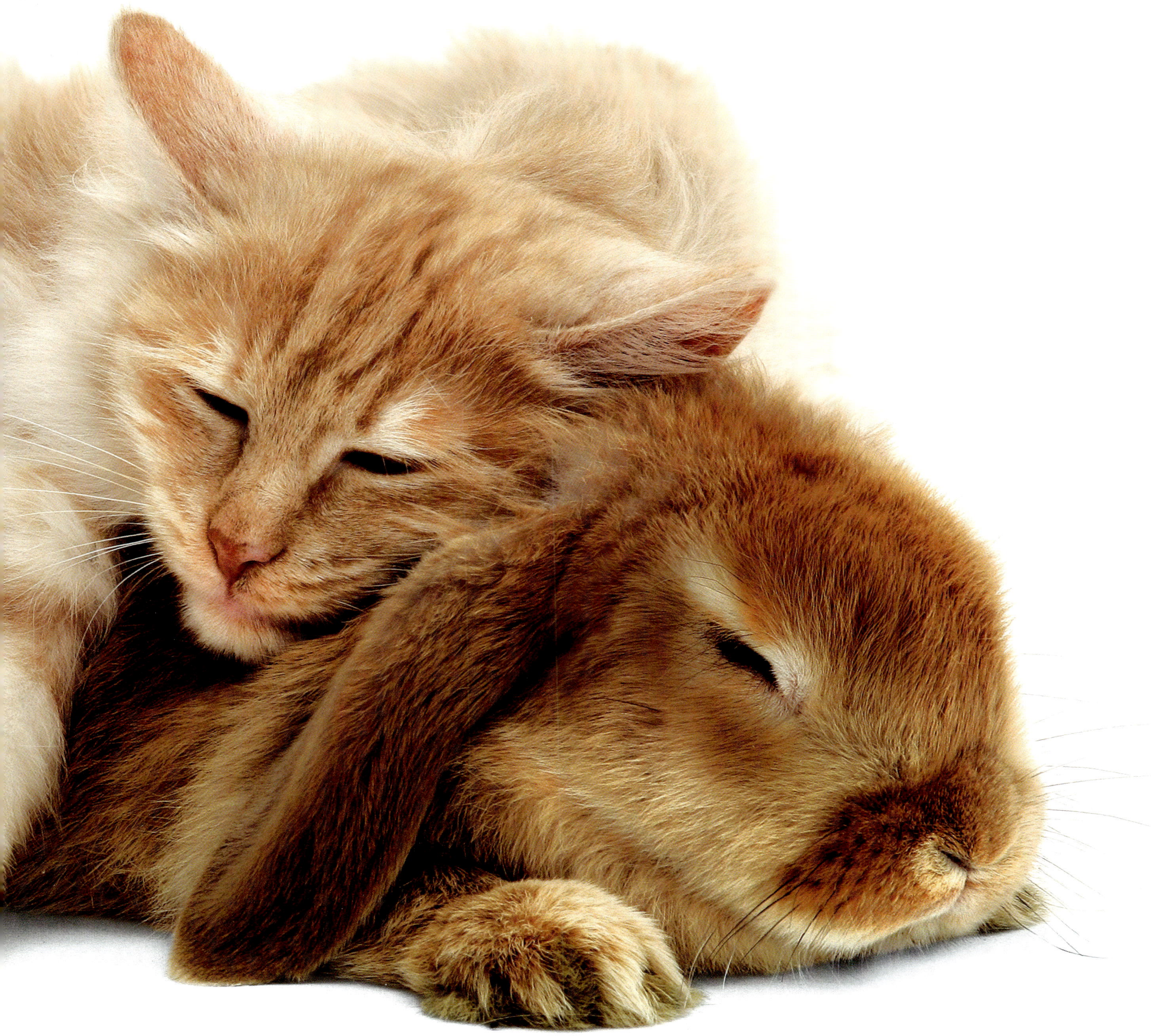

A DAY TO FORGET

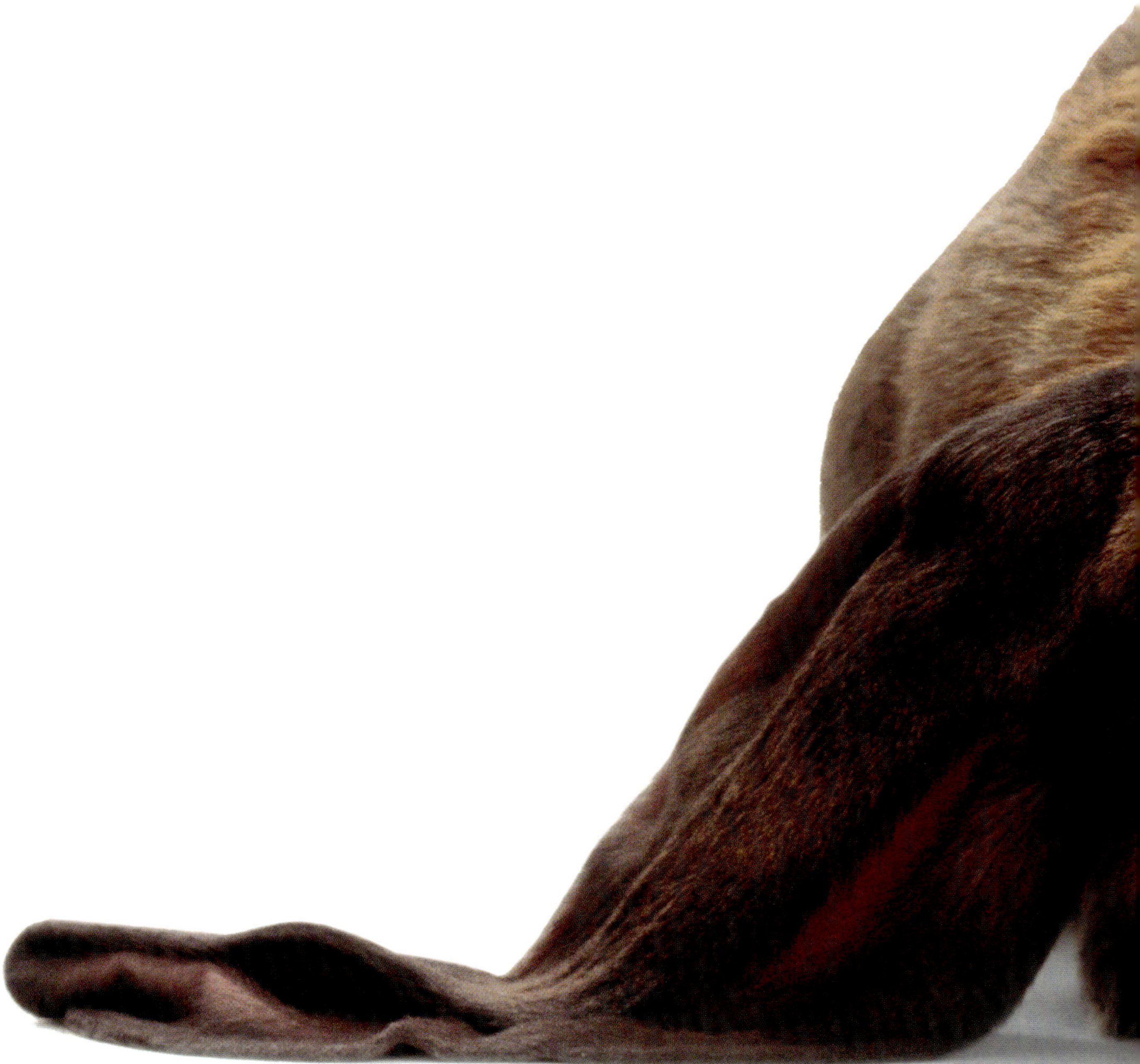

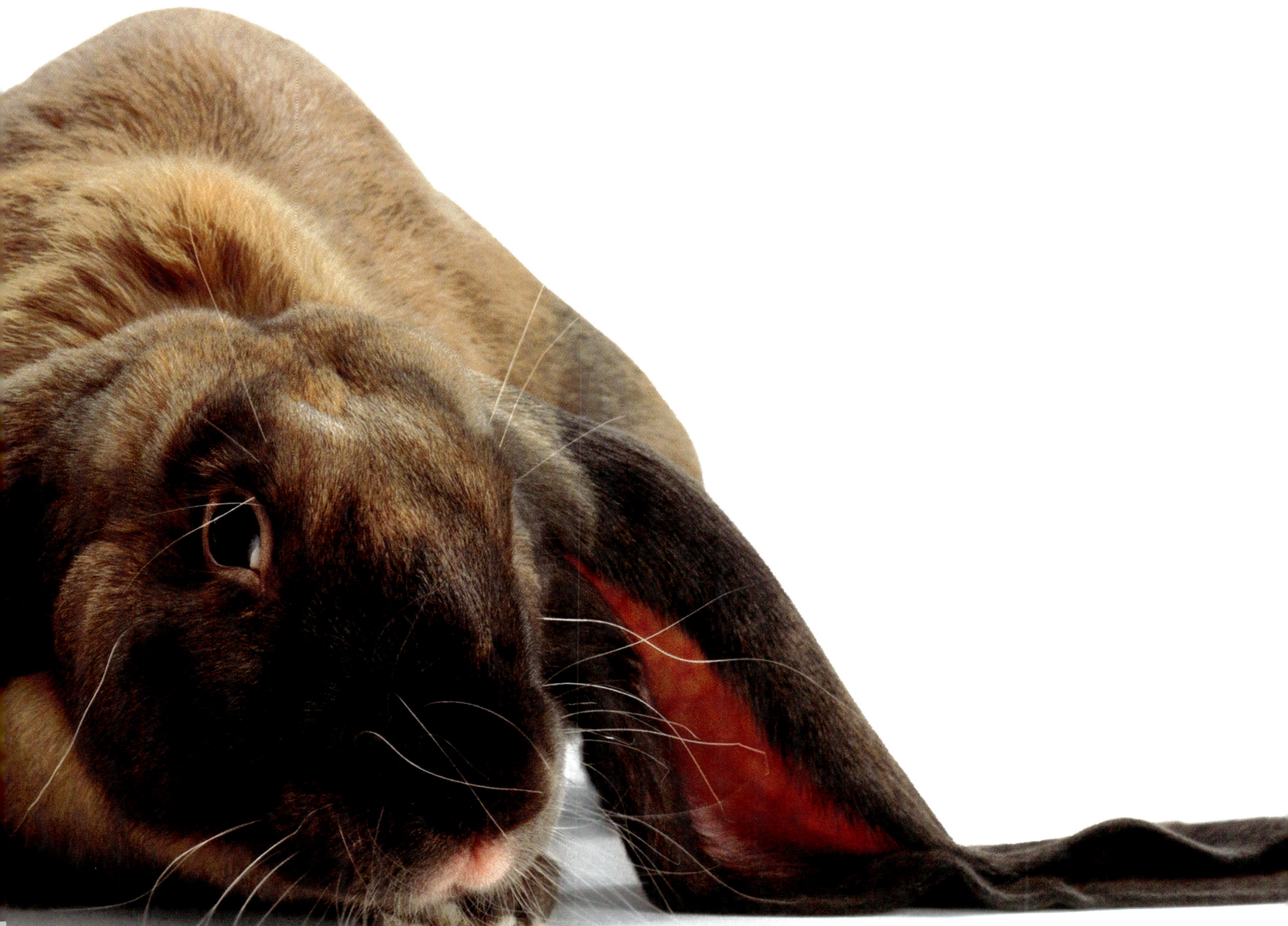

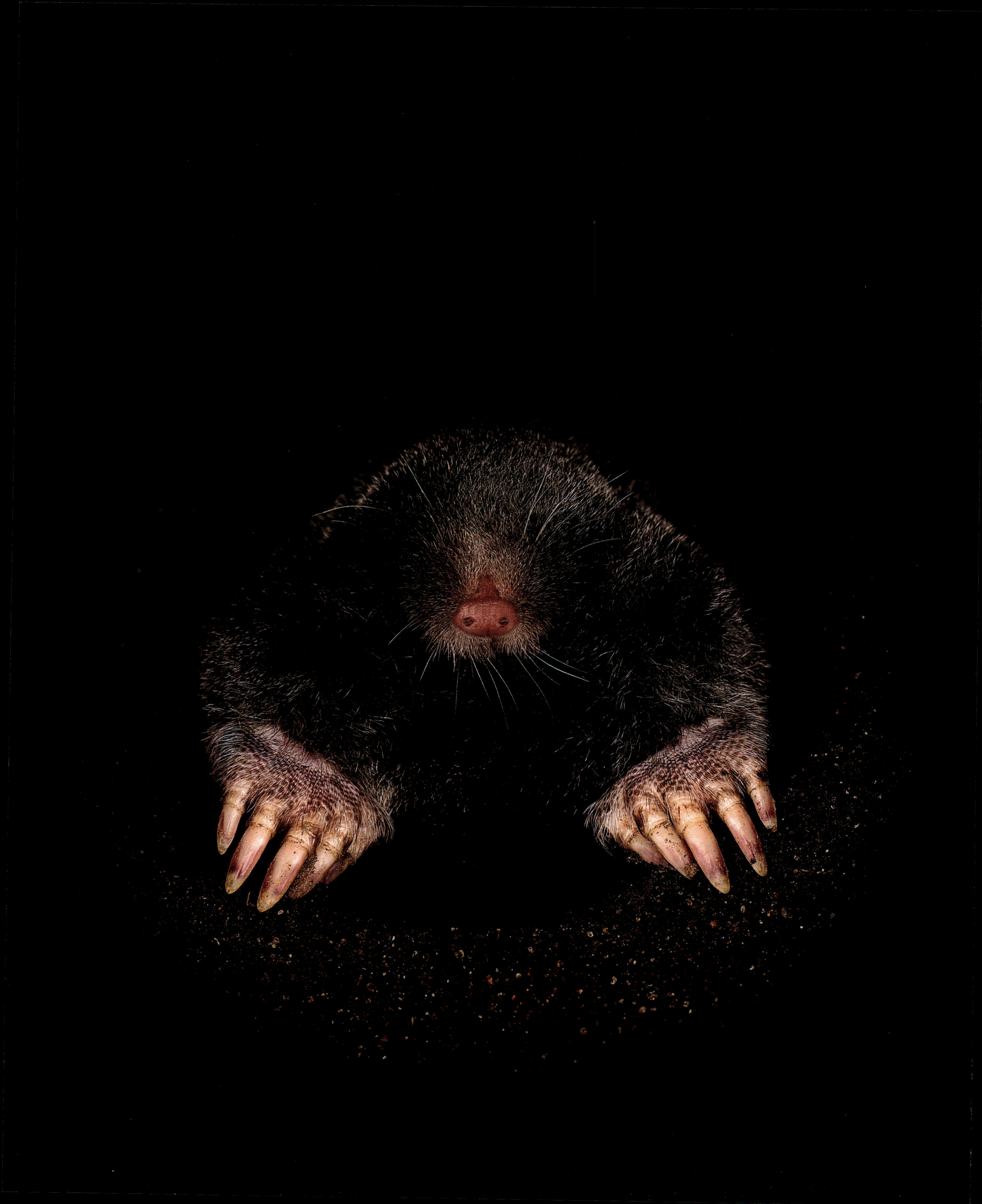

“If you know a better hole, go to it.” (Bruce Bairnsfather)

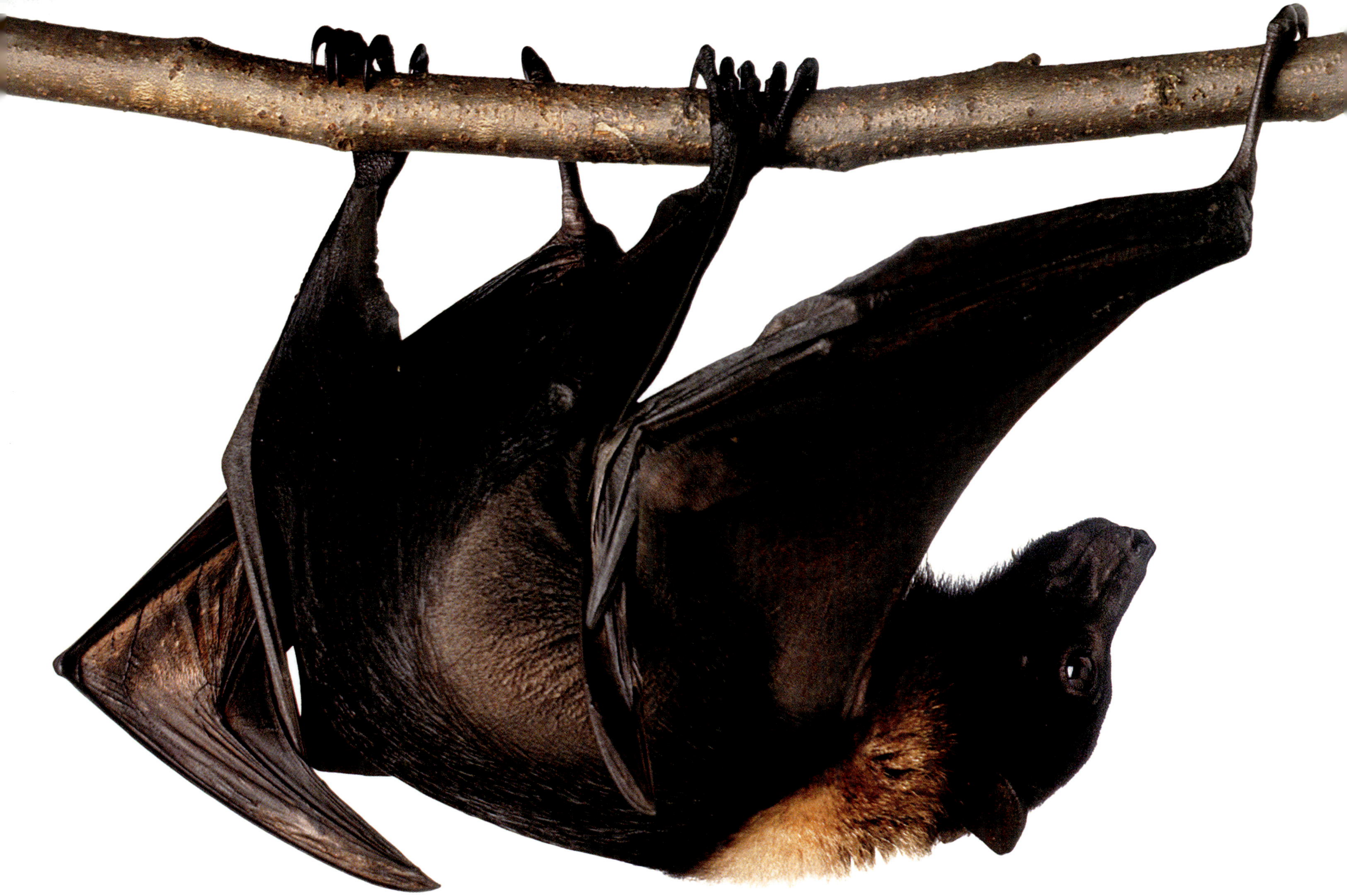

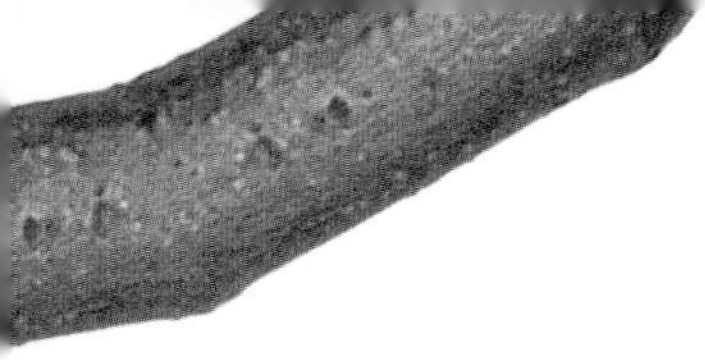

HALLOWE'EN NIGHT

“I’m Dracula, I bid you welcome.” (Bram Stoker)

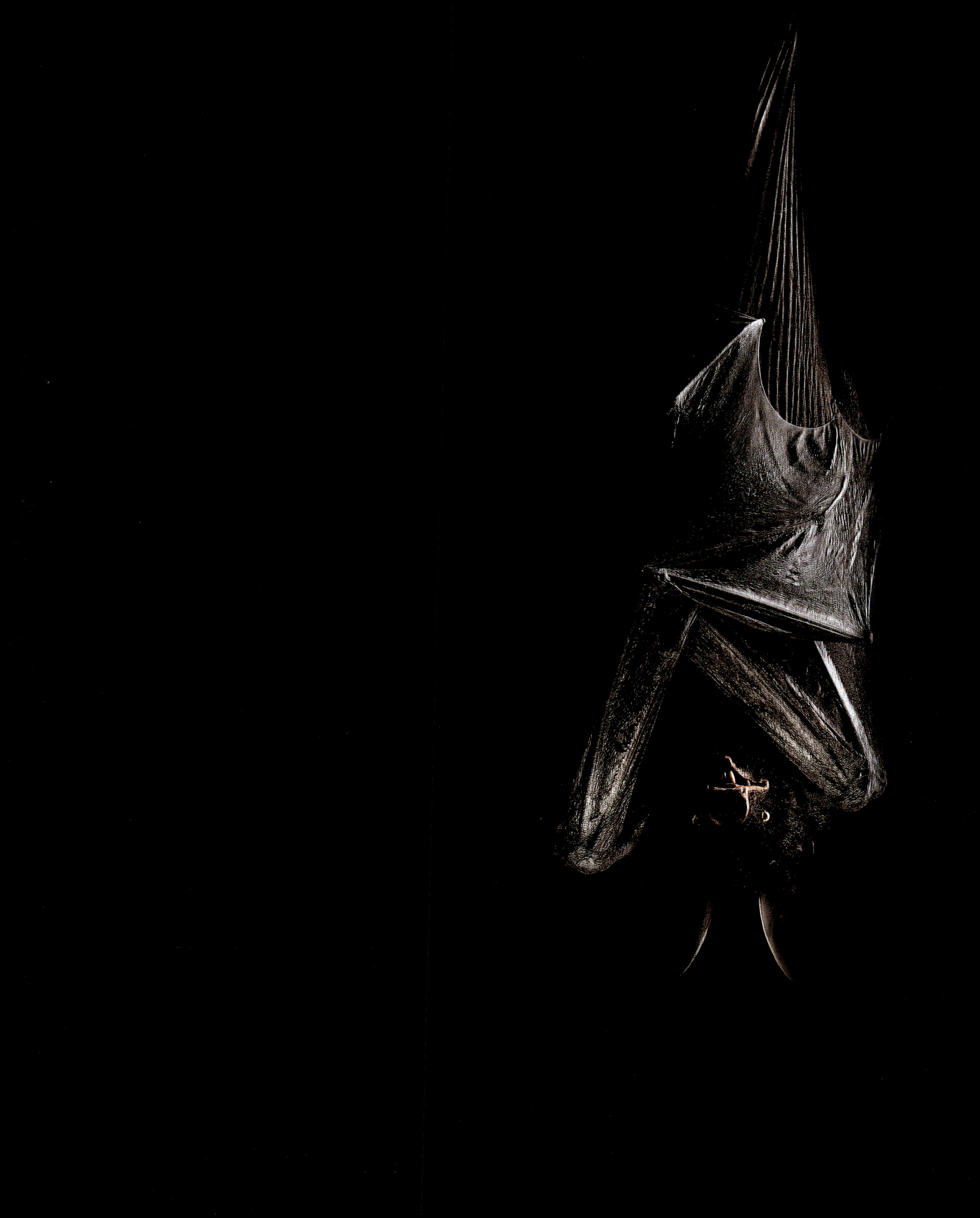

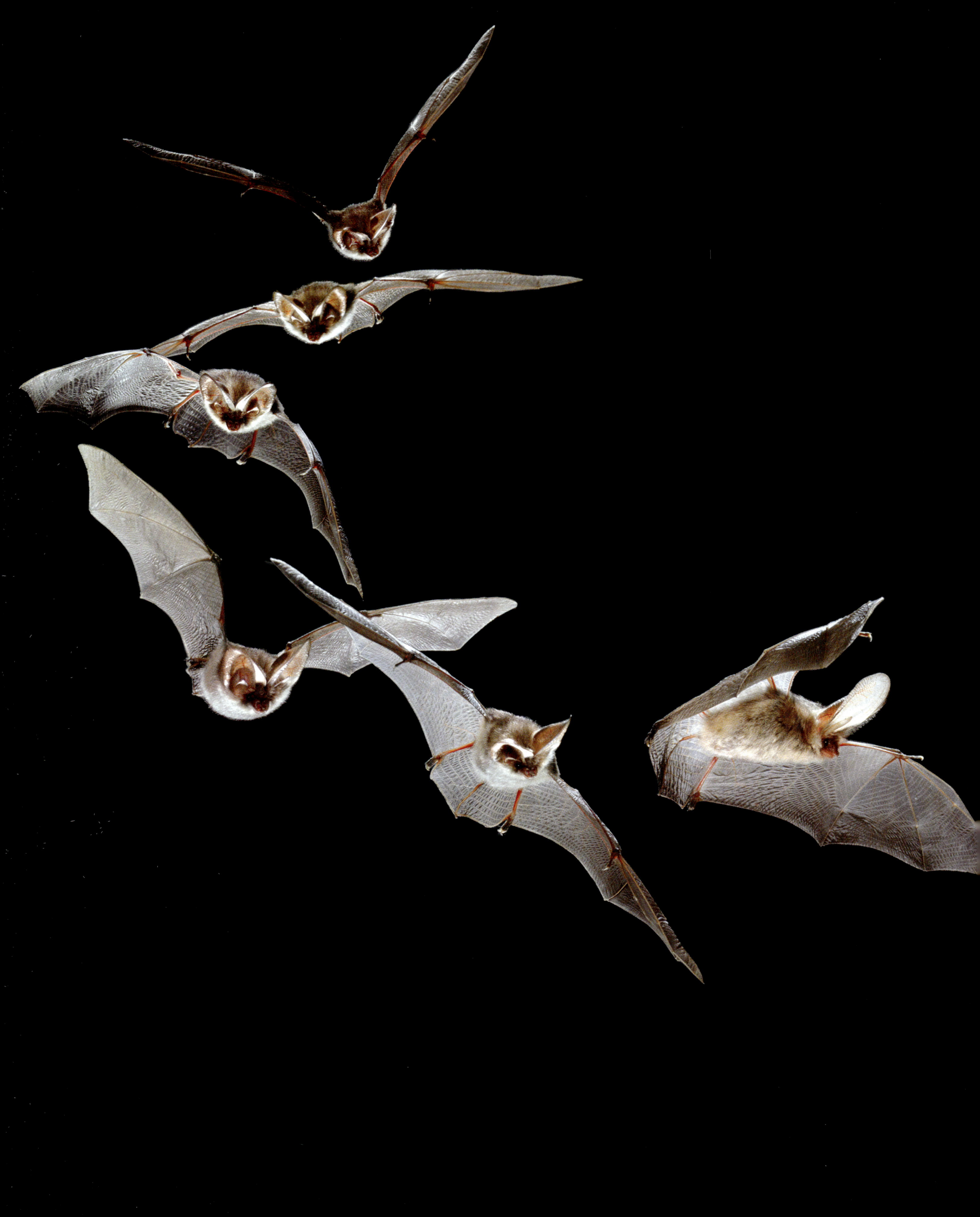

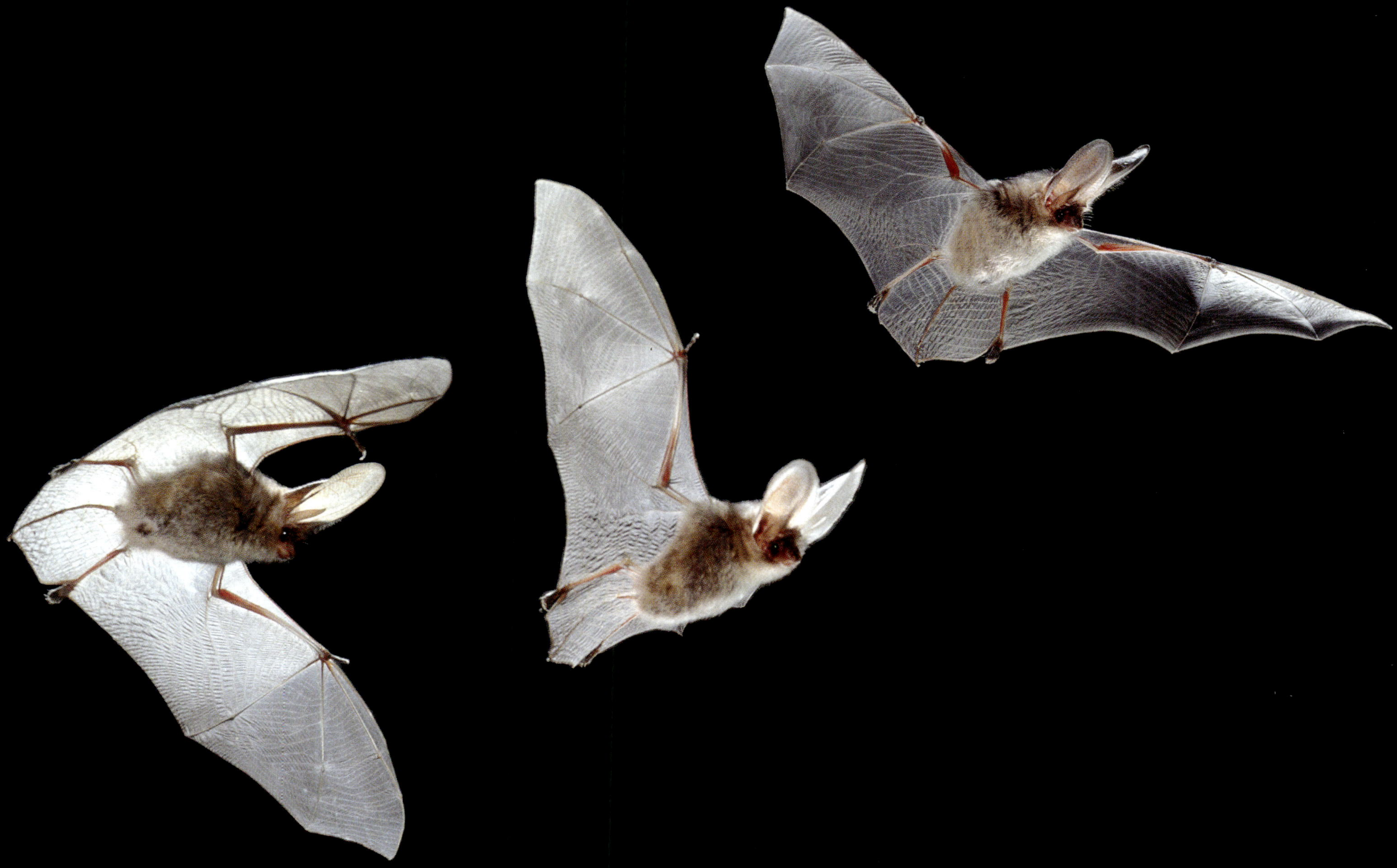

“Suspicions amongst thoughts are like bats amongst birds,
they ever fly by twilight.”

(Francis Bacon)

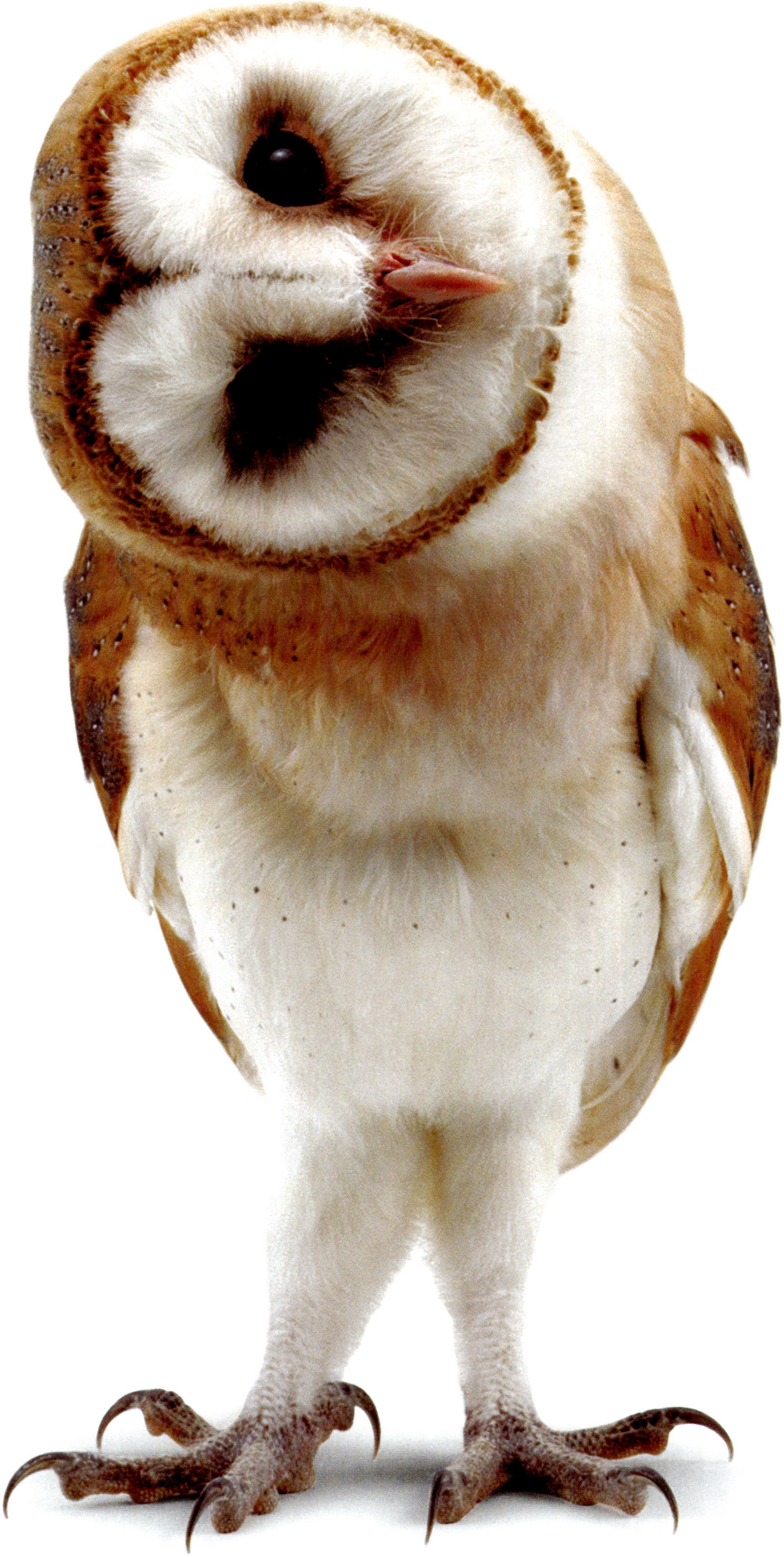

RASH ACTS

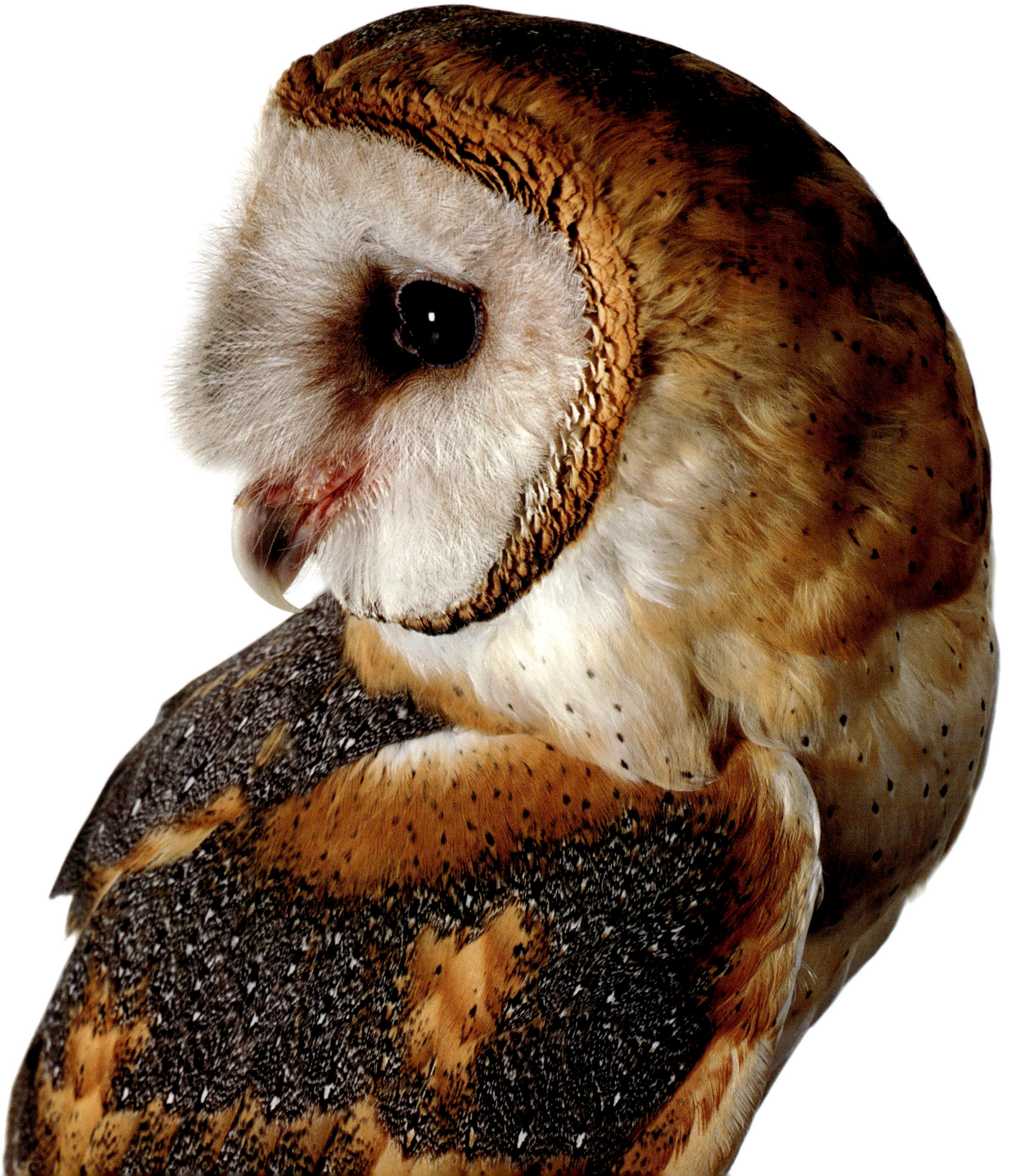

“There is none so blind as they that won't see.”
(Jonathan Swift)

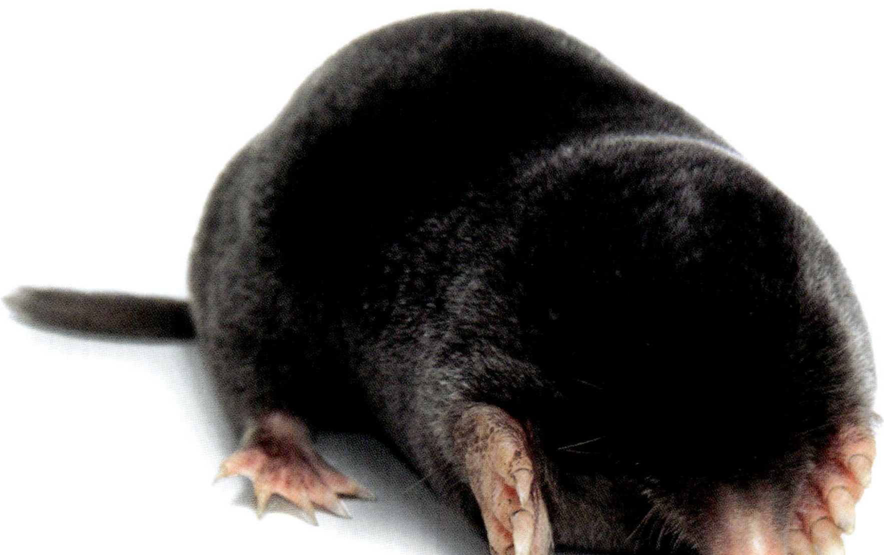

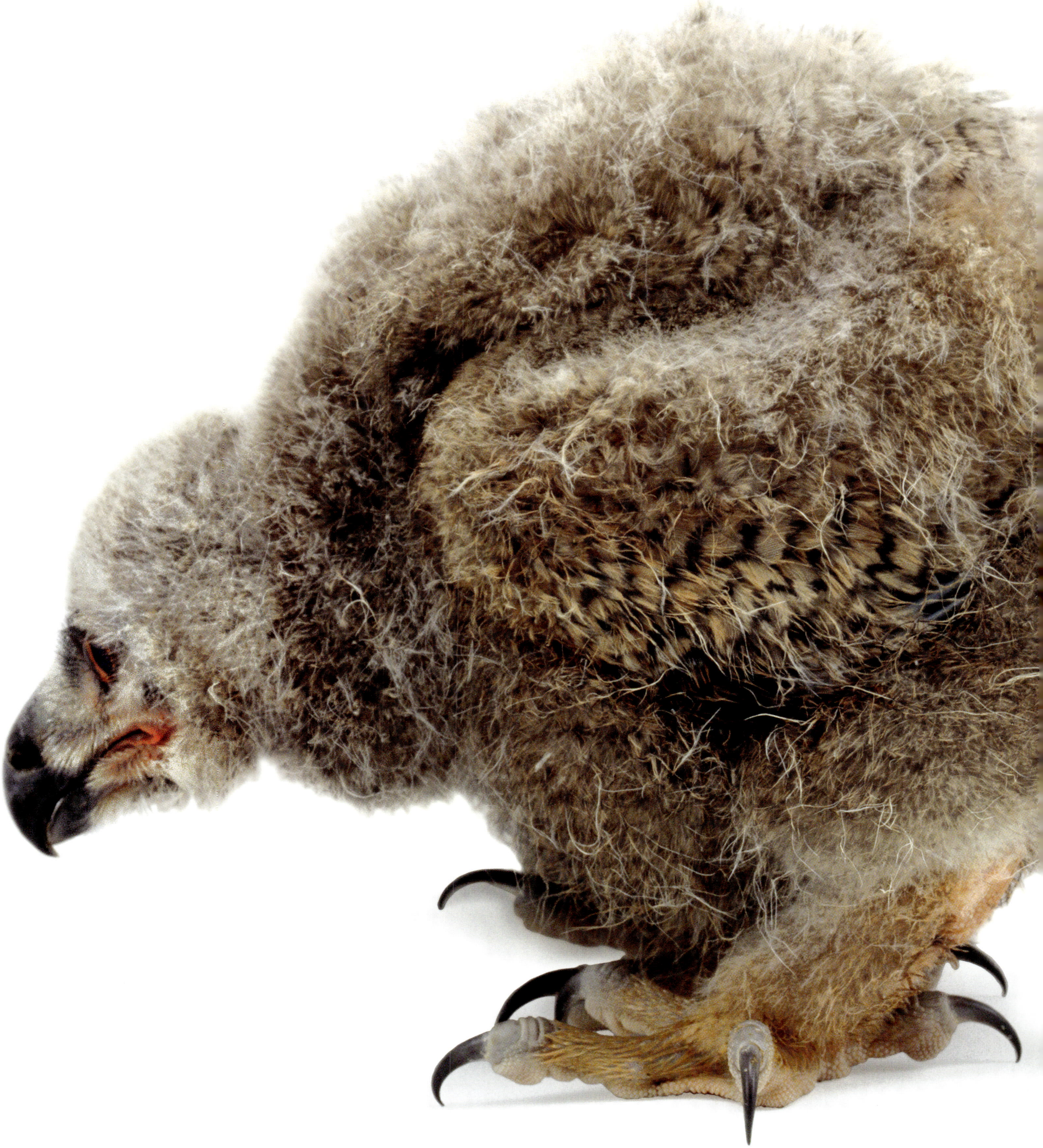

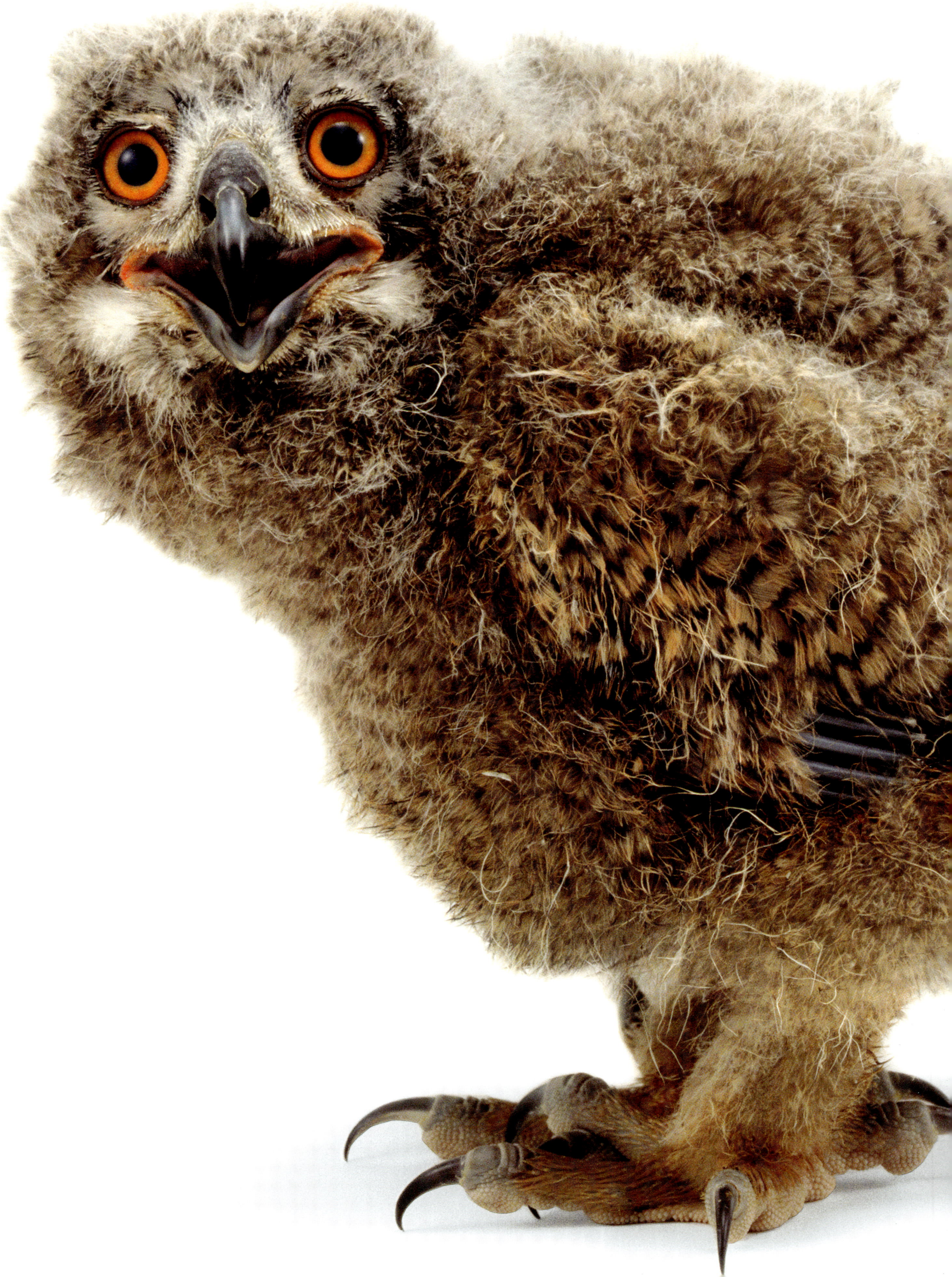

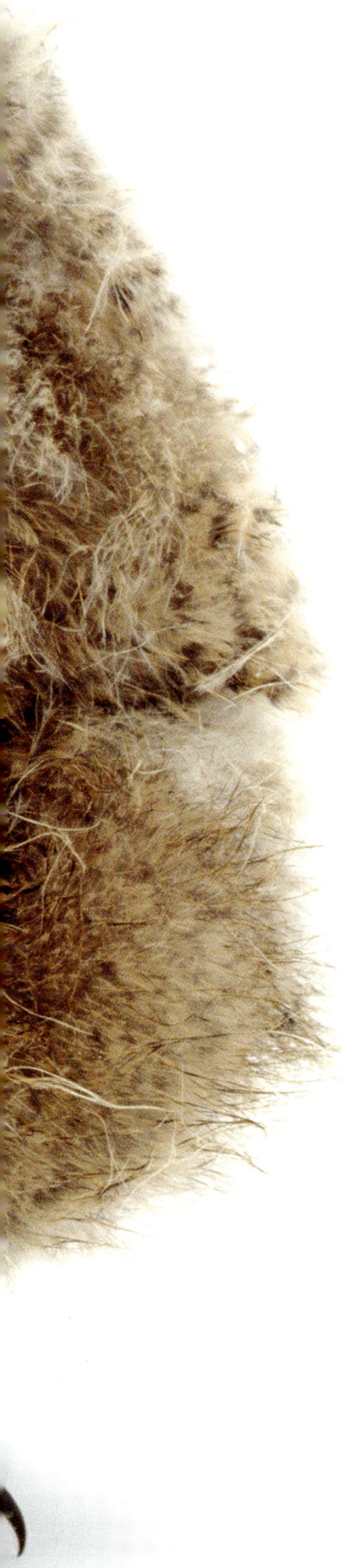

The Day of the Owl

(Leonardo Sciascia)

“Everything has rhythm. Everything dances.” (Maya Angelou)

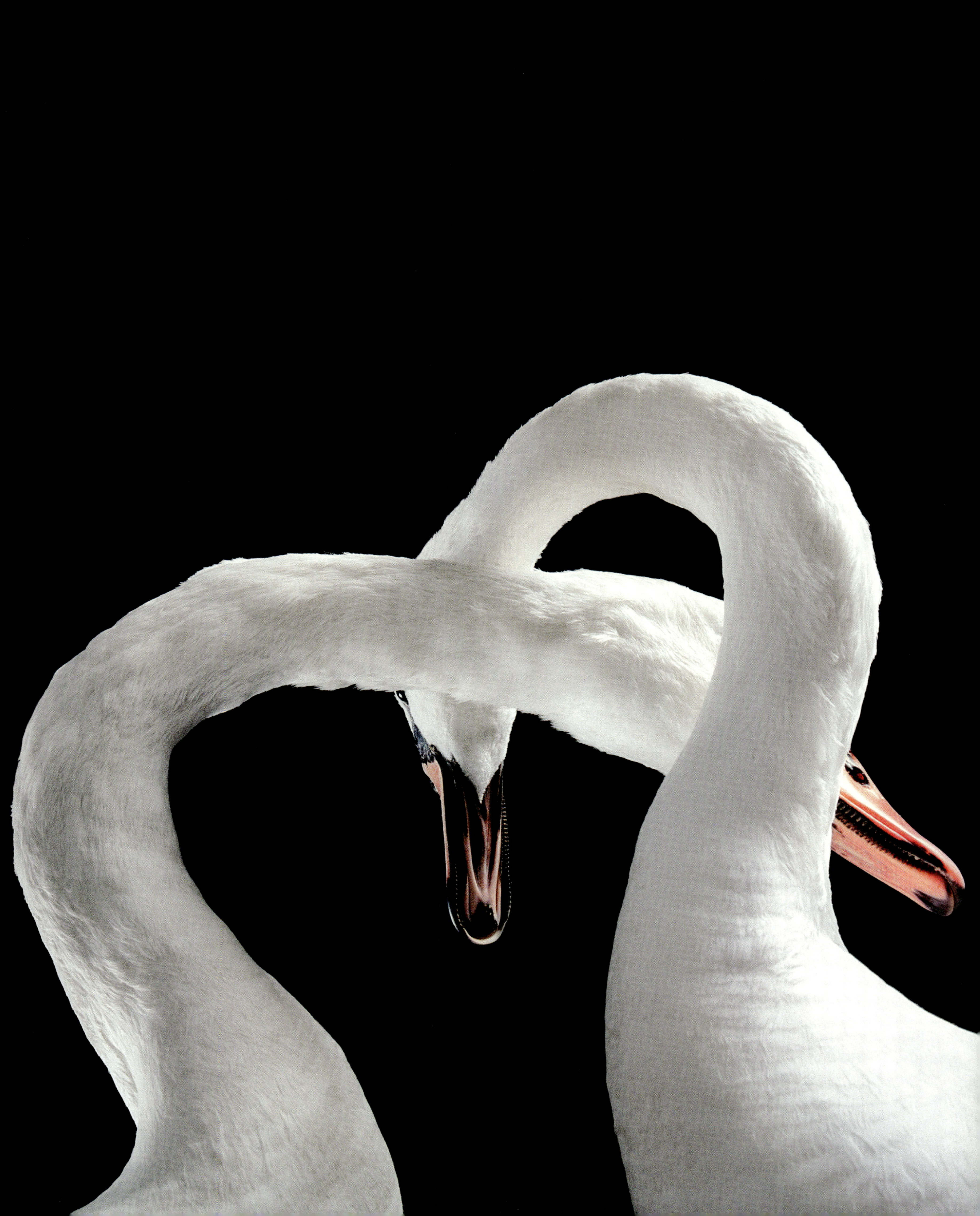

SWAN LAKE

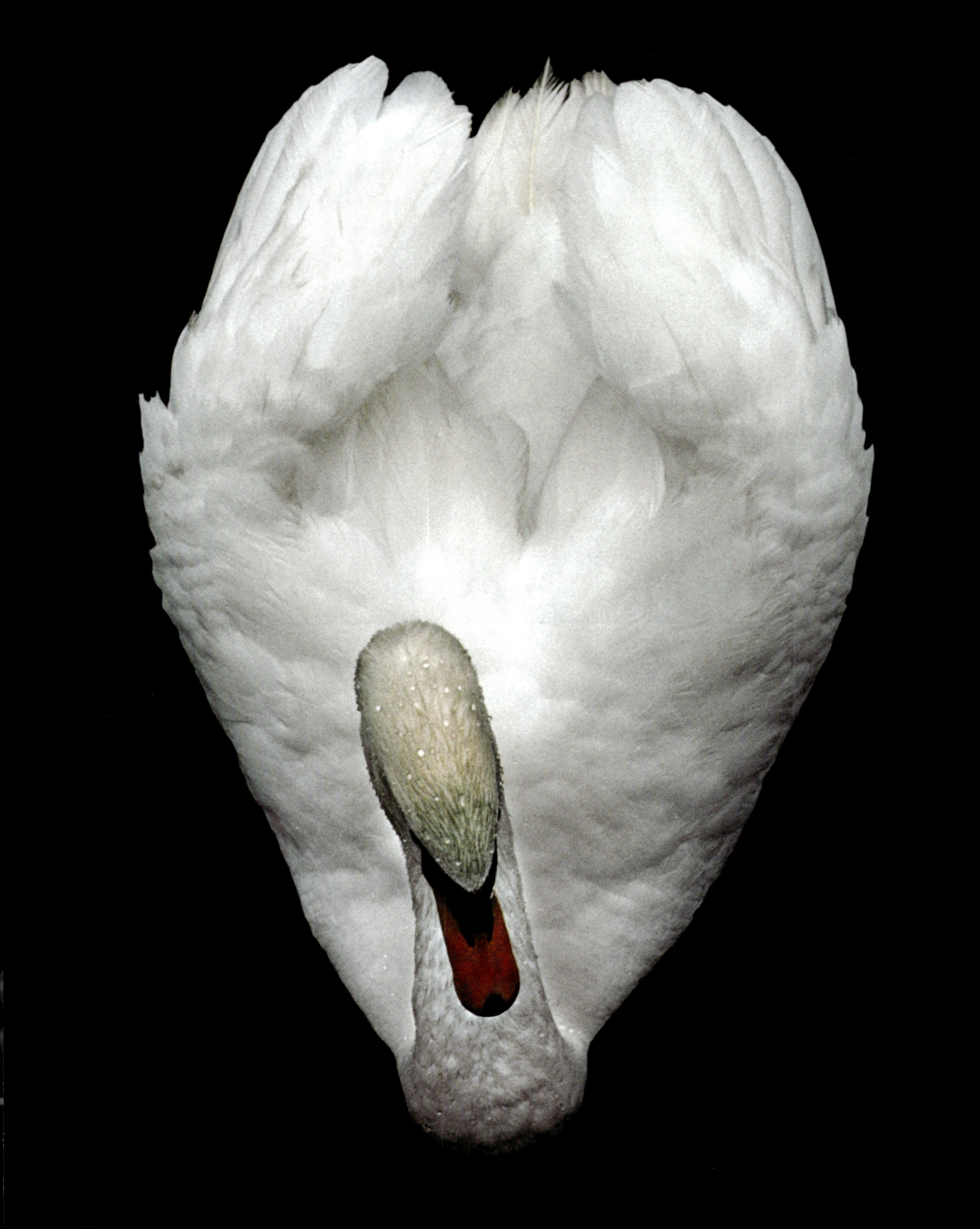

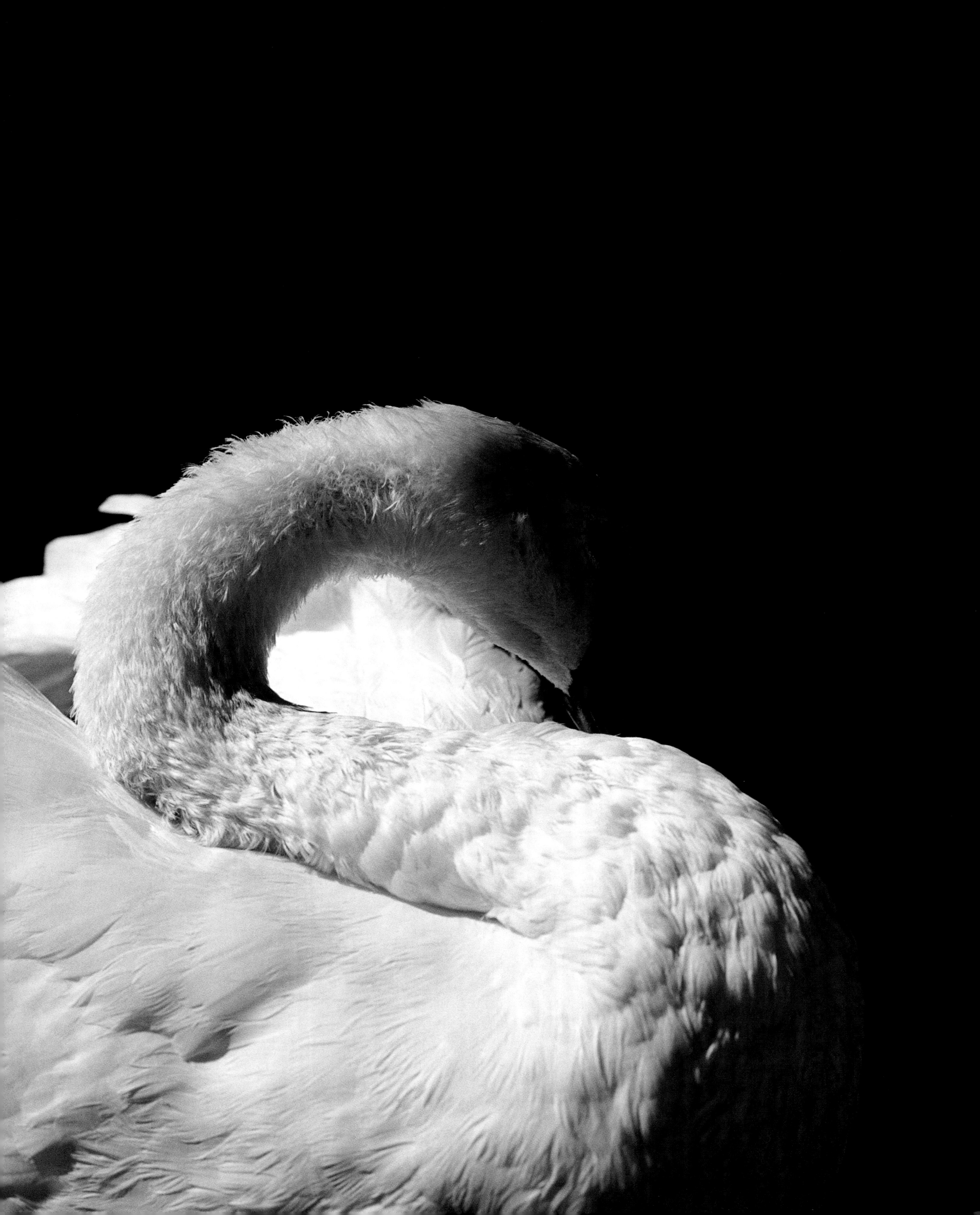

“Now wakes the owl. Now sleeps the swan. Behold a dream, the dream is gone.” (Pink Floyd – A Pillow of Winds)

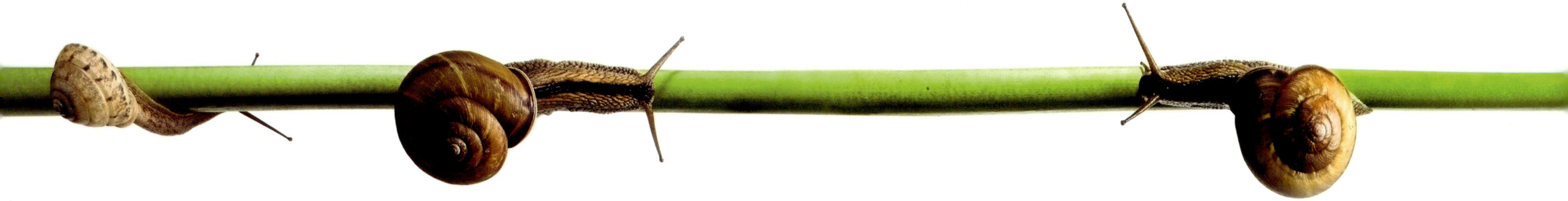

“Many are stubborn in pursuit of the path they have chosen, few in pursuit of the goal.” (Friedrich Nietzsche)

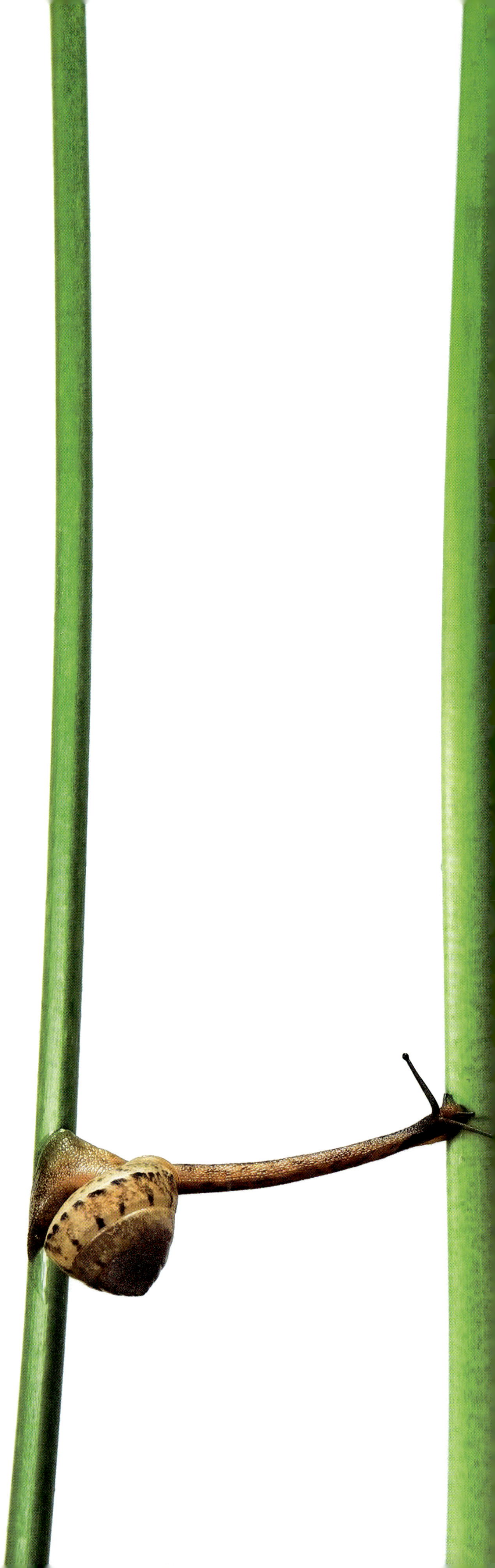

“Be slow in choosing a friend, slower in changing.”
(Benjamin Franklin)

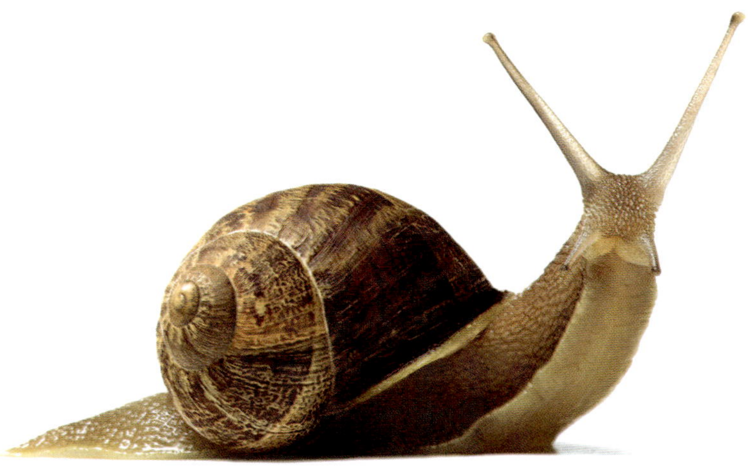

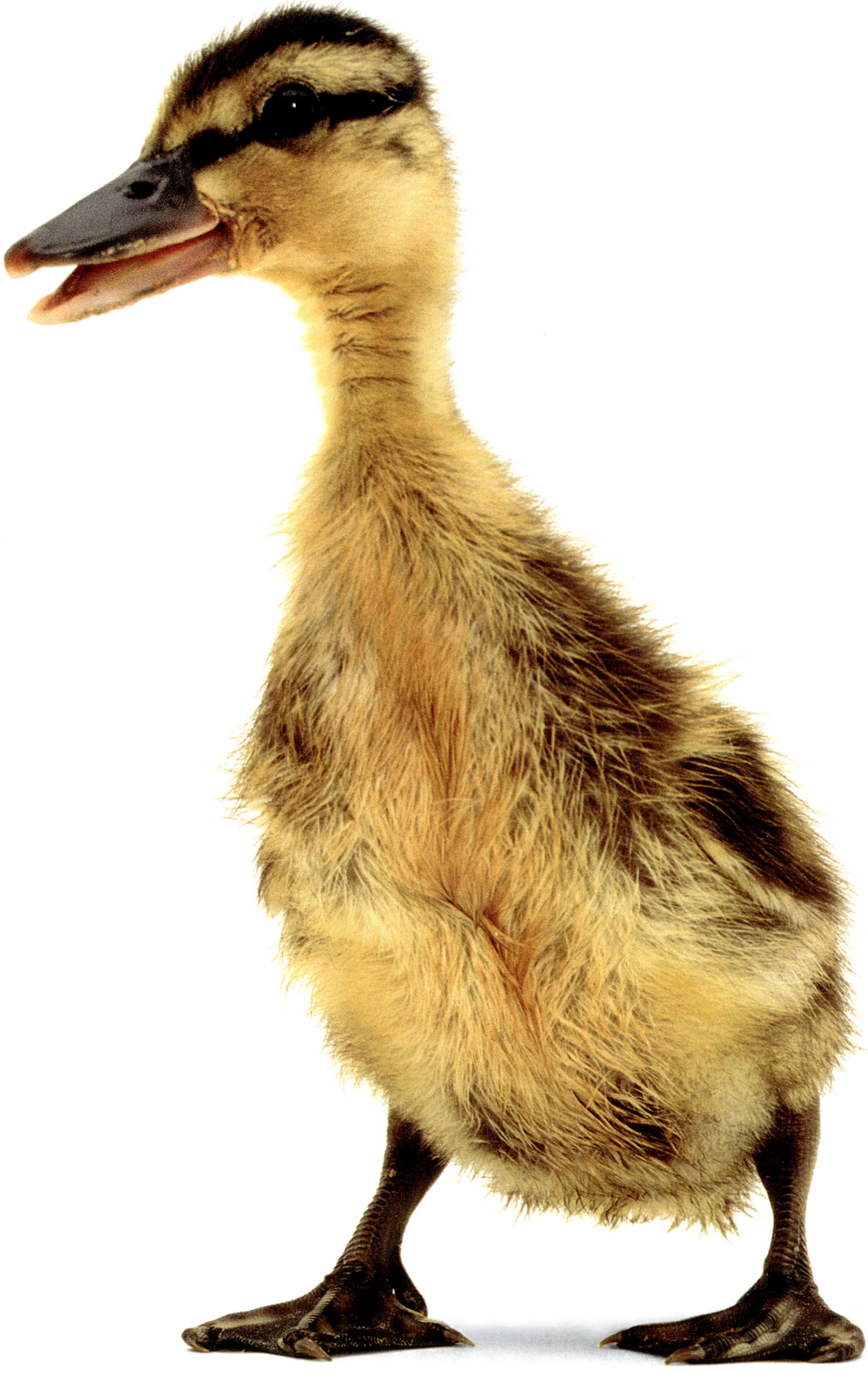

ROYAL PARADE

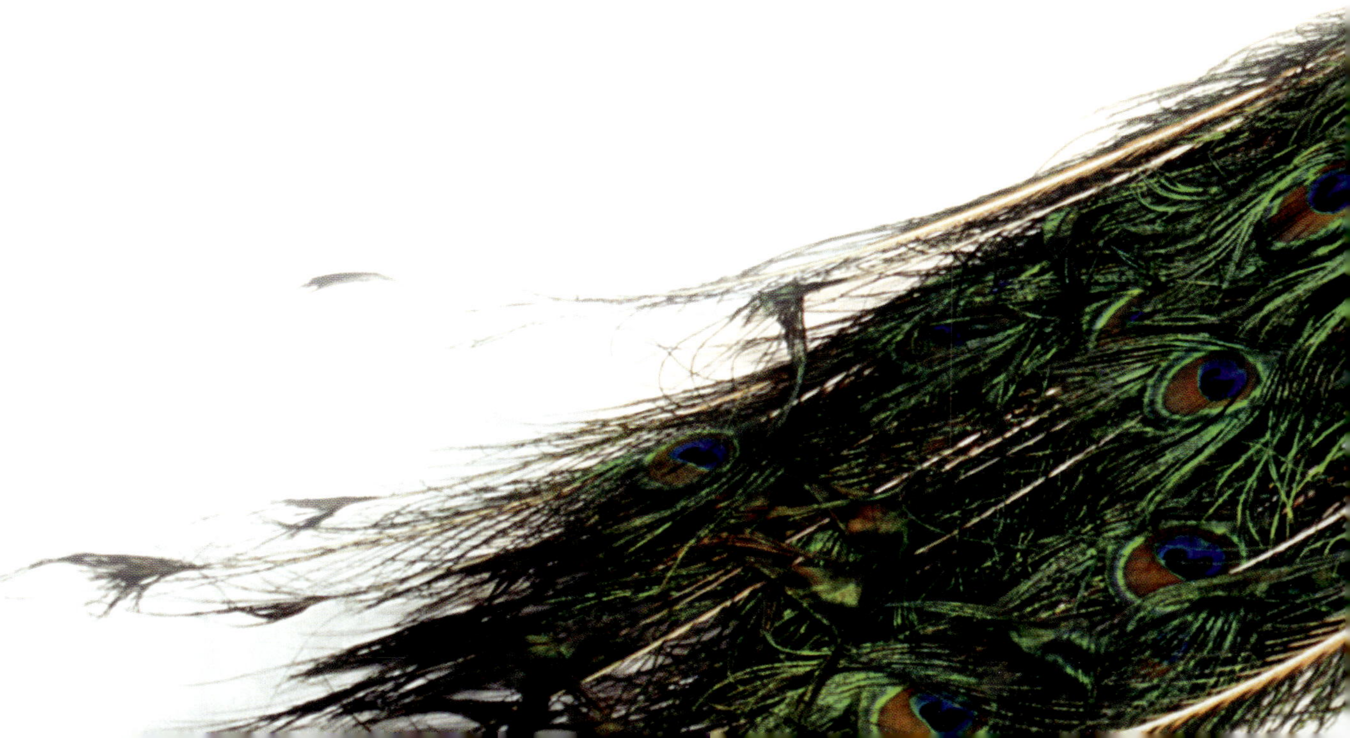

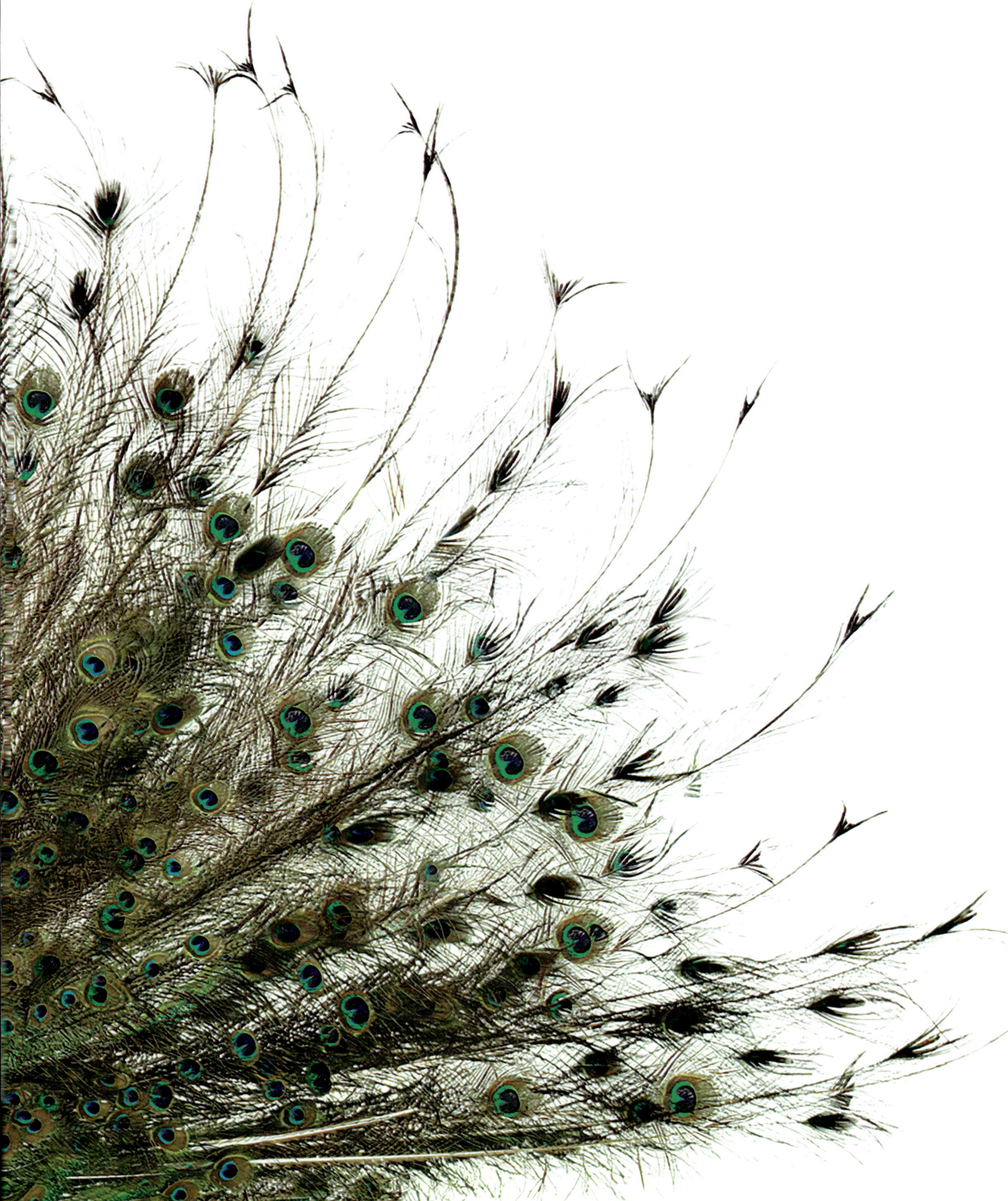

BELLE ÉPOQUE

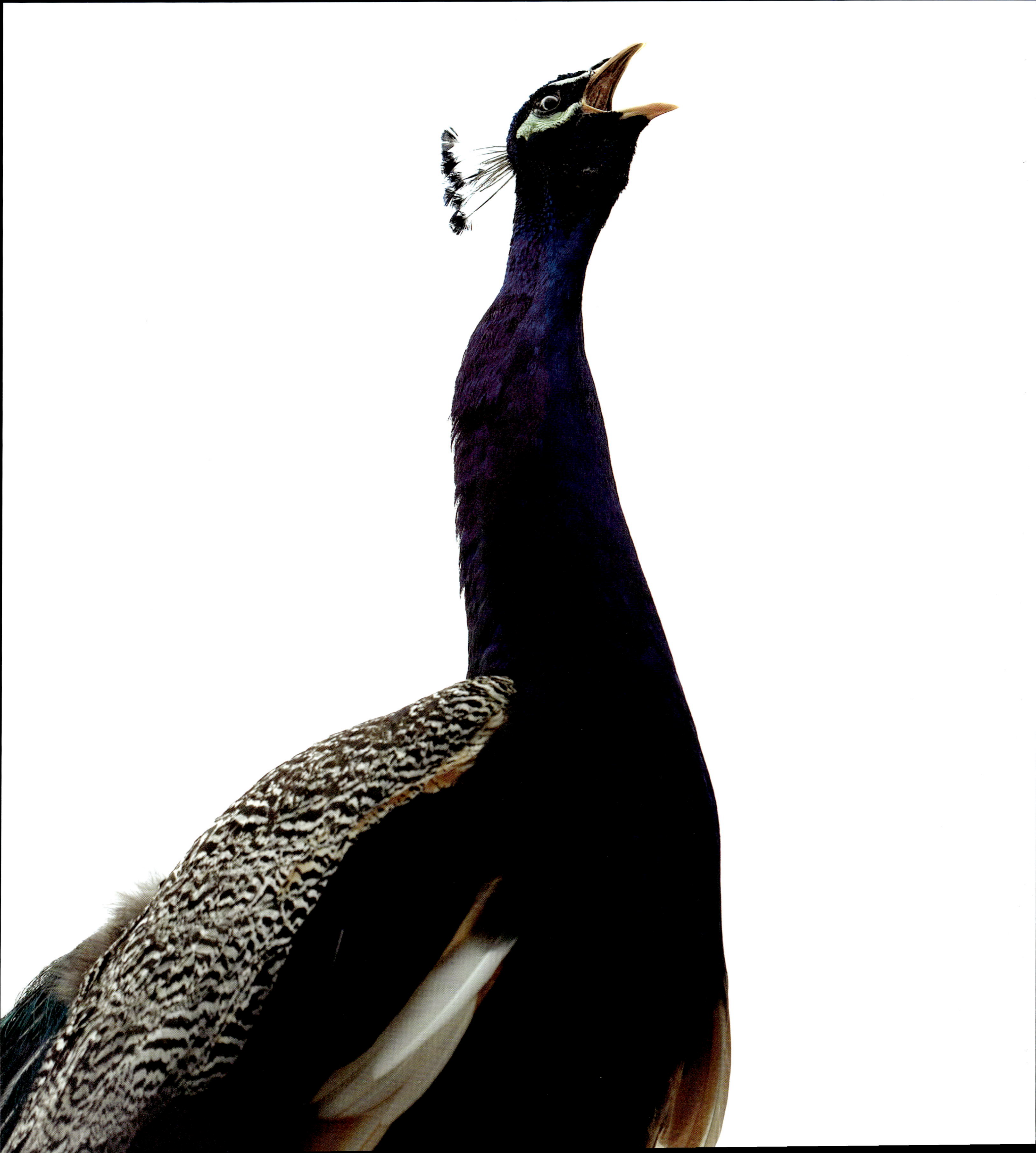

"Hope is the thing with feathers, that perches in the soul,
and sings the tune without words, and never stops at all."

(Emily Dickinson)

THE LIFE SEASONS

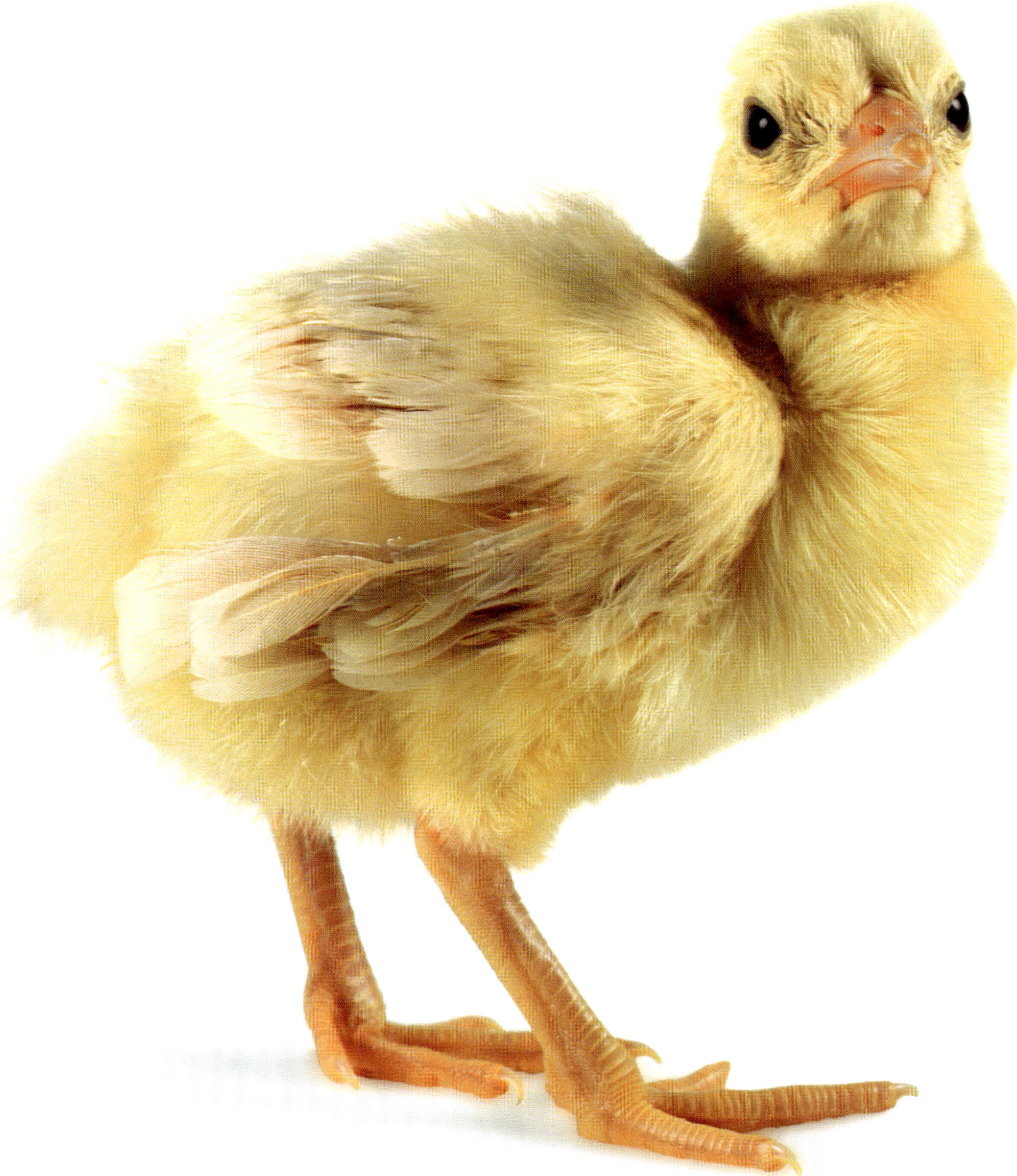

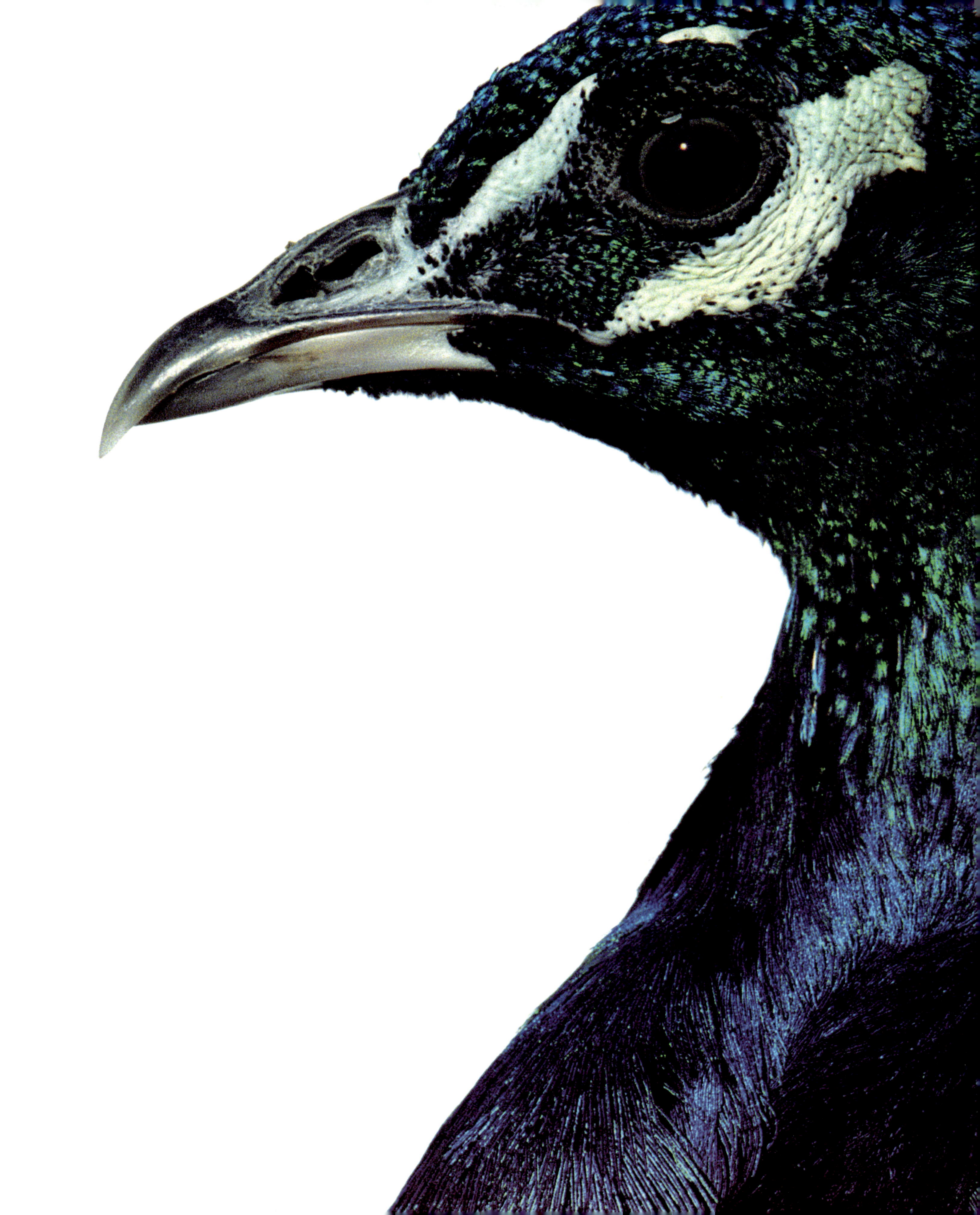

“To see what is in front of one's nose needs a constant struggle.”

(George Orwell)

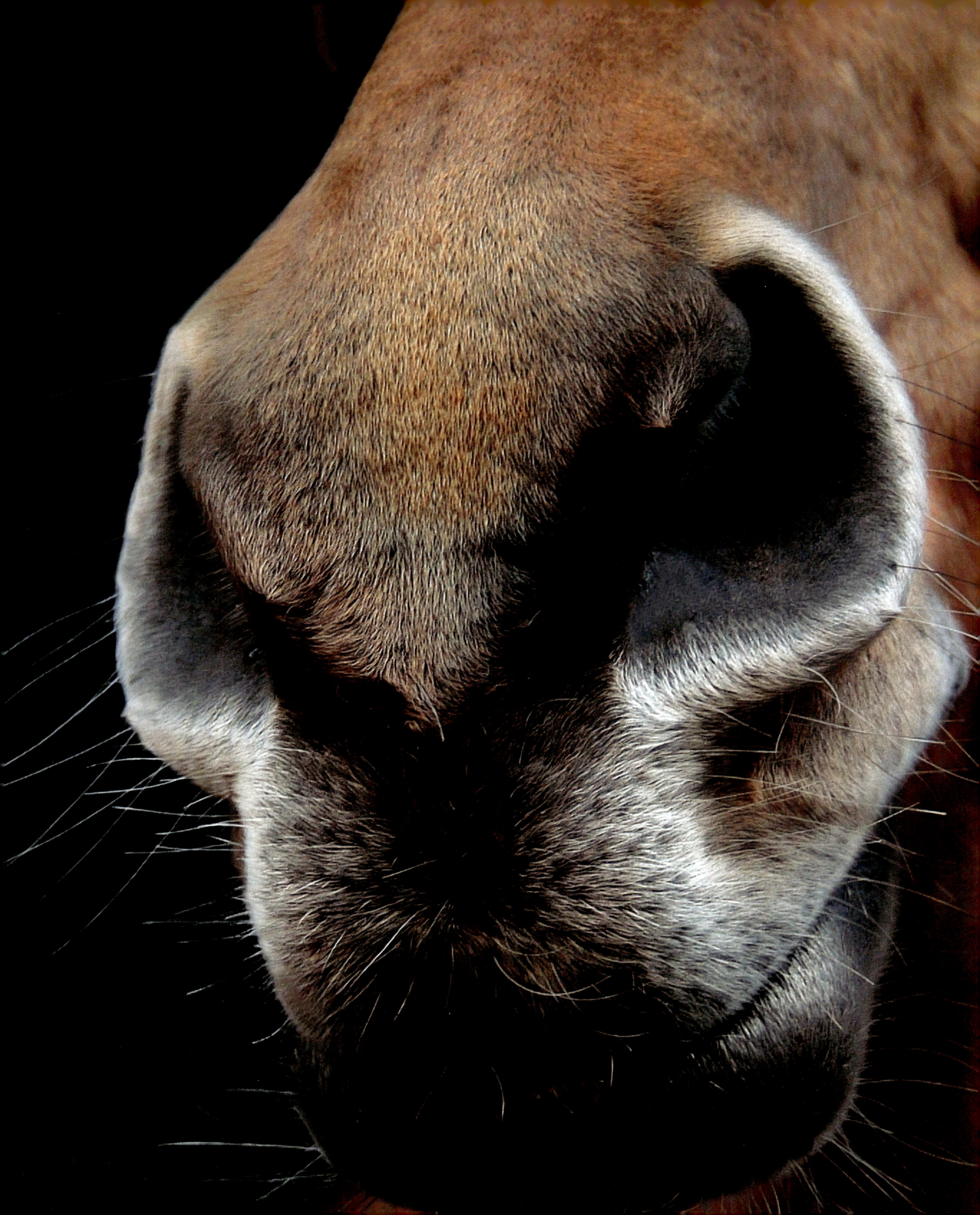

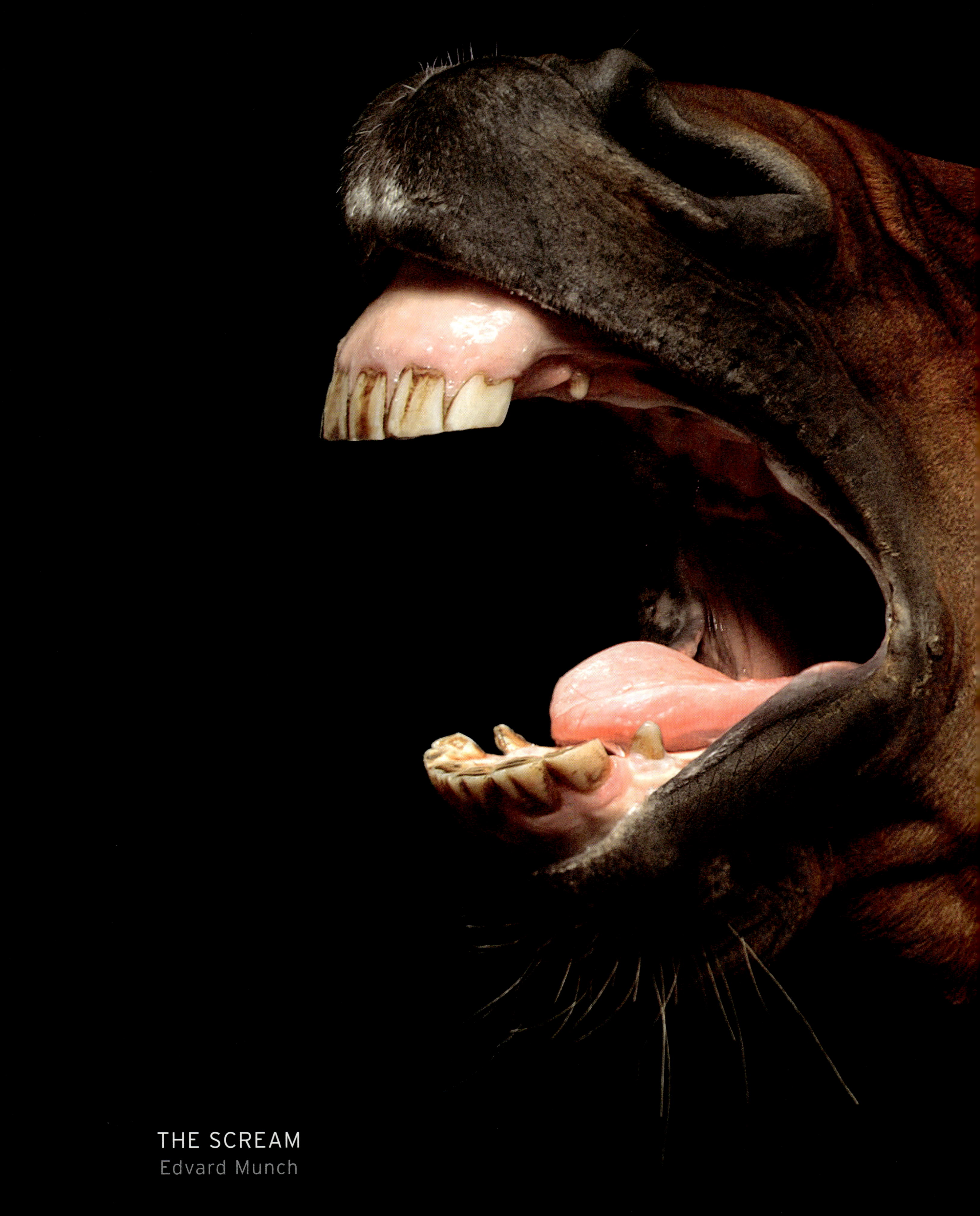

THE SCREAM
Edvard Munch

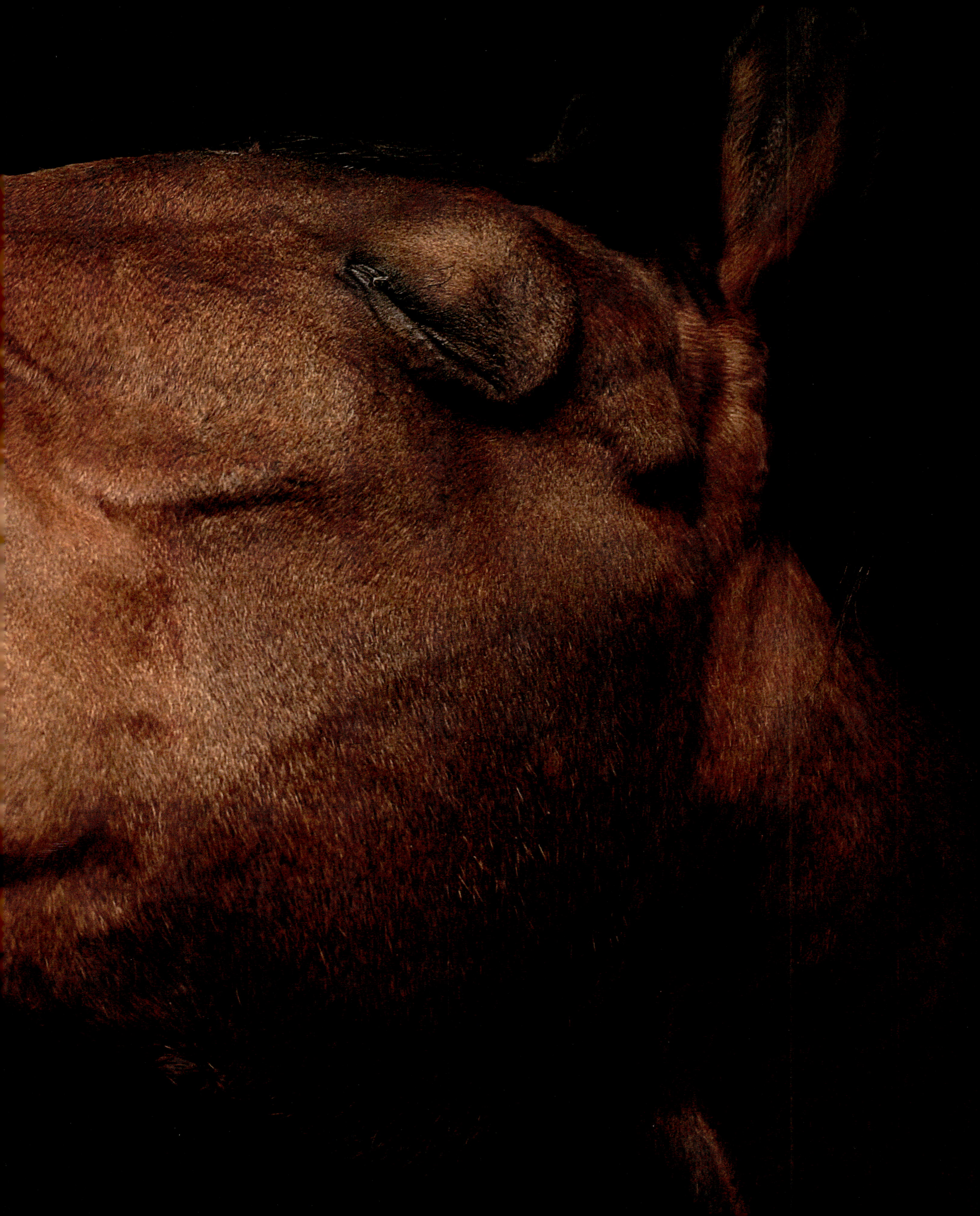

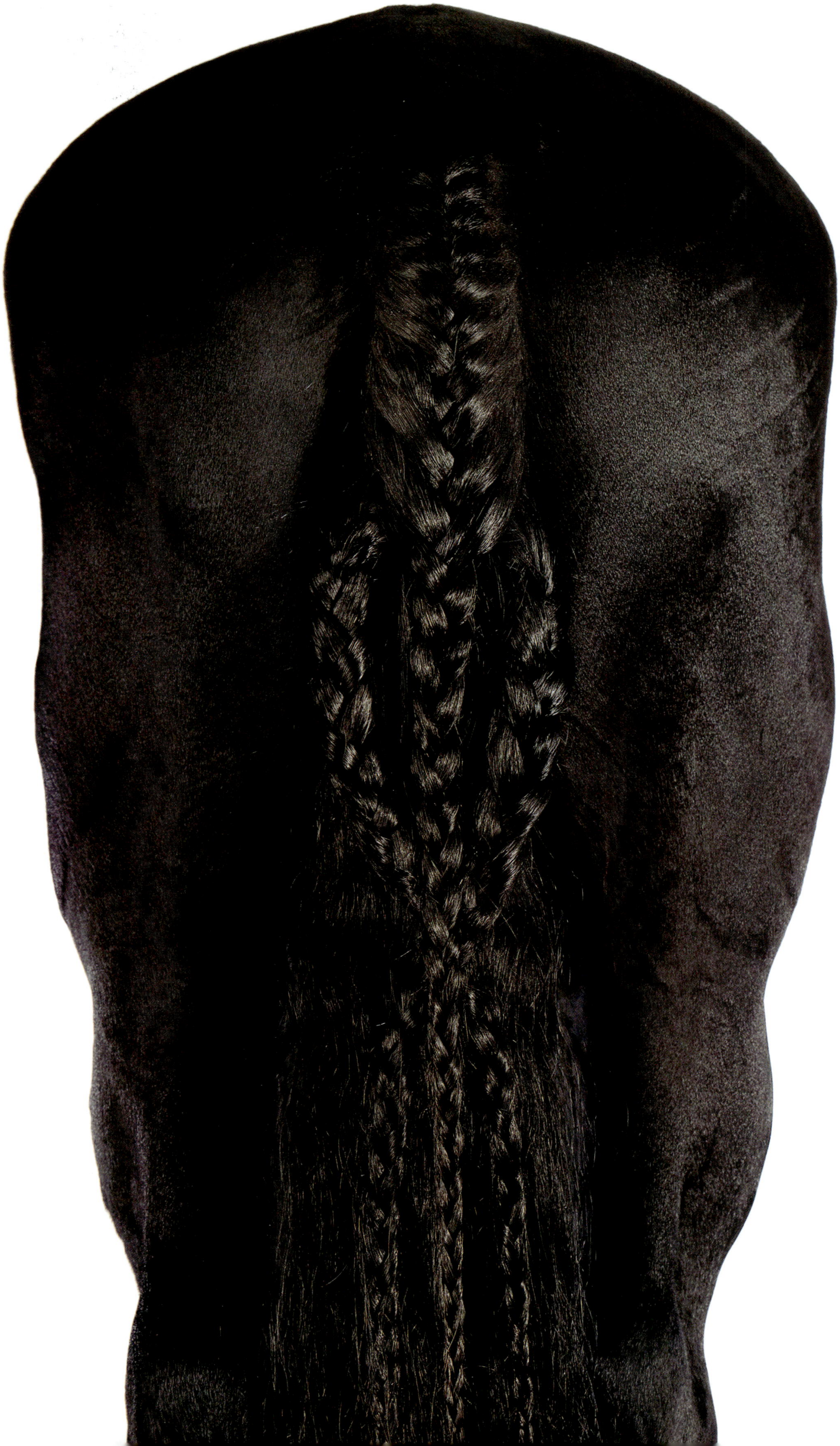

HAIR-STYLE

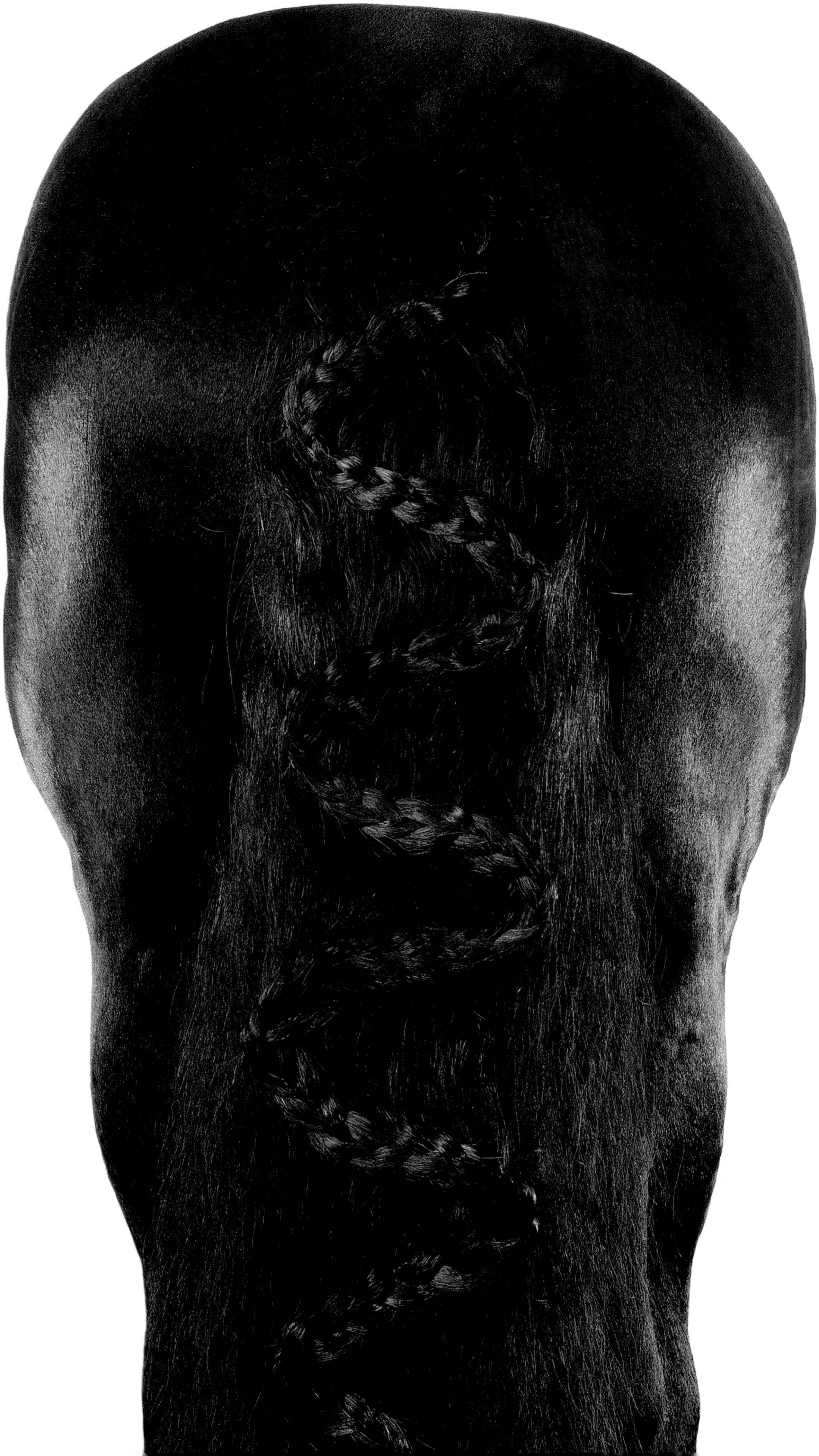

"A horse is a thing of beauty. . ."
(Xenophon)

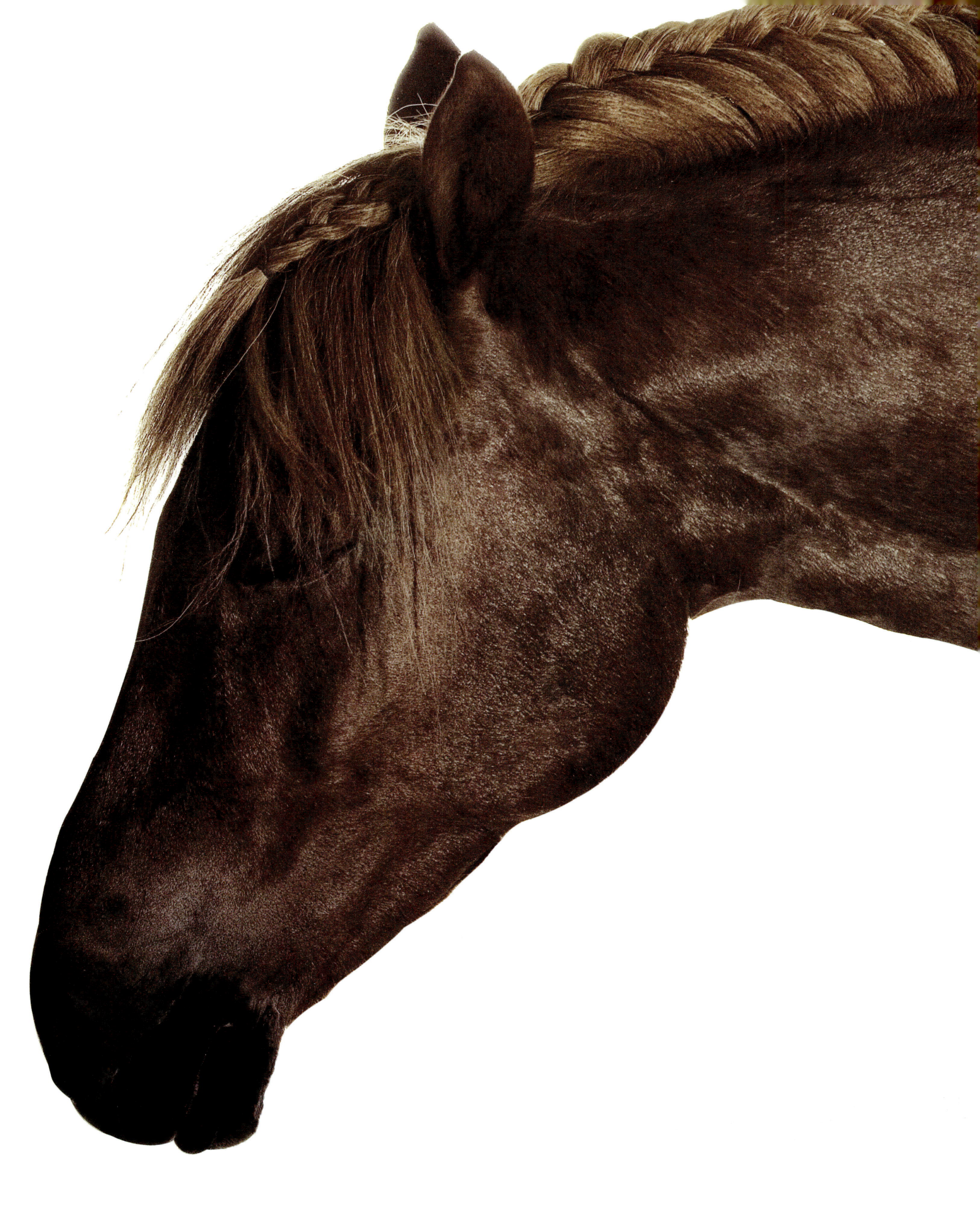

“A horse is the projection of peoples' dreams about themselves – strong, powerful, beautiful.”
(Pam Brown)

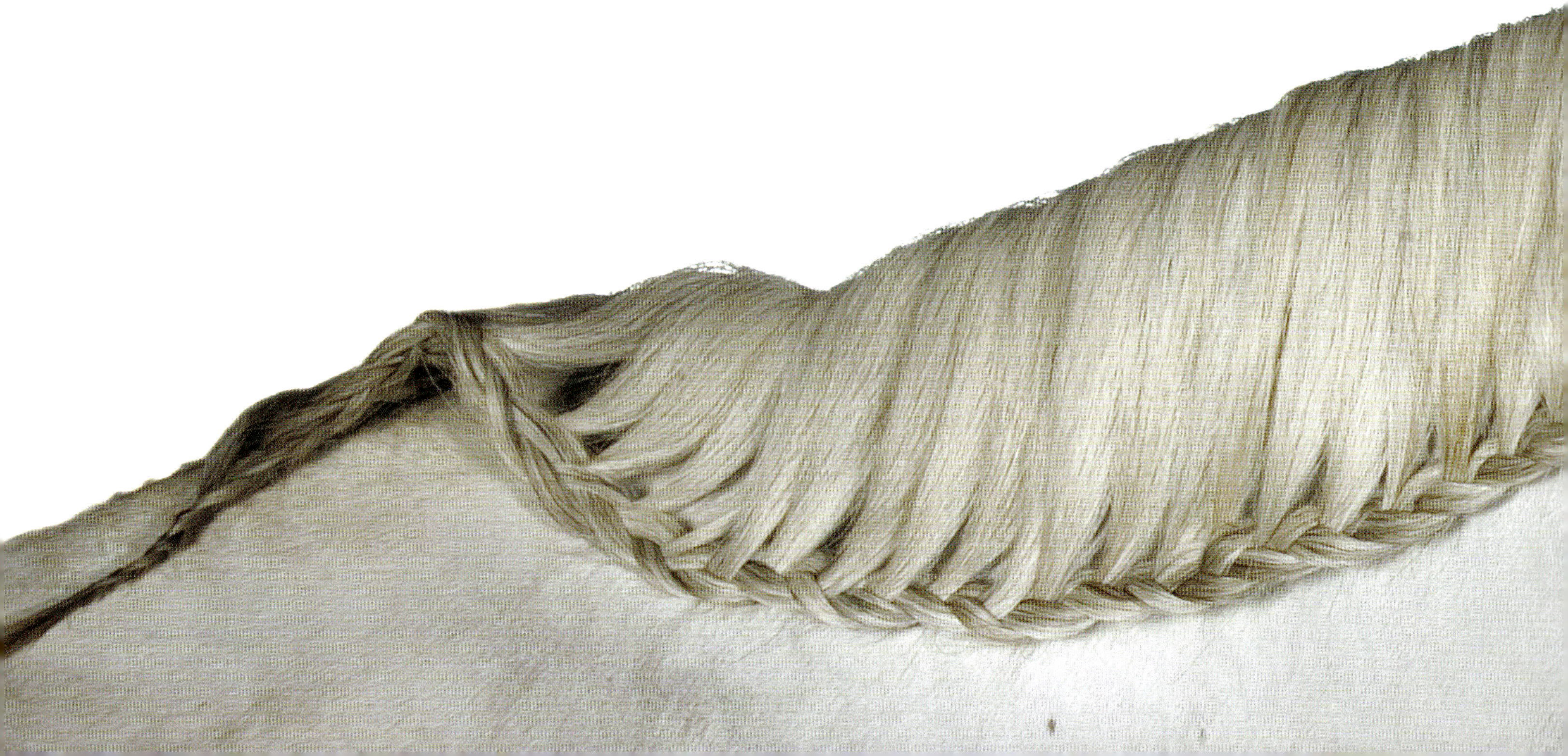

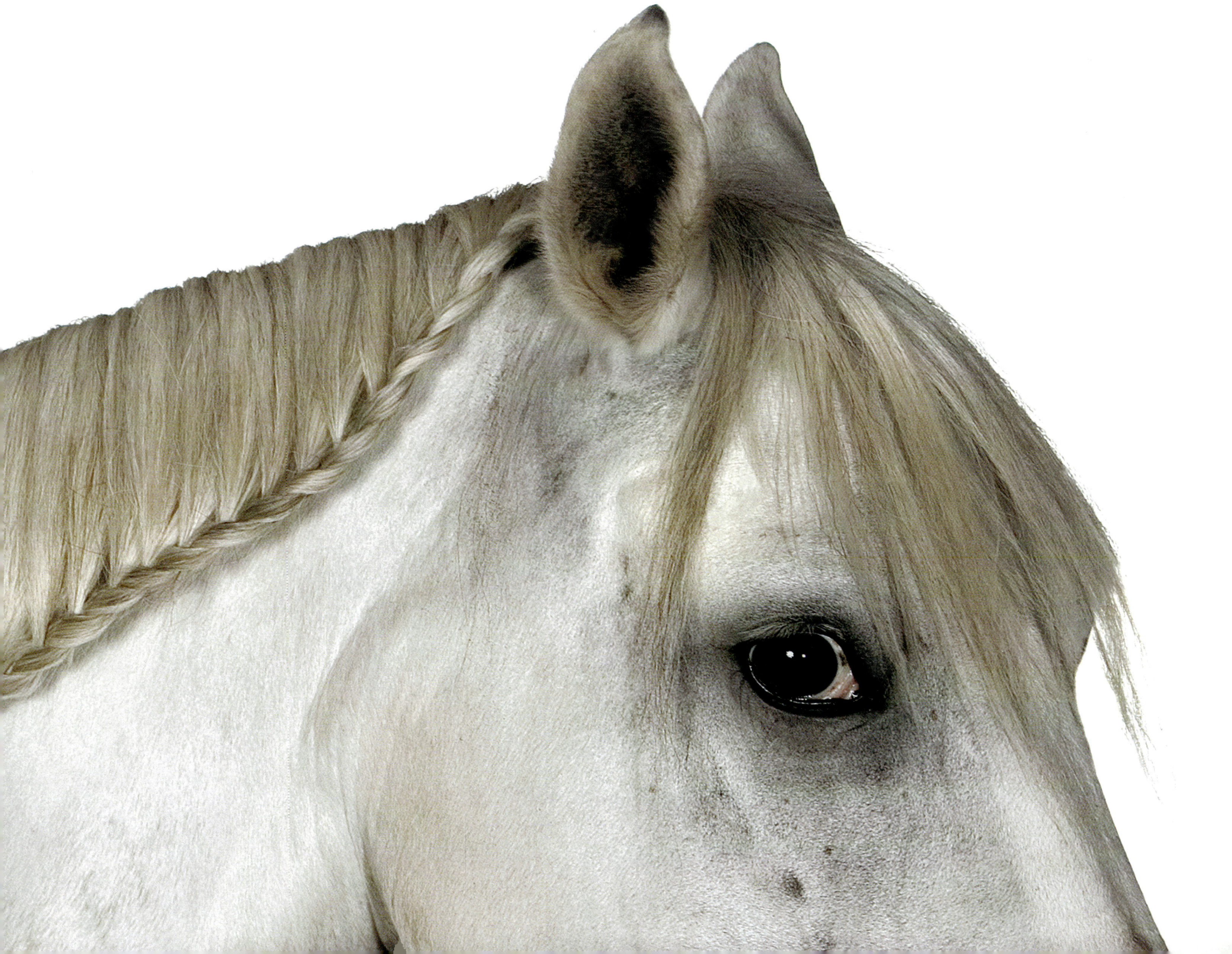

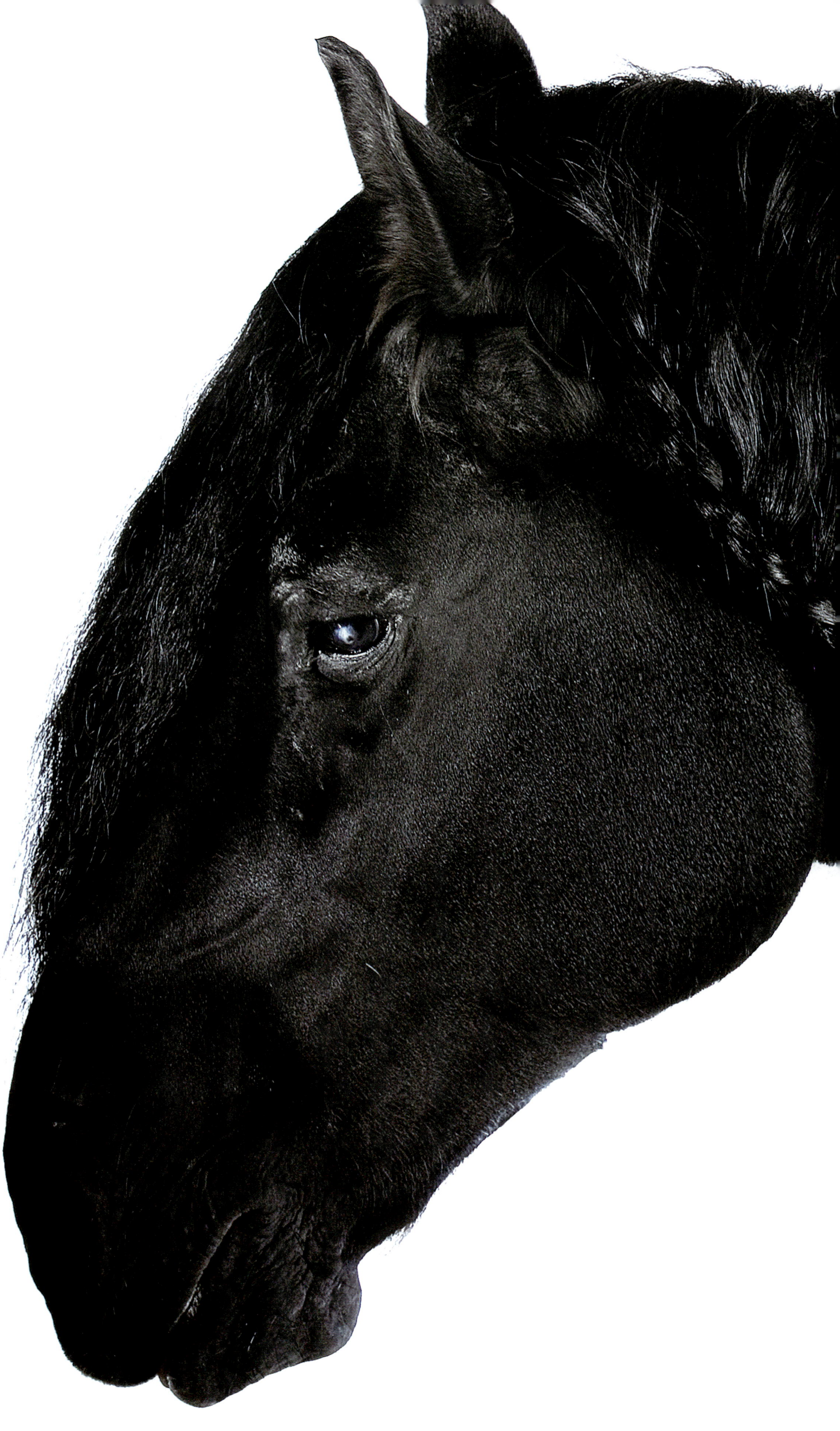

“My kingdom for a horse!”

(Richard III - William Shakespeare)

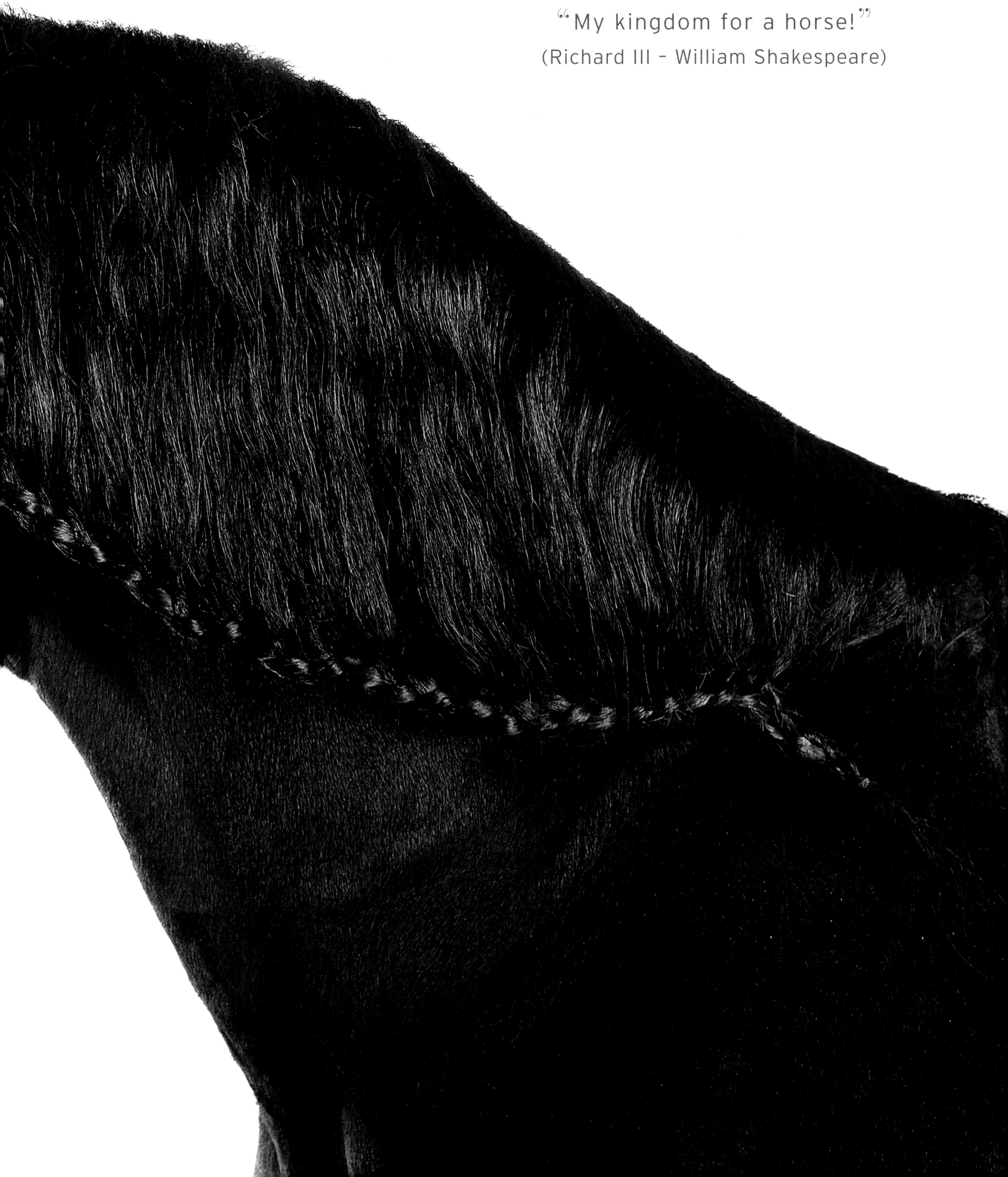

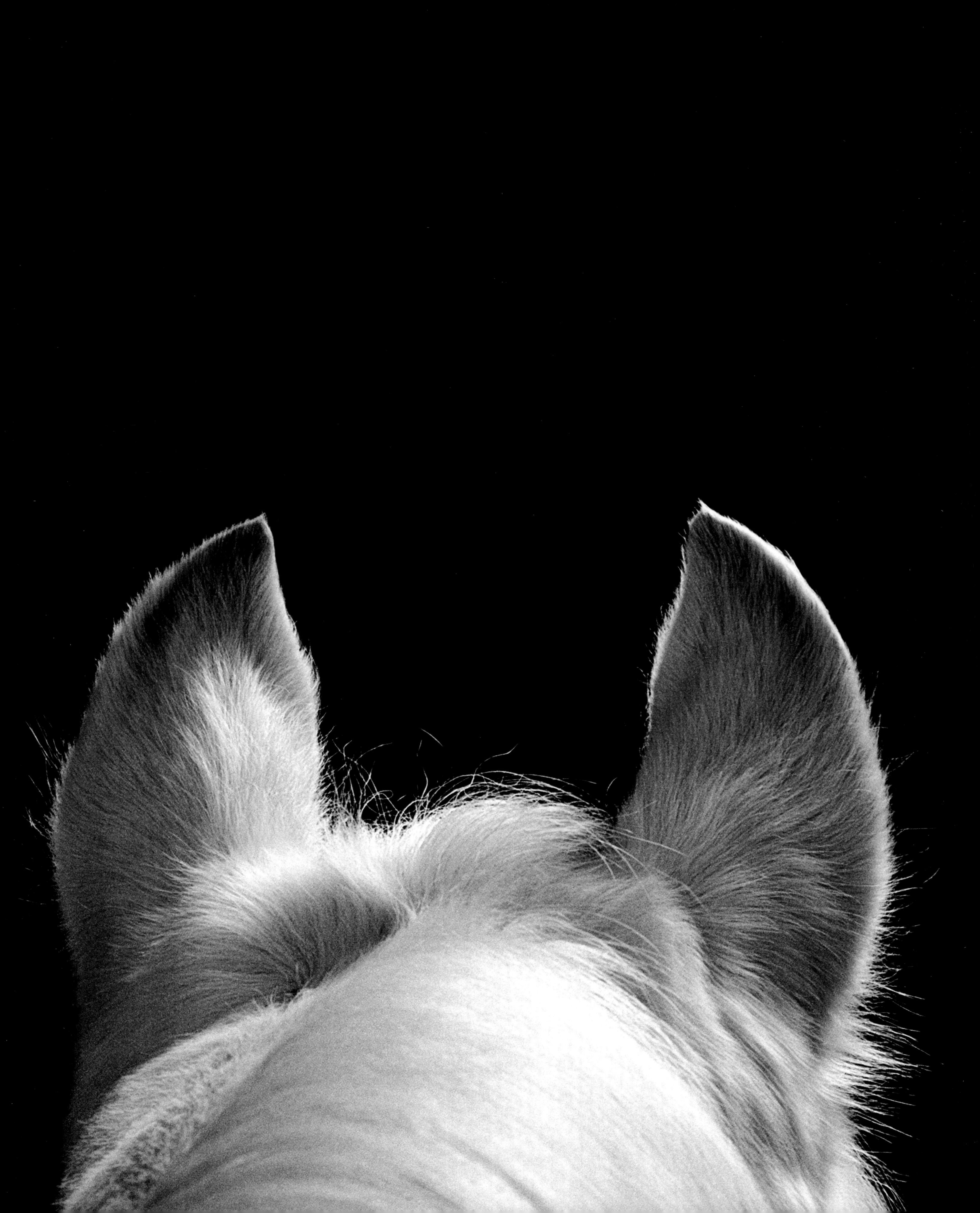

"The world seems better
seen through a horse's ears."
(Anonymous)

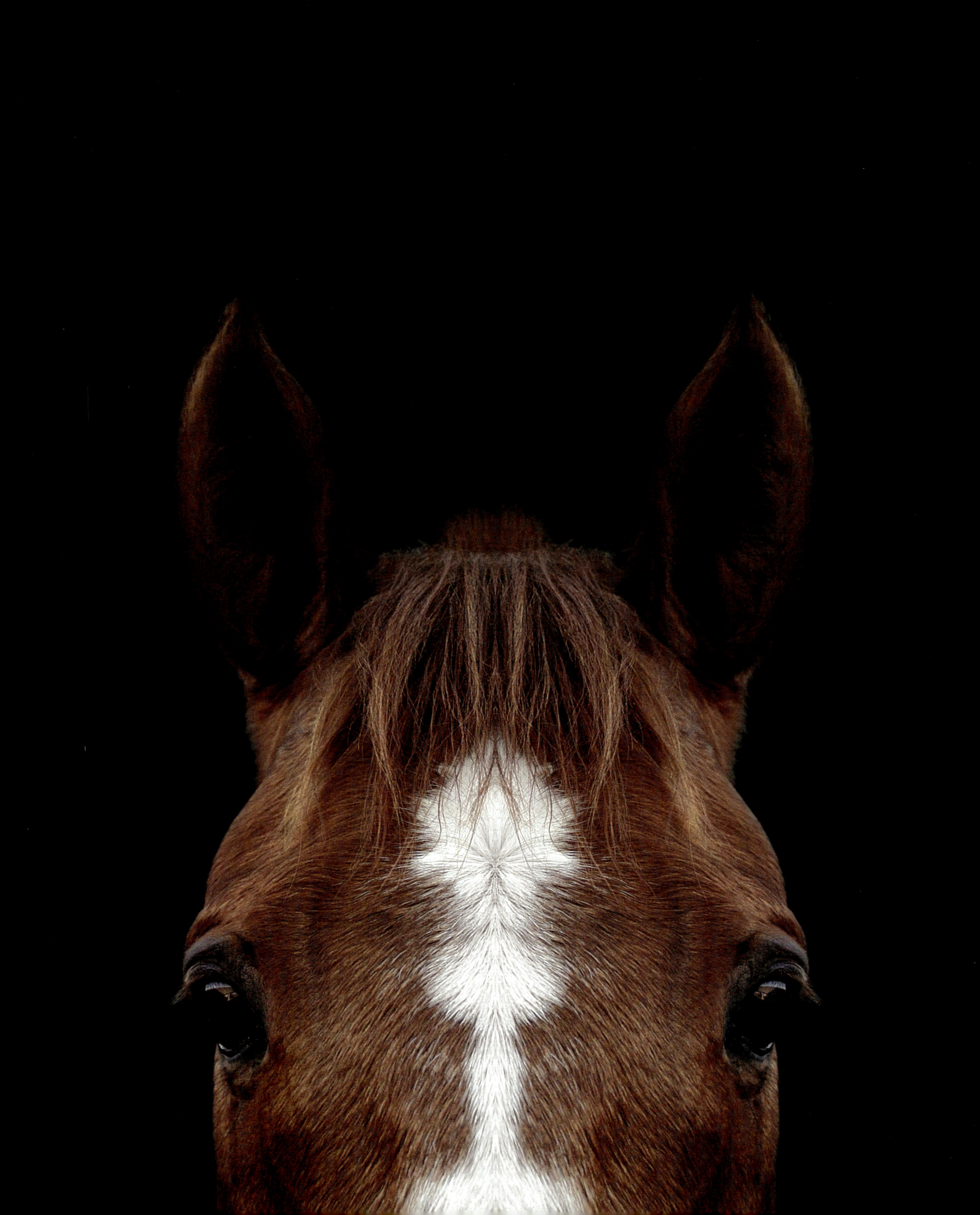

“Not all horses were born equal.
A few were born to win.”

(Mark Twain)

THE BEGINNING OF A STORY

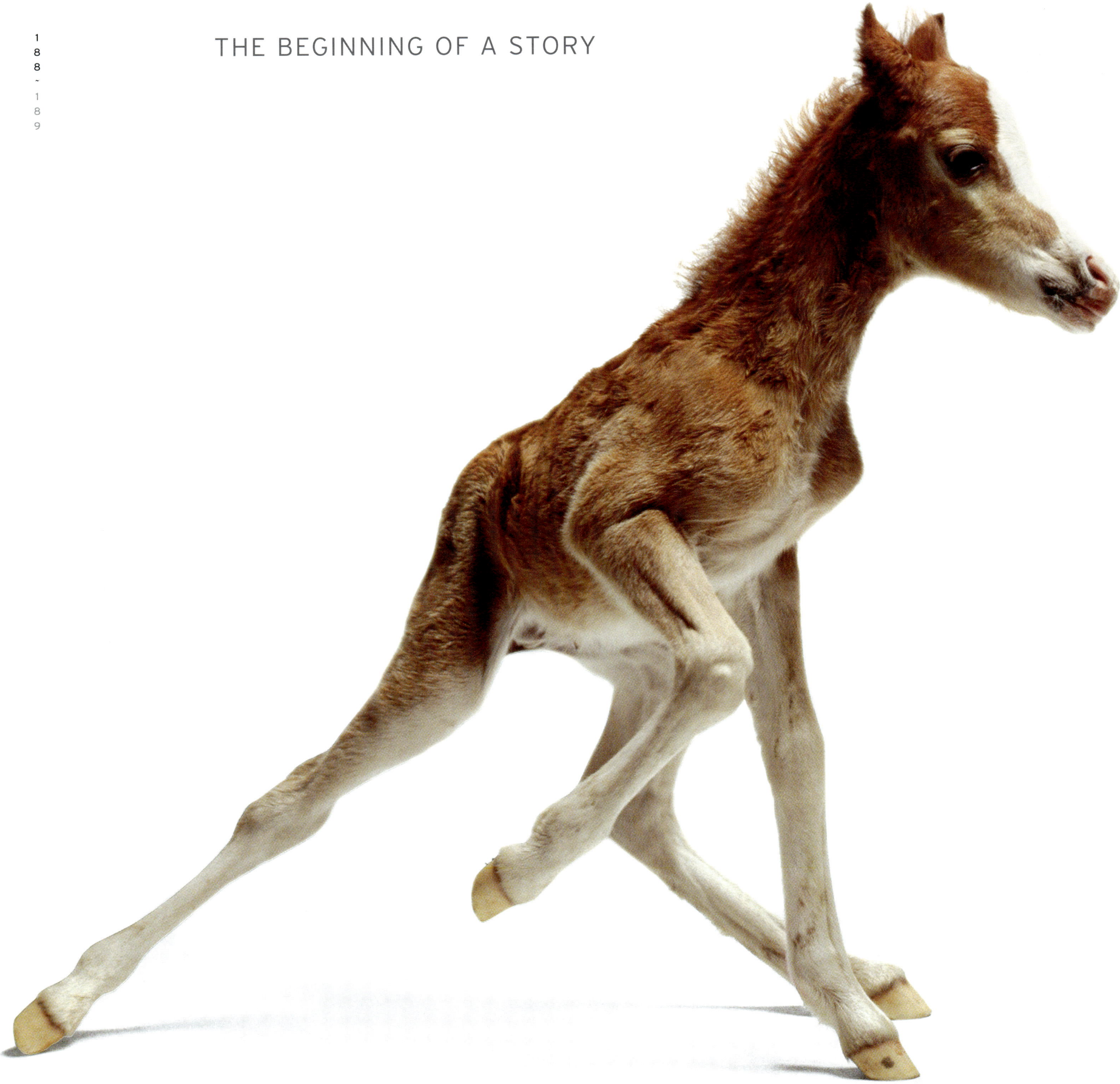

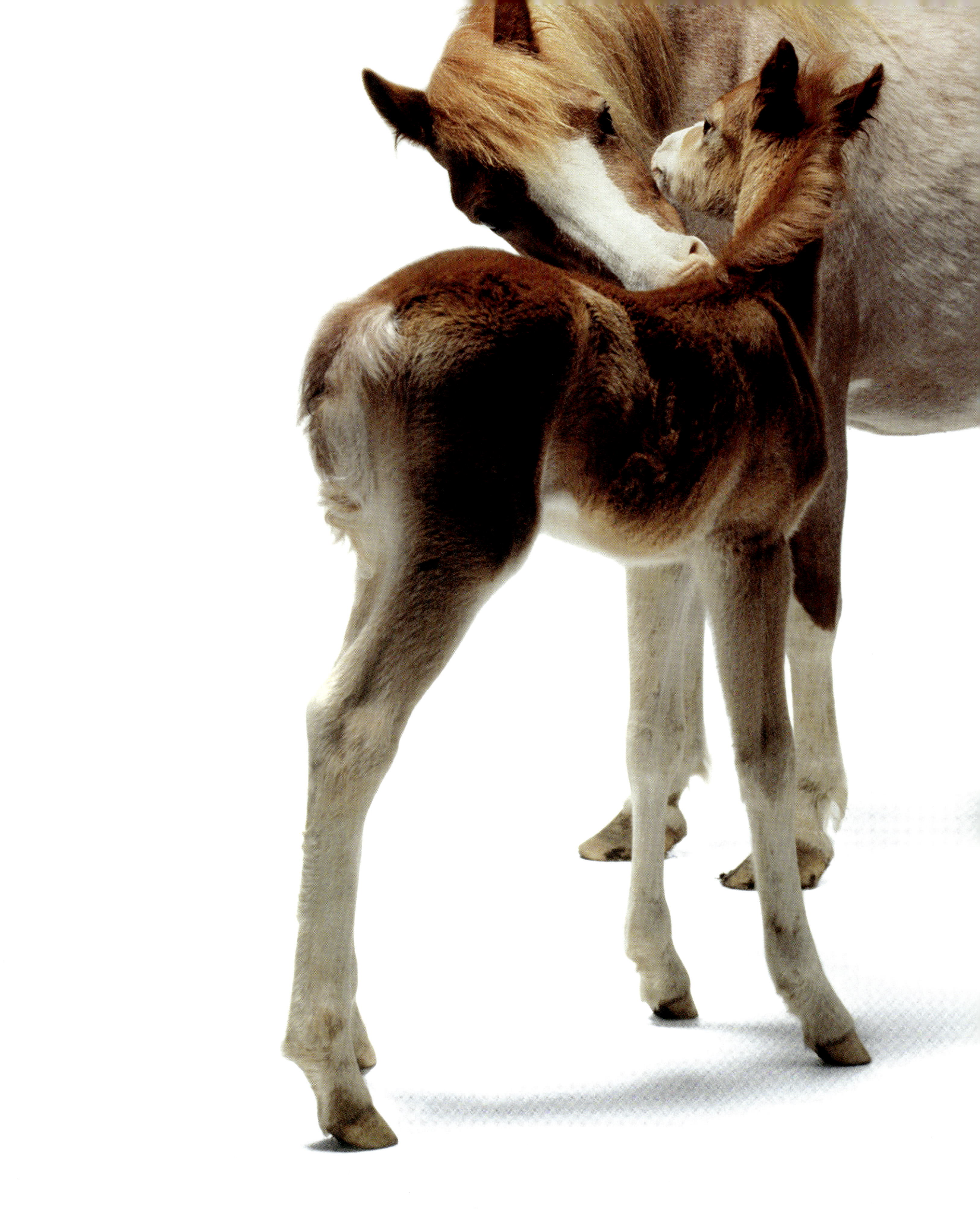

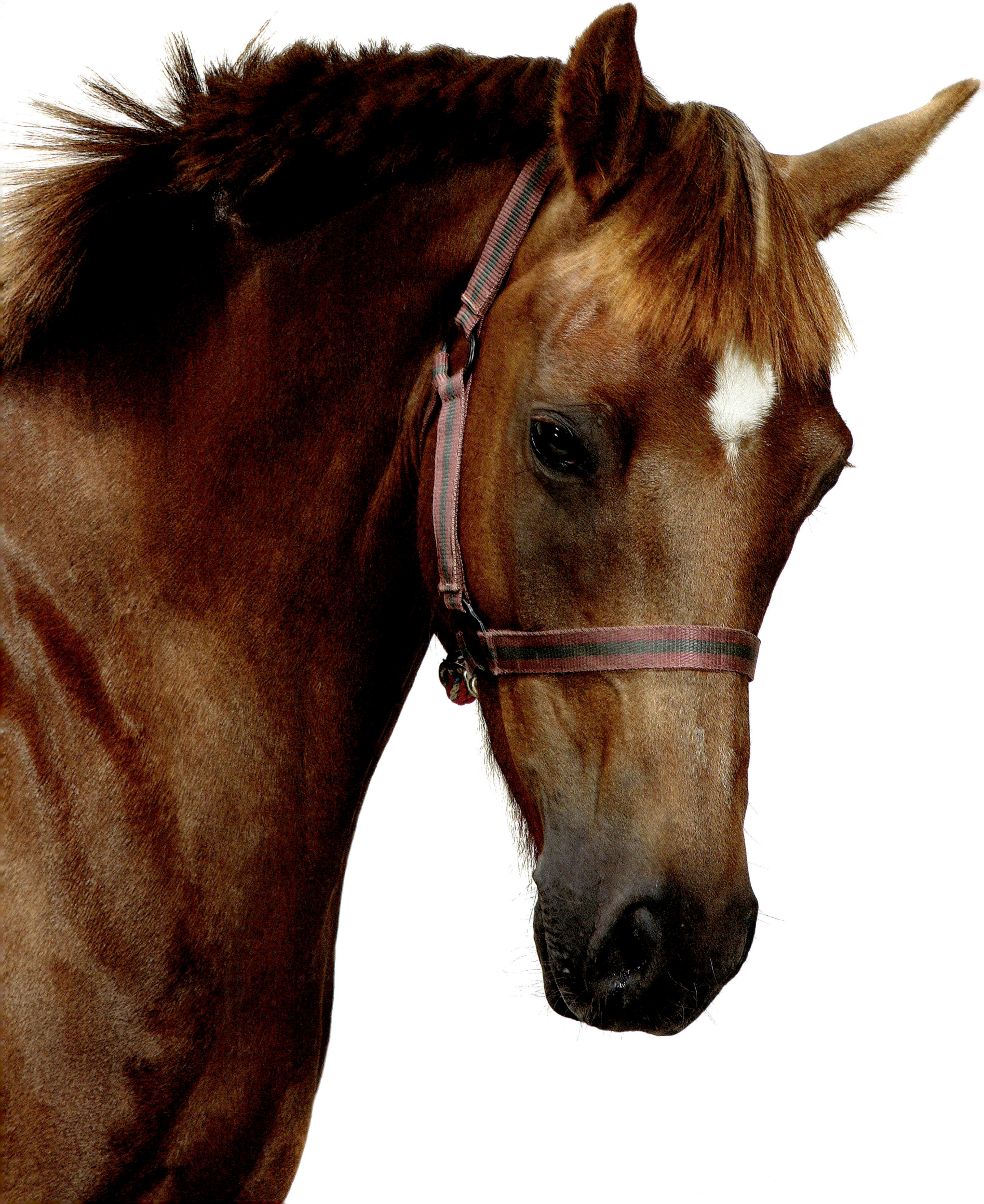

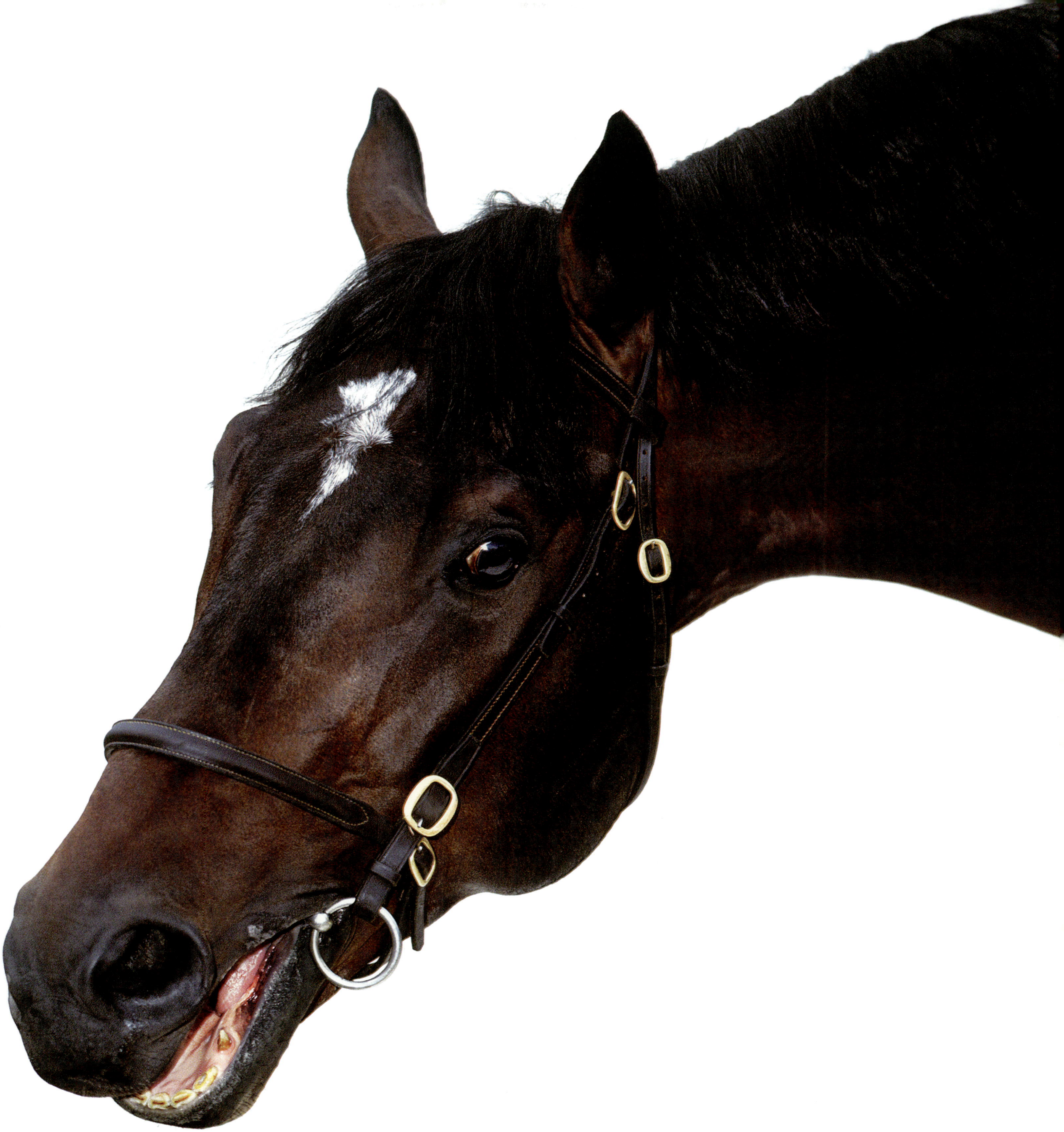

SHOW ME YOUR HORSE AND I'LL TELL YOU WHO YOU ARE

CLASS CONFLICTS

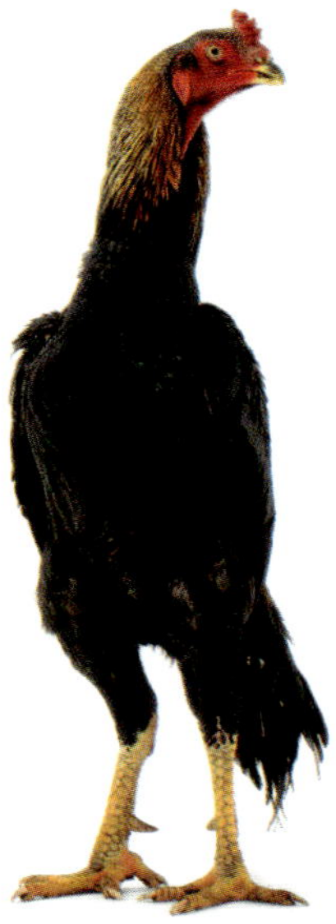

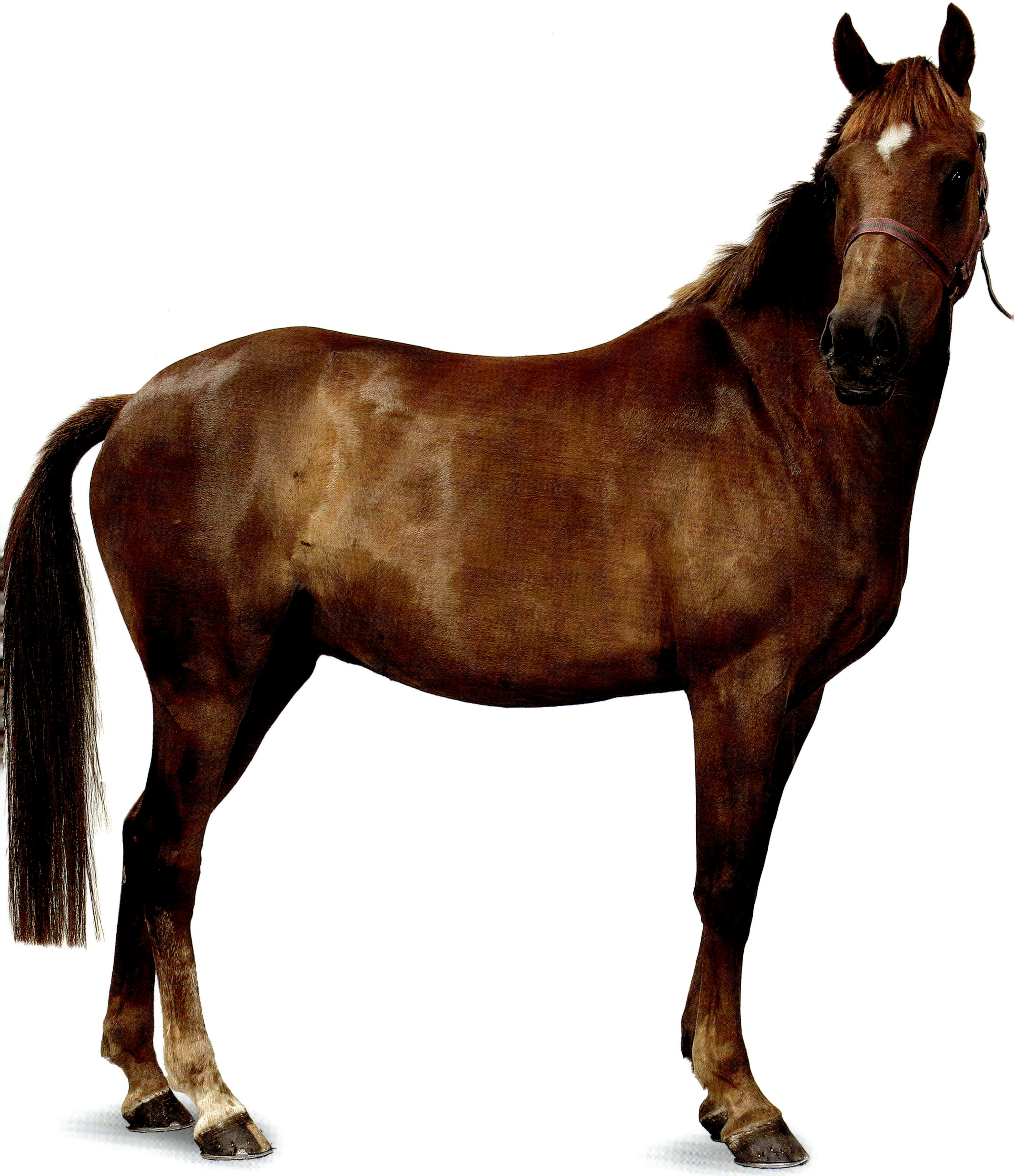

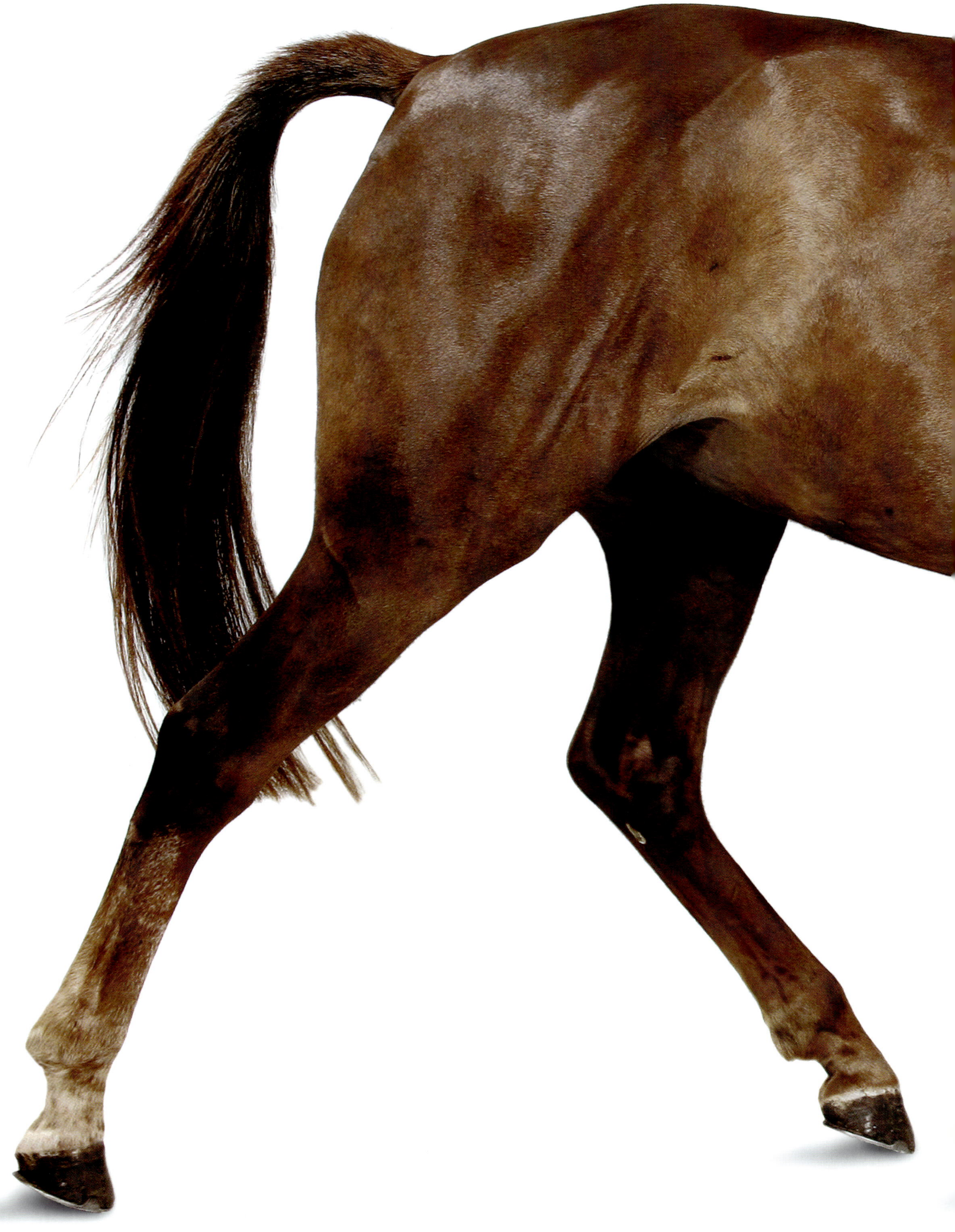

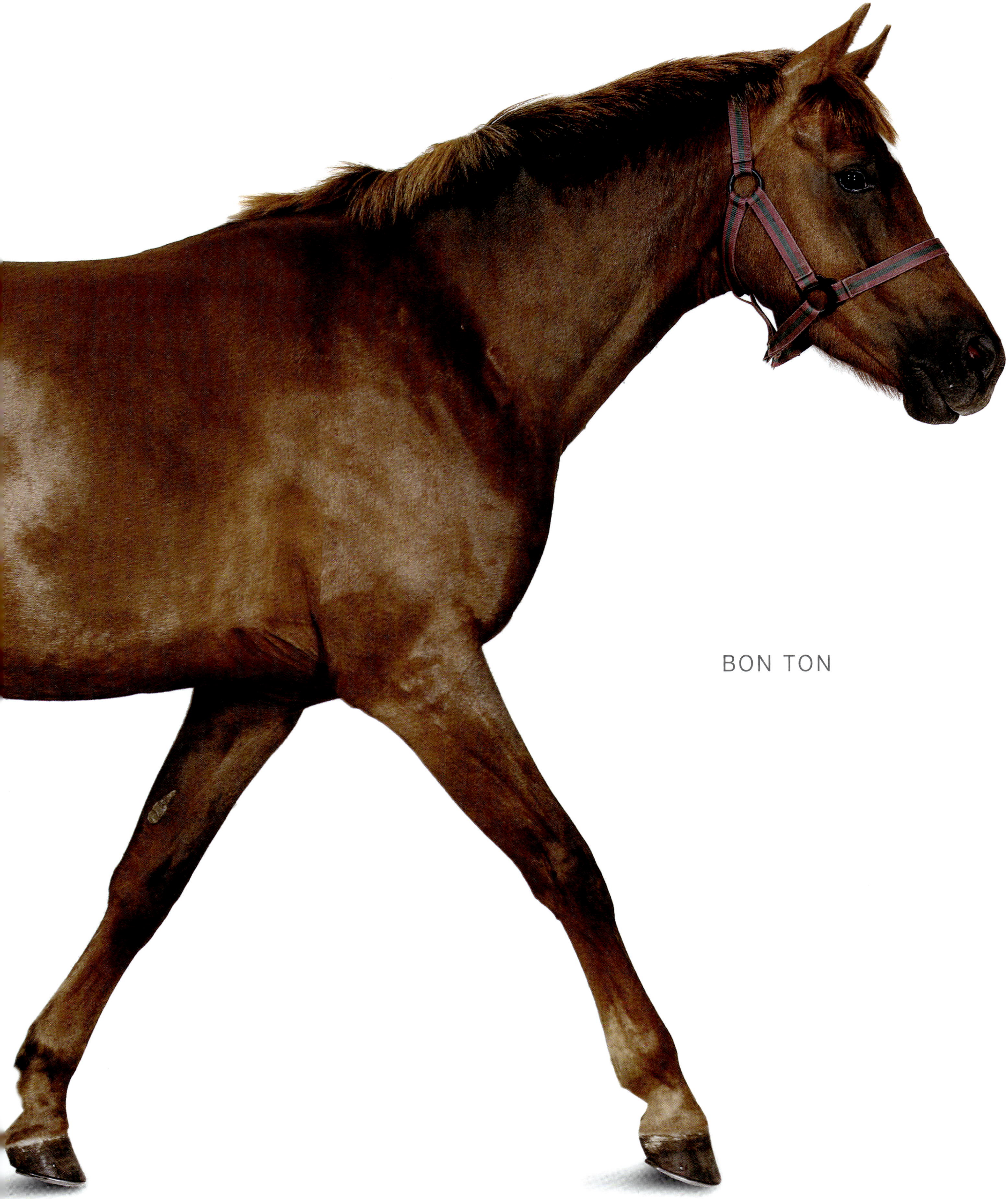

BON TON

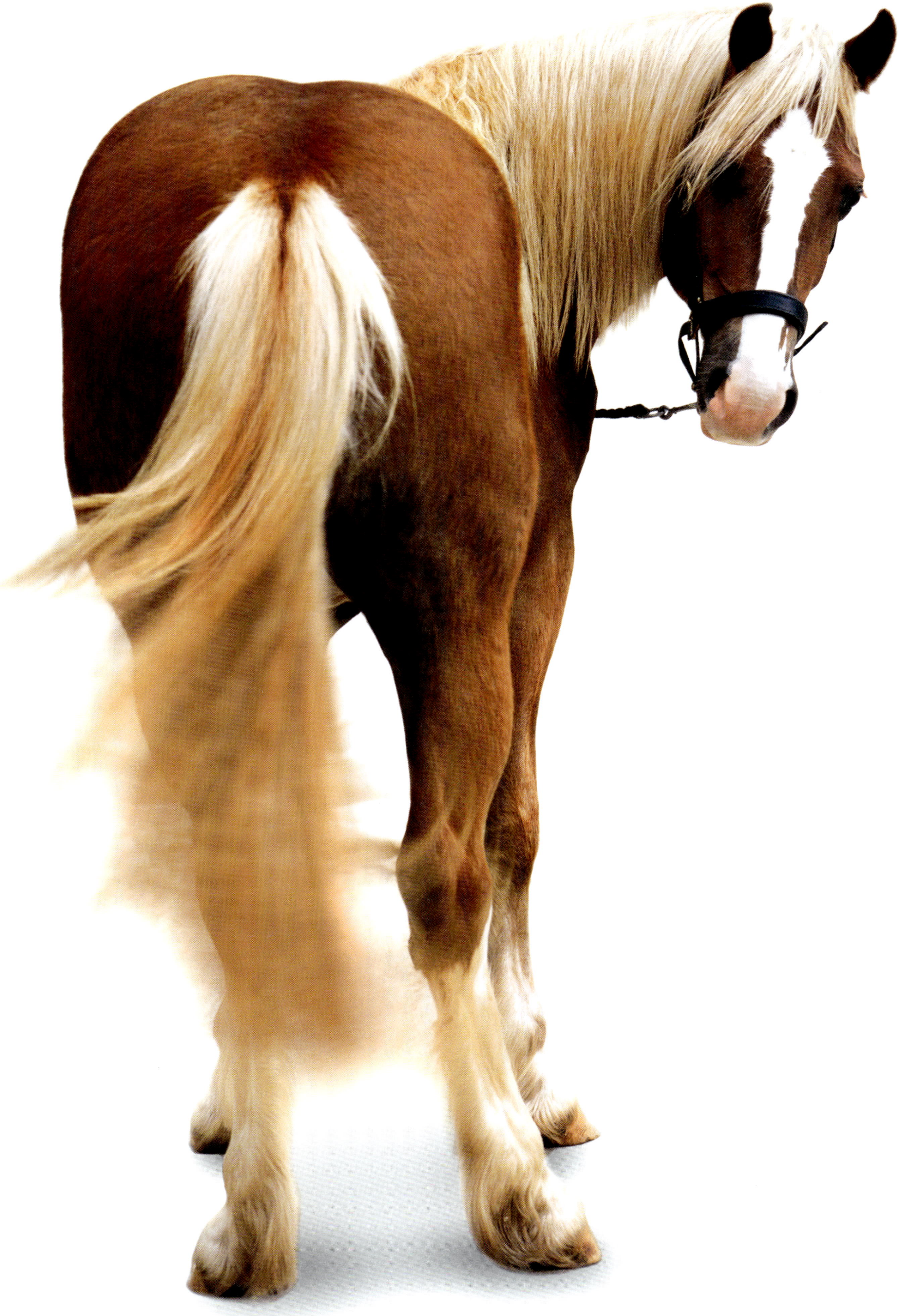

“If you’re looking for a lasting friendship,
you’d better choose a horse.”

(Anonymous)

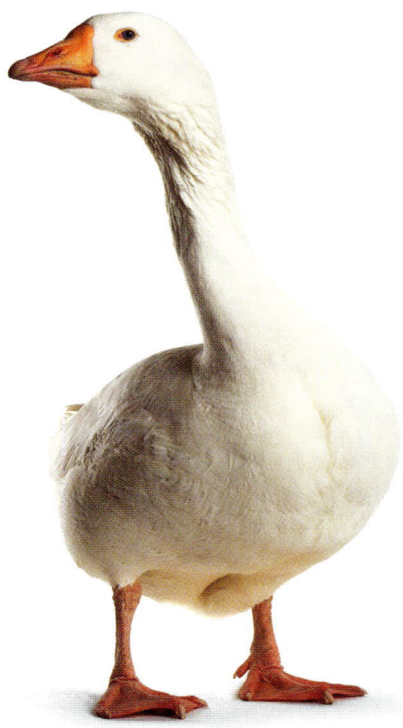

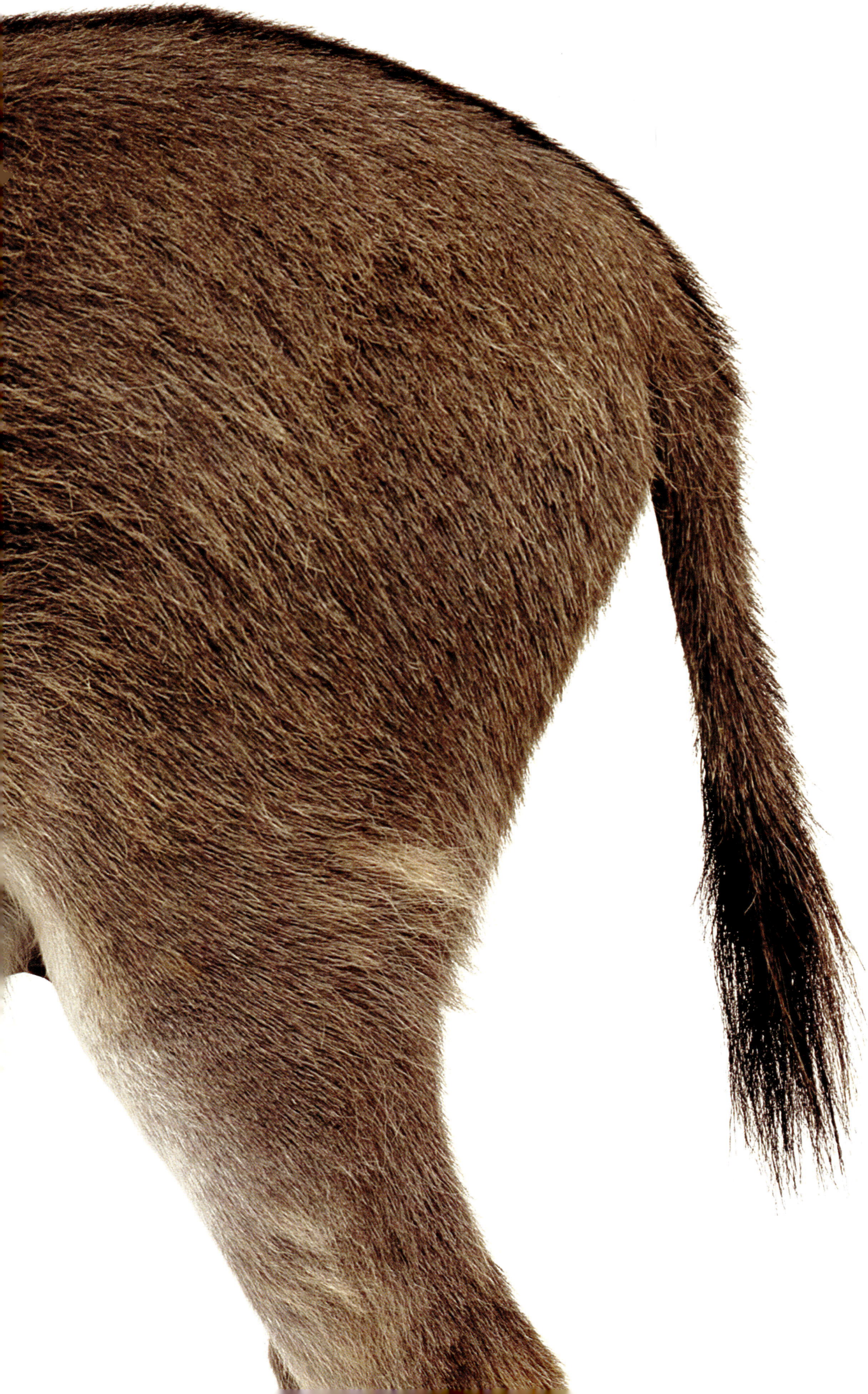

IRREMEDIABLE DIFFERENCES

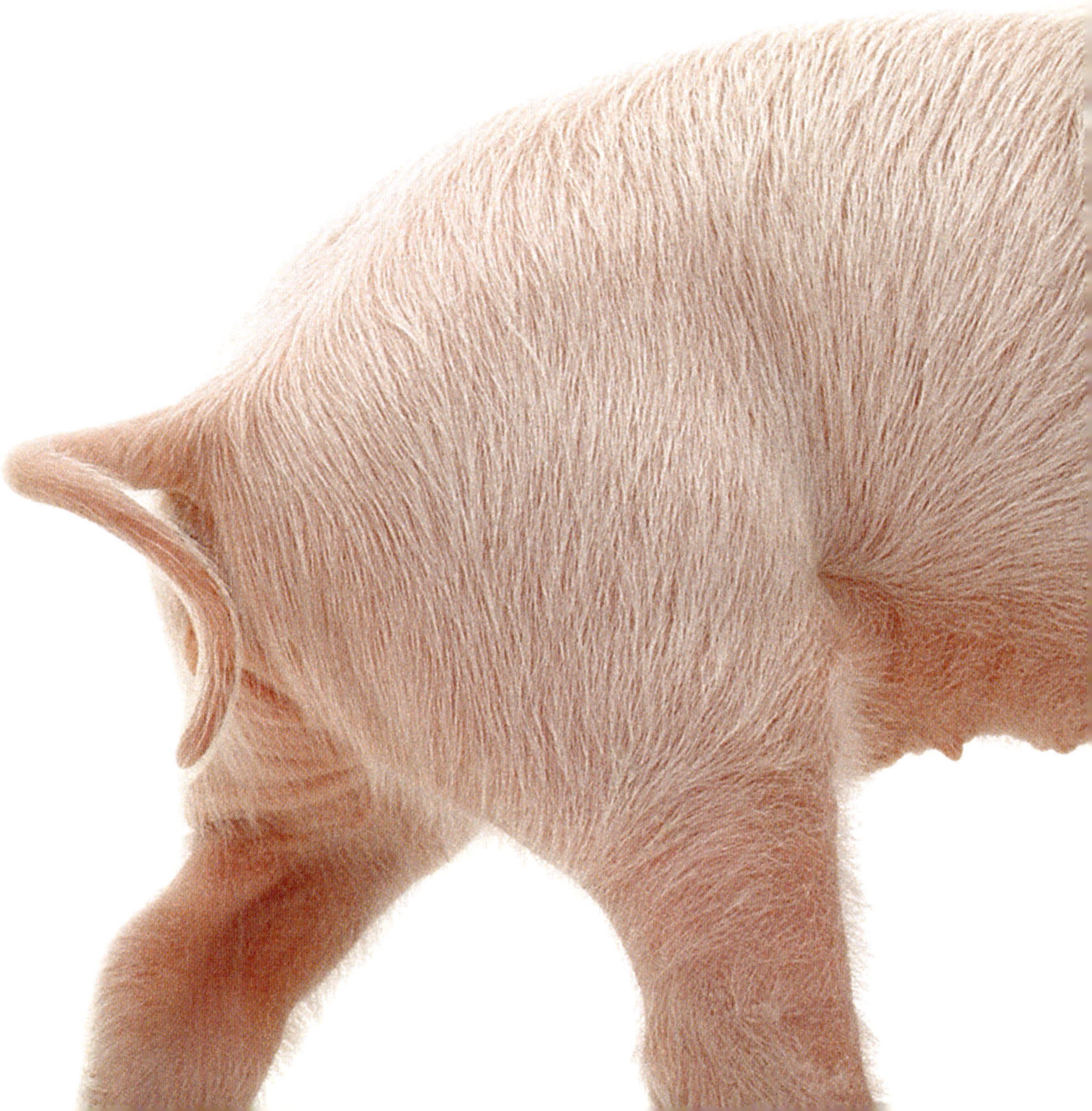

LAZY STUDENT

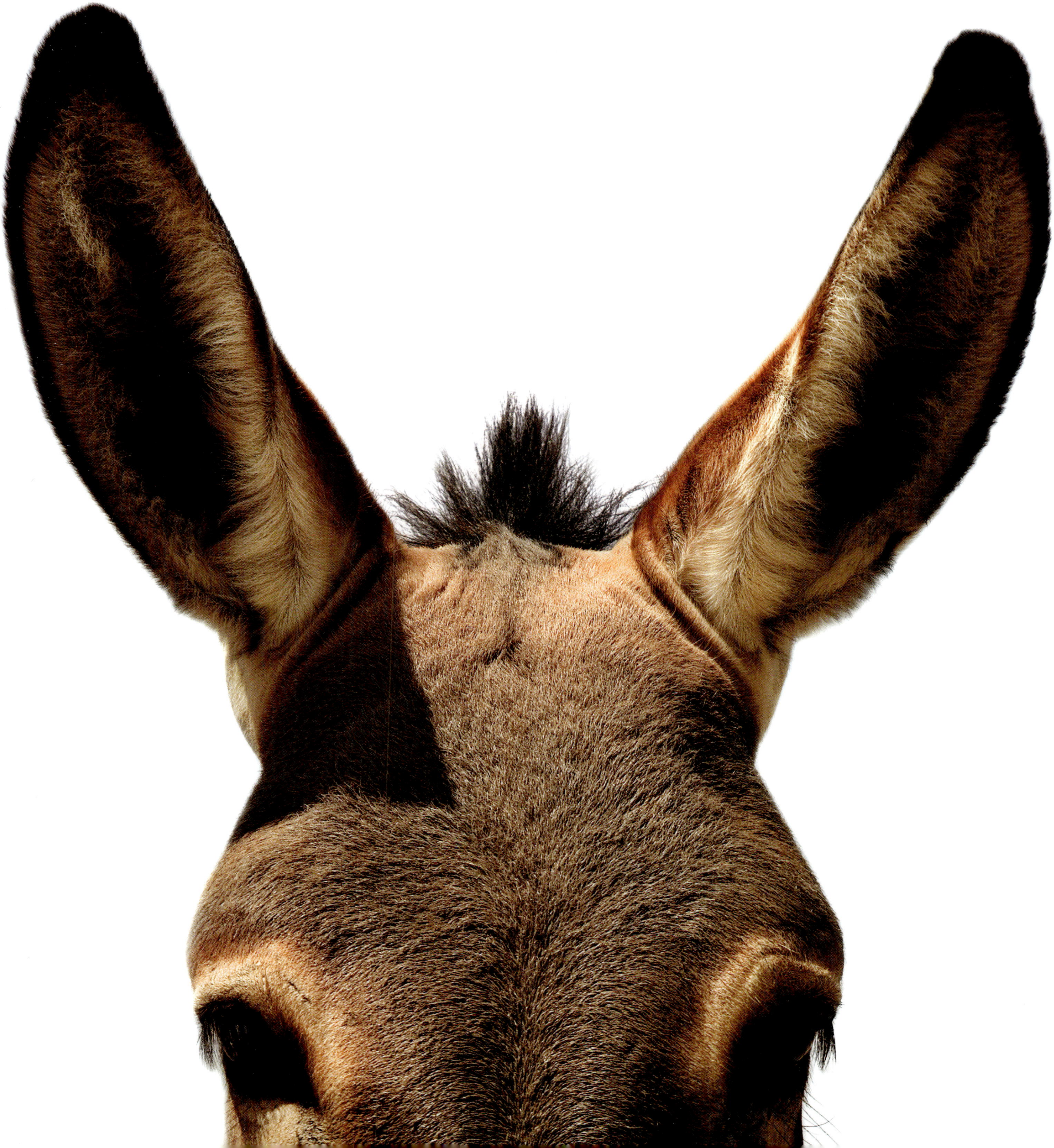

“He who conquers himself is the mightiest warrior.” (Confucius)

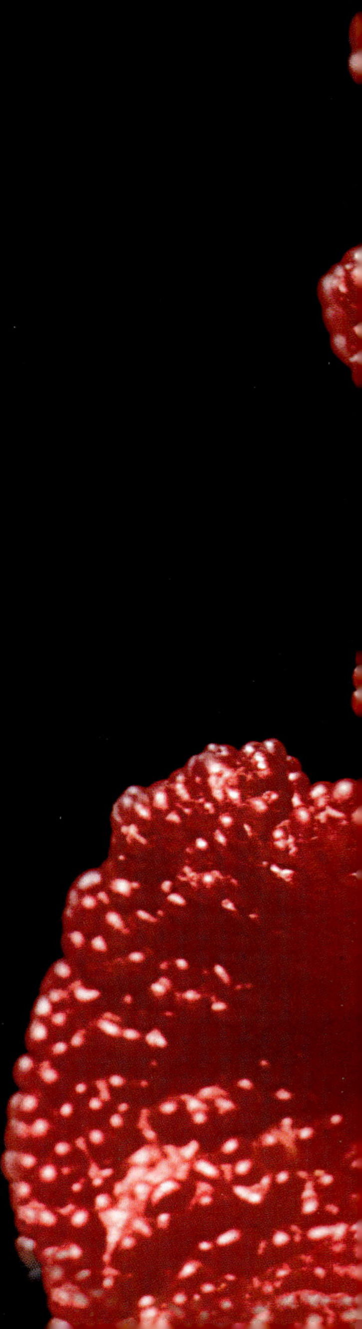

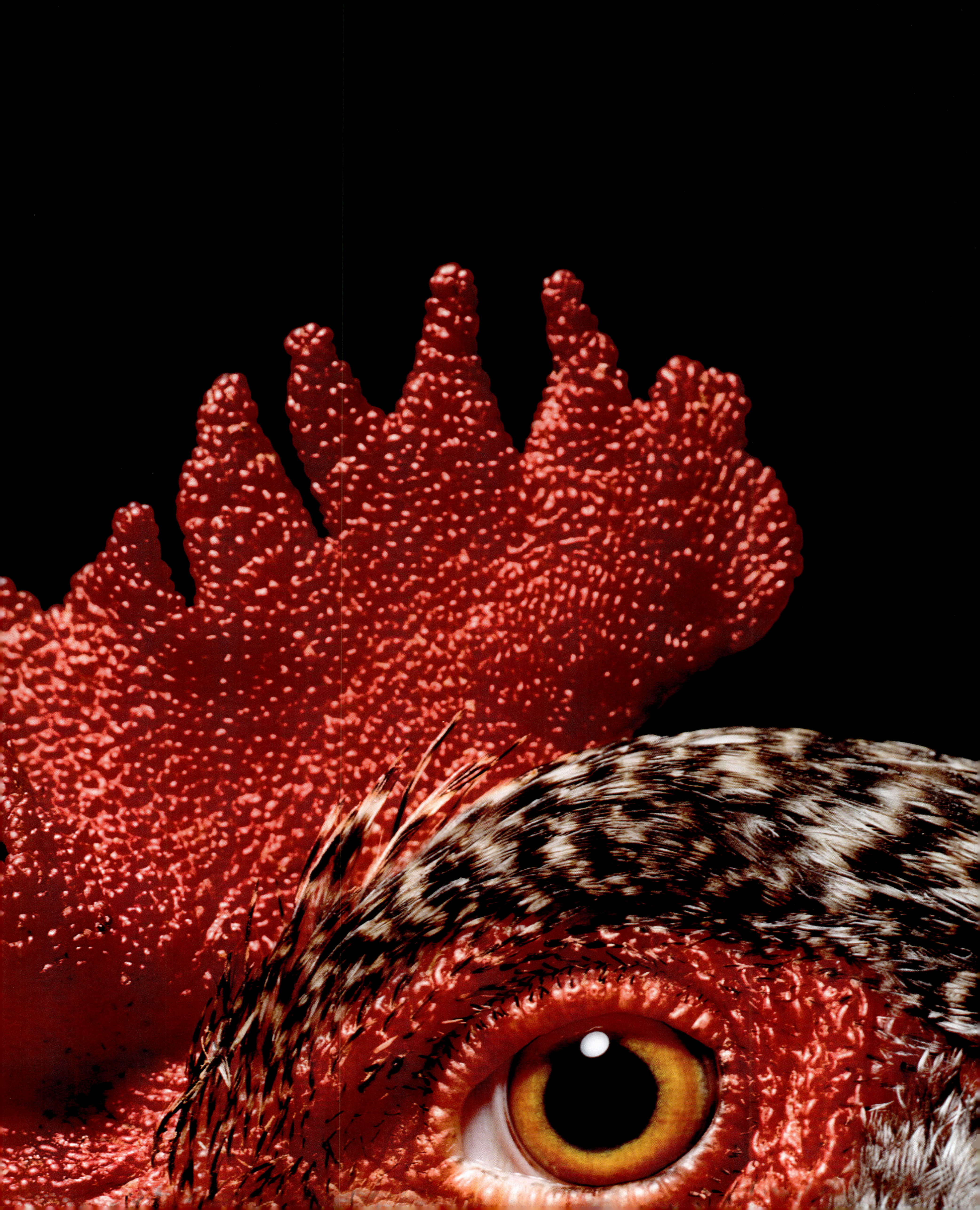

EVERYONE LINED UP!

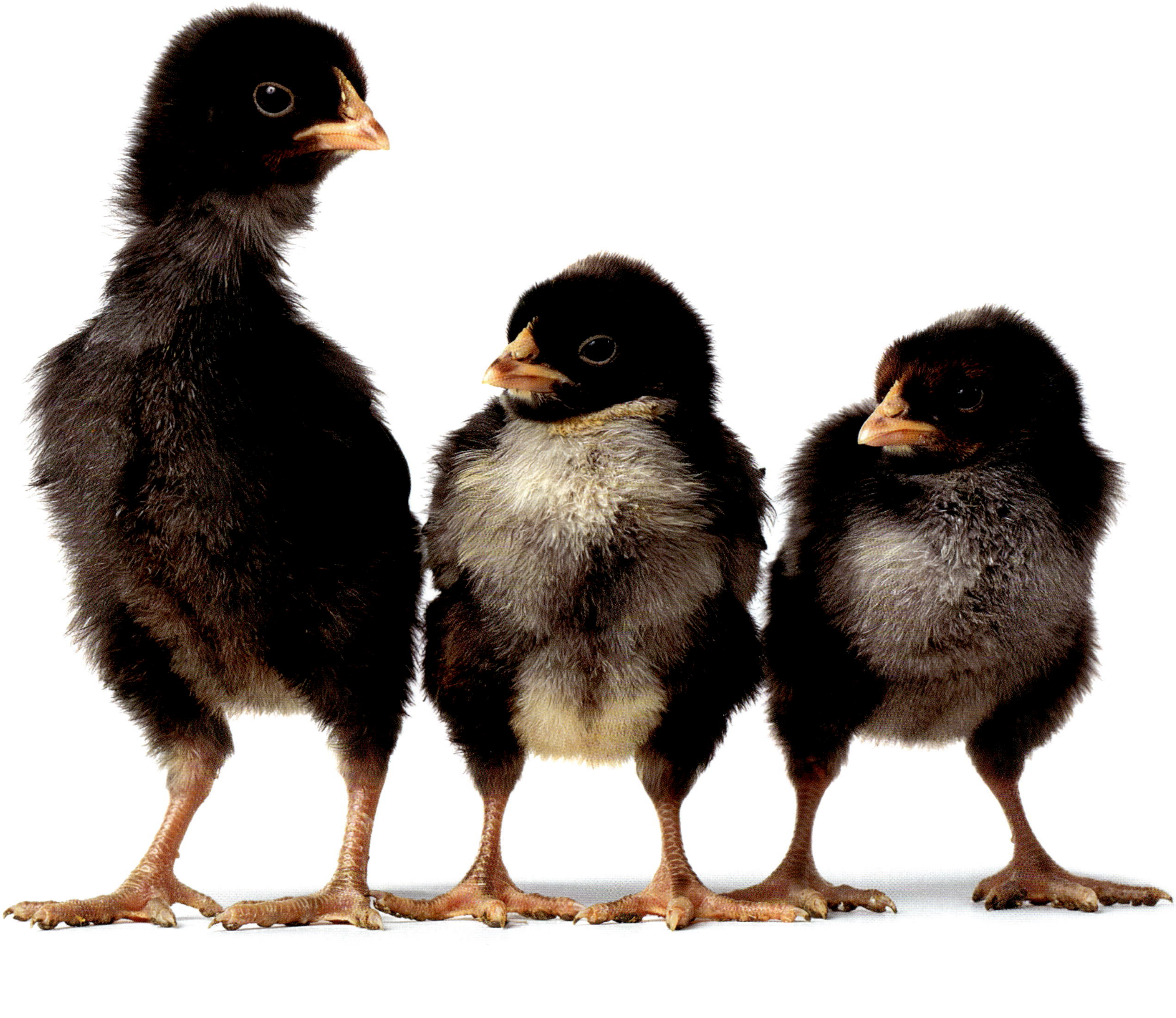

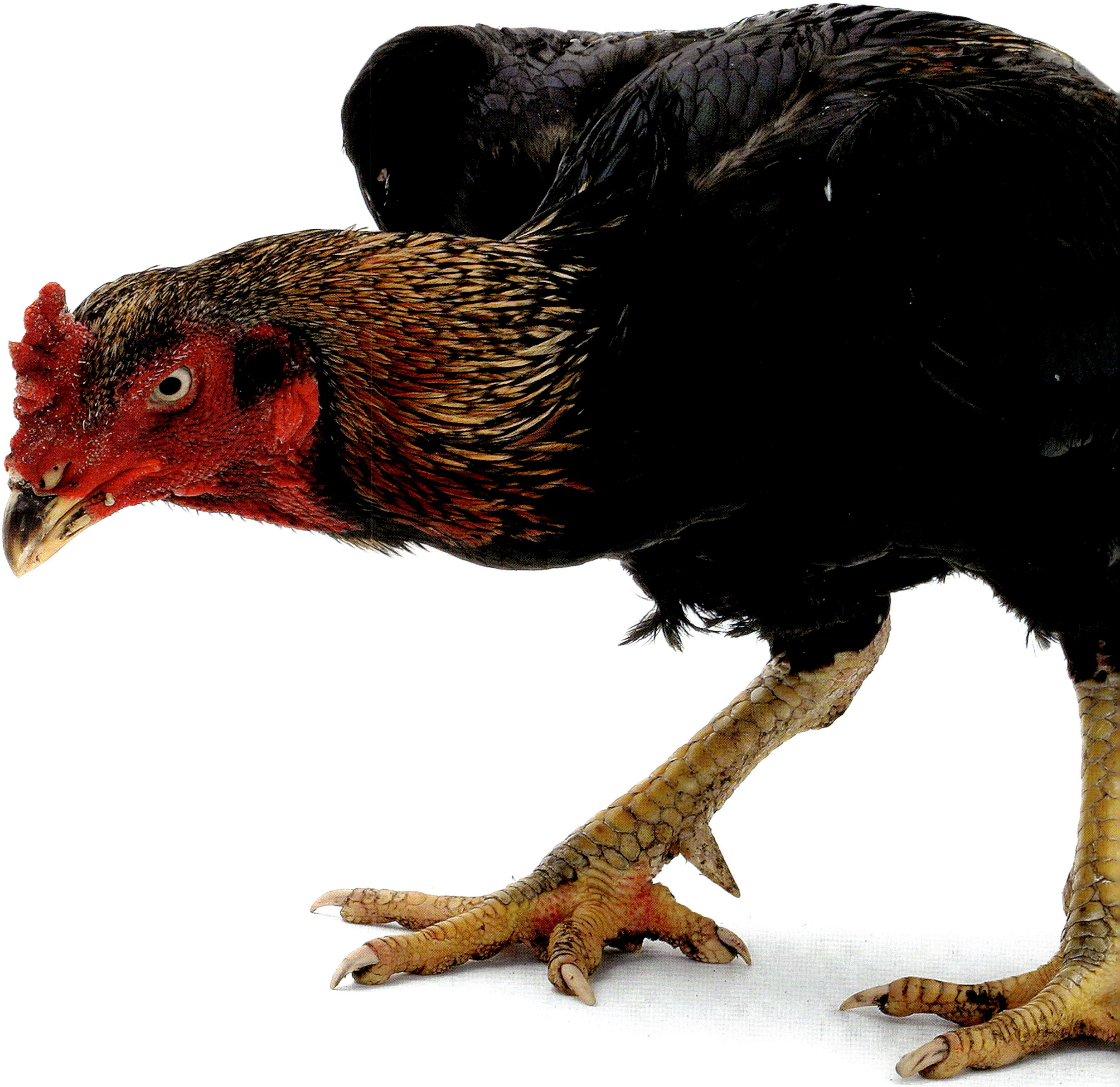

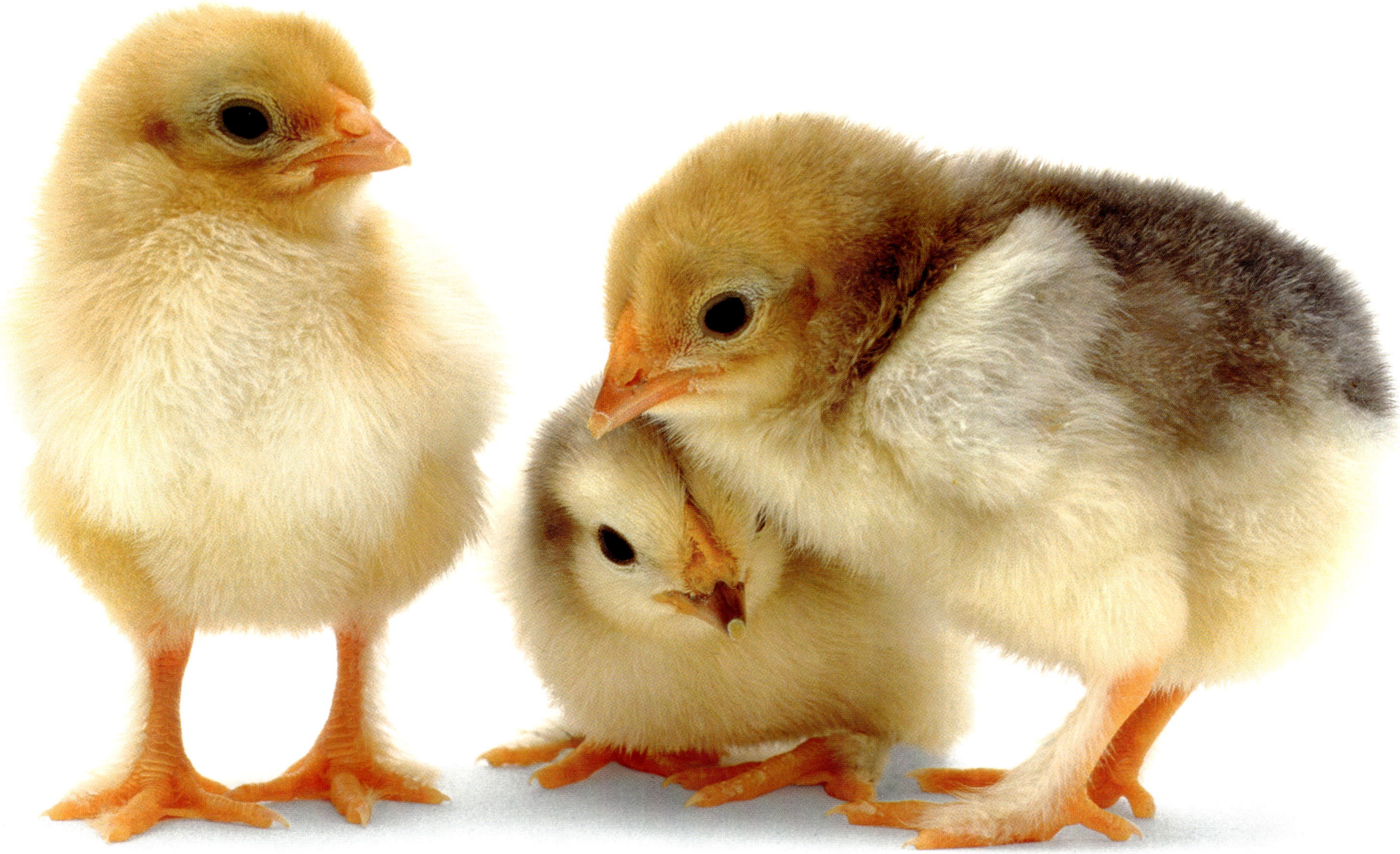

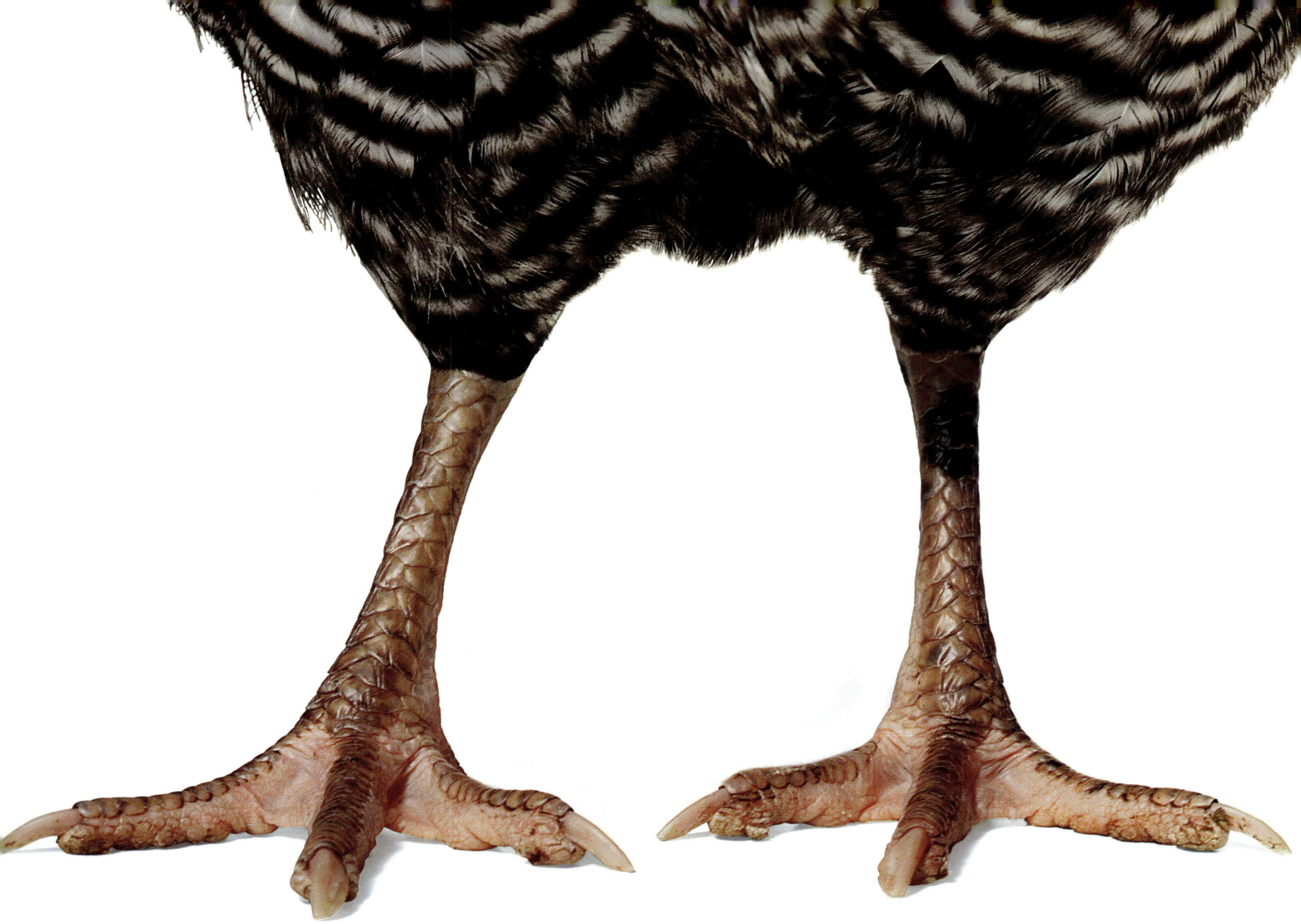

A SENSE OF PROPORTION

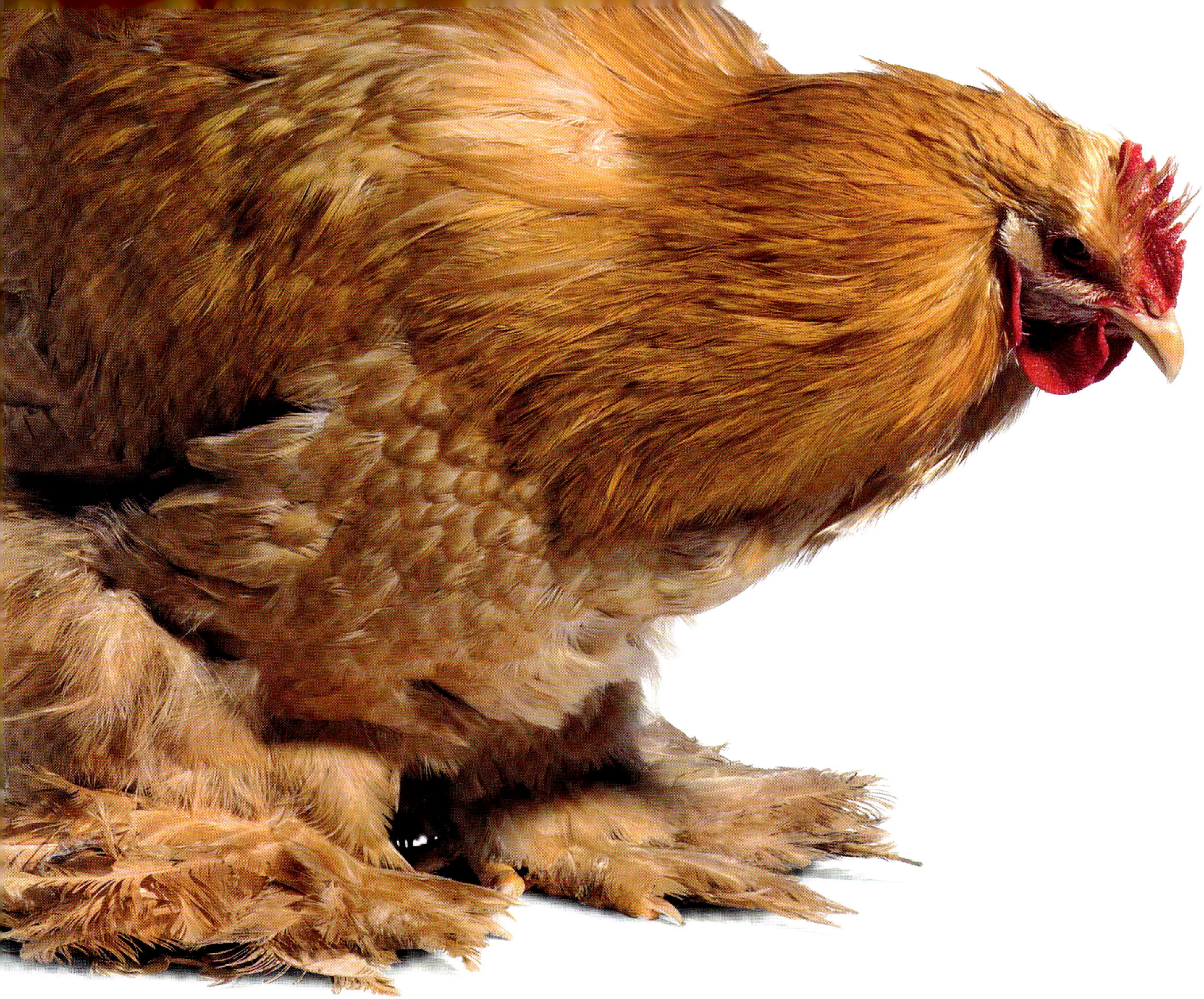

THE FIRST DAY OF SCHOOL

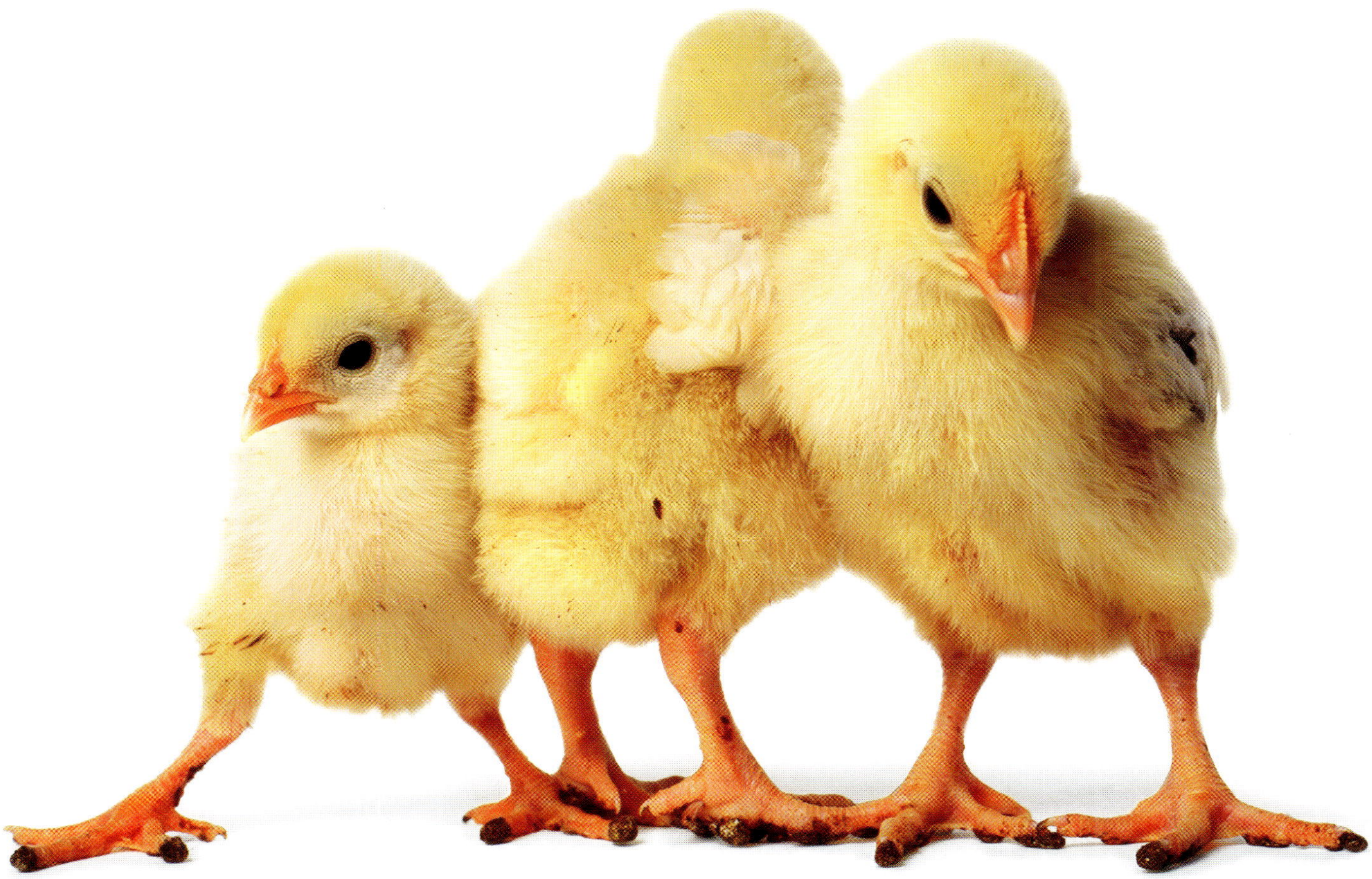

NURSERY

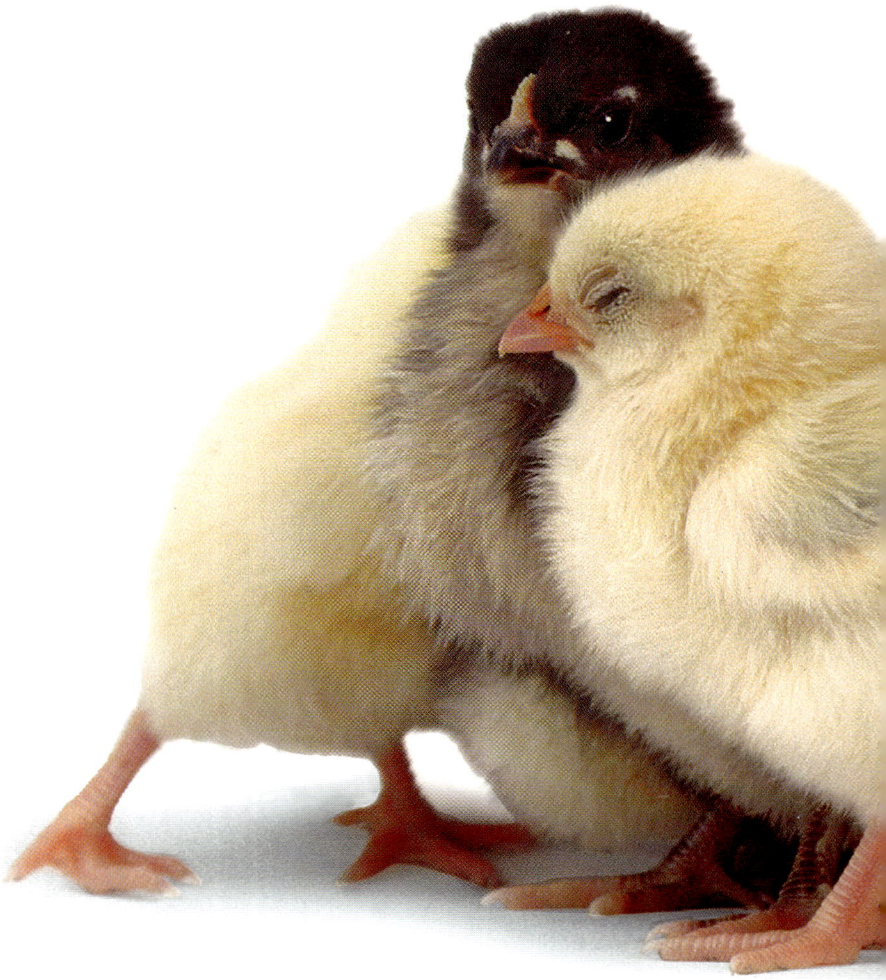

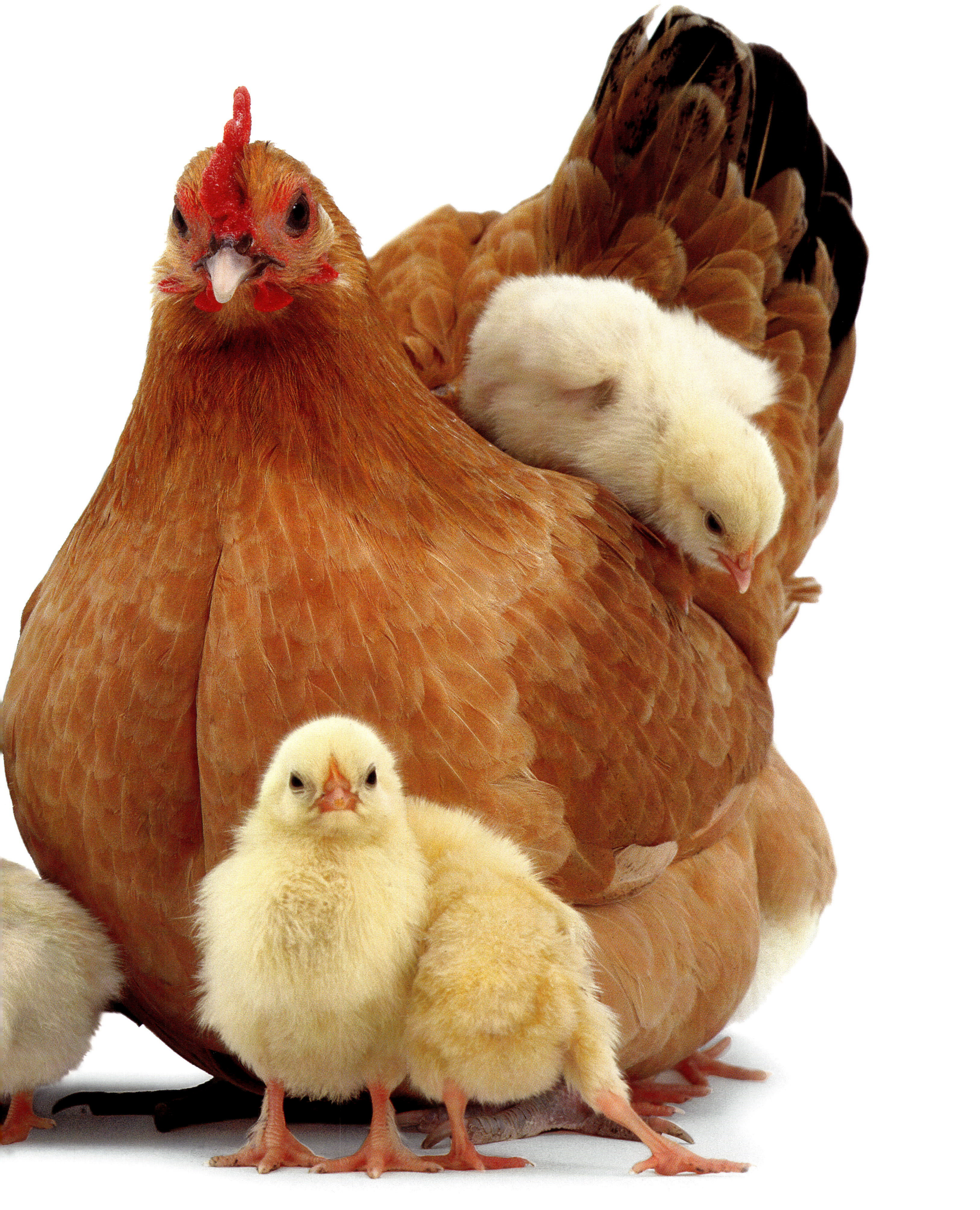

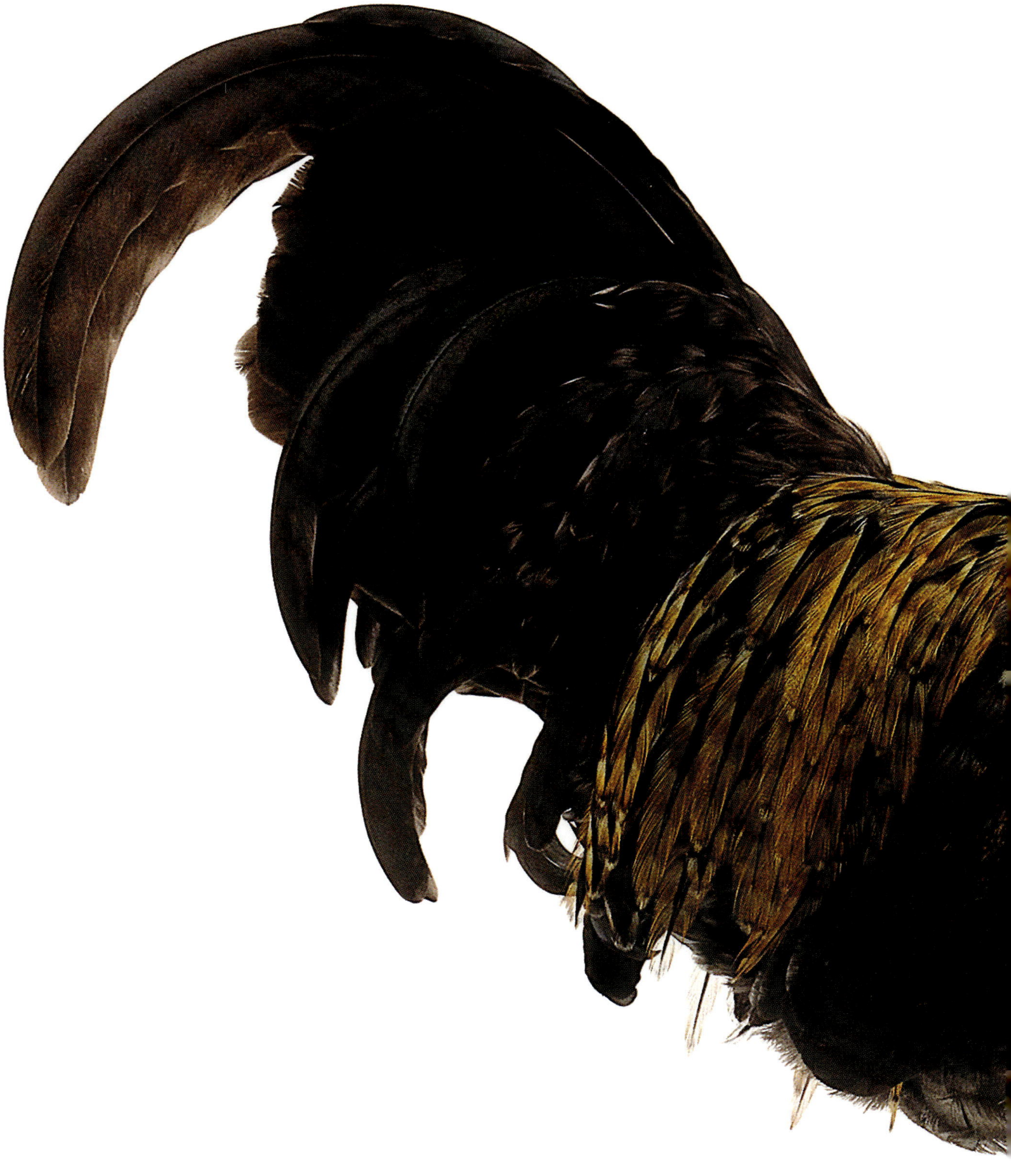

CRESTS . . . IN PROPORTION

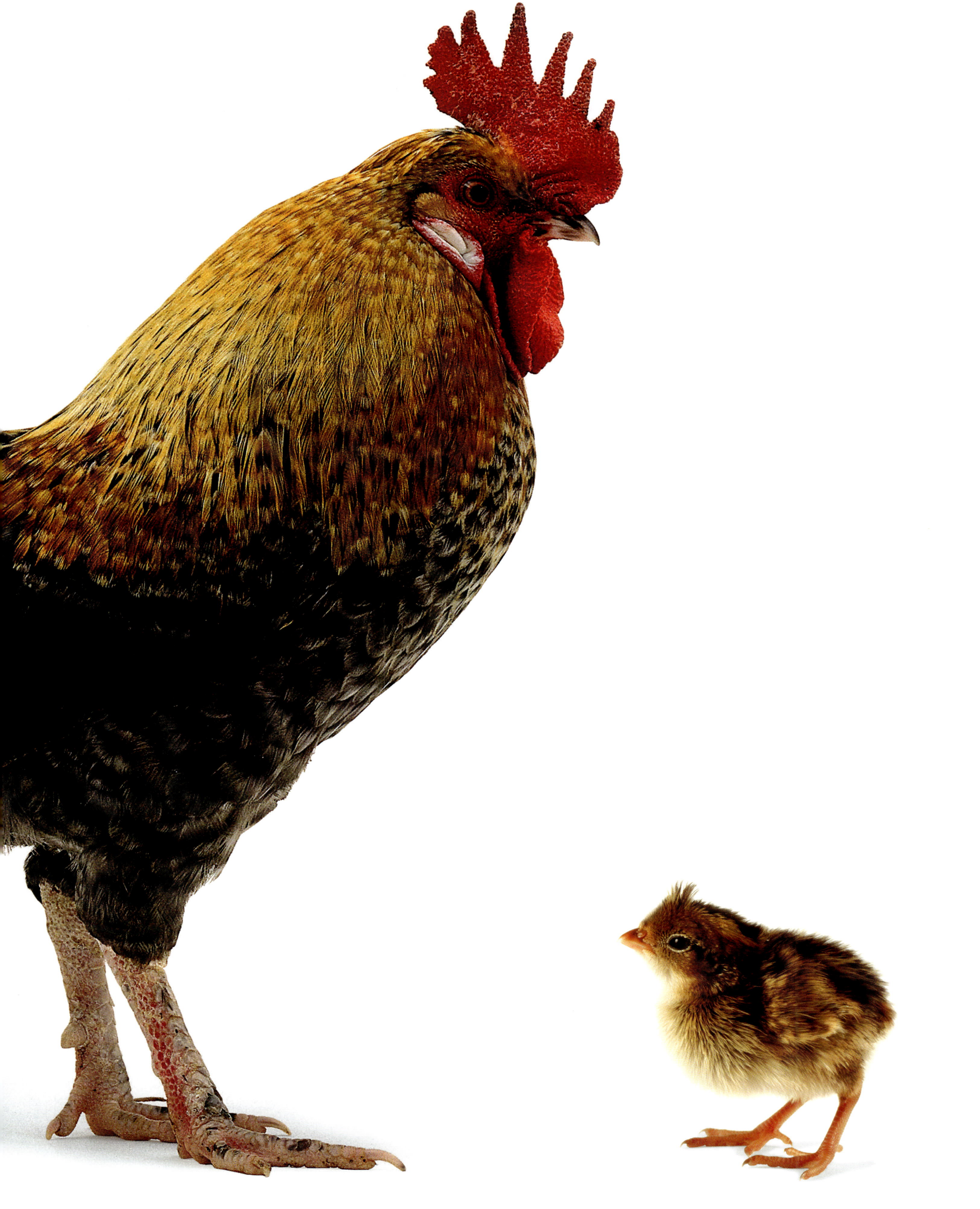

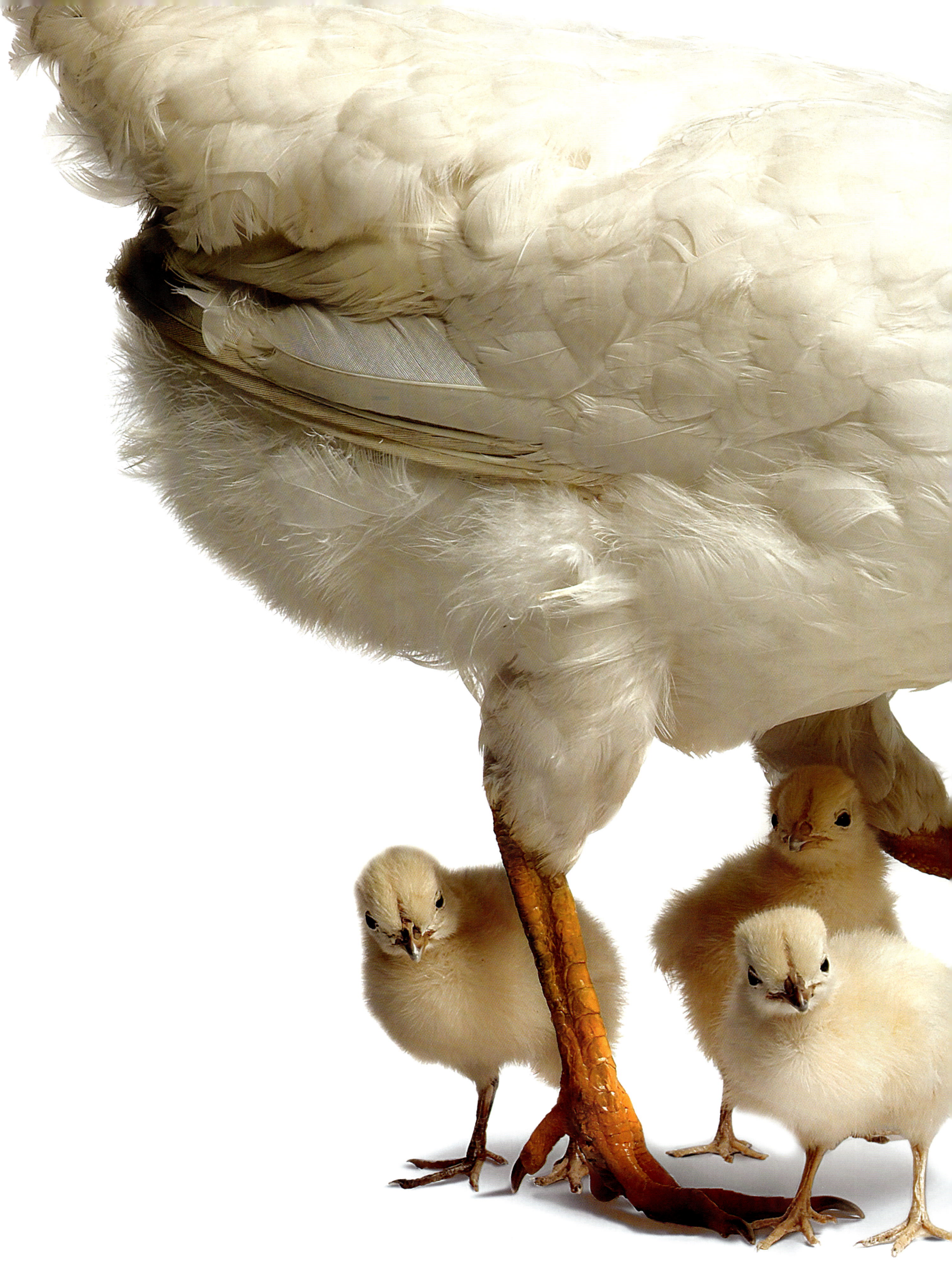

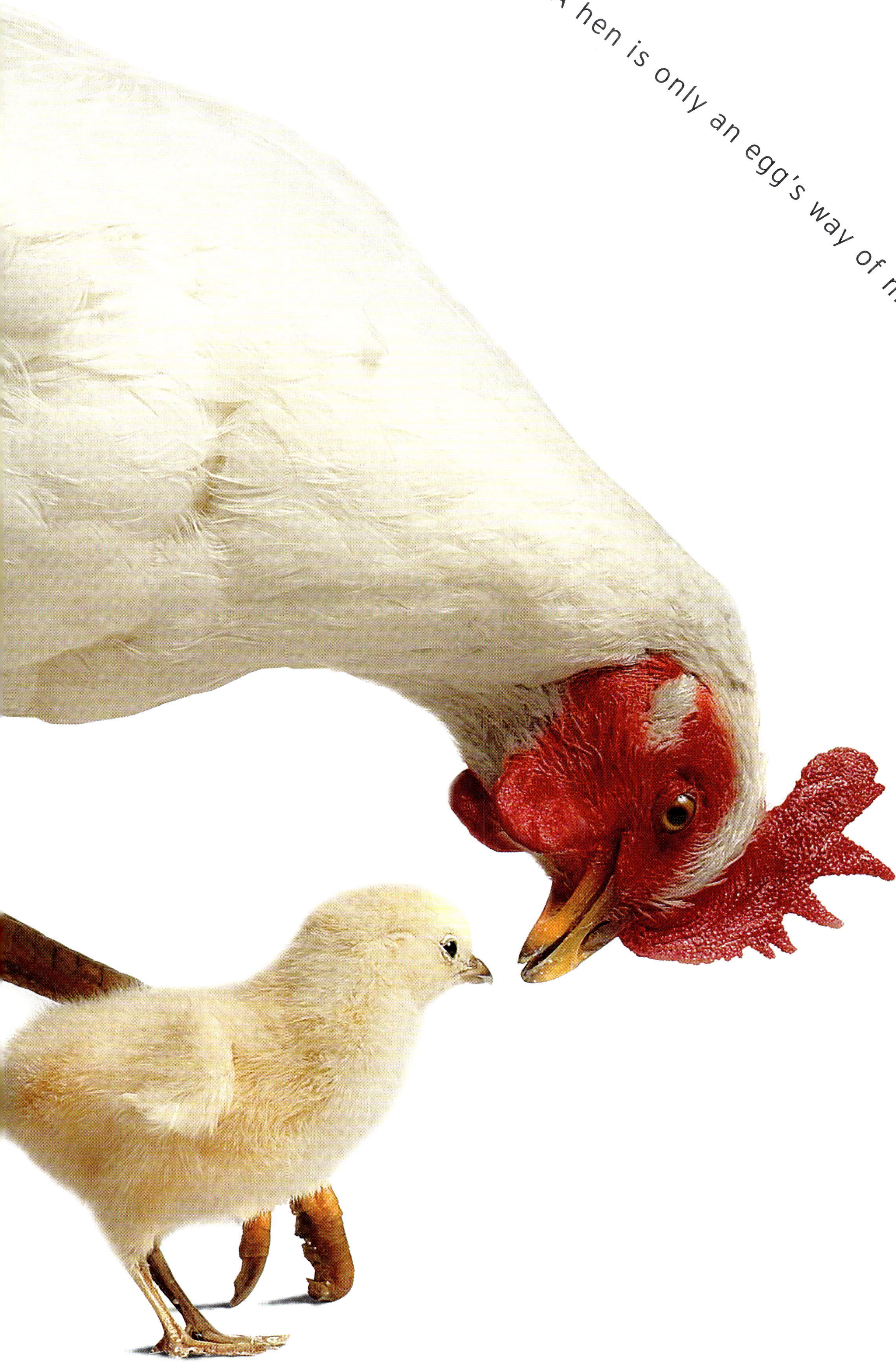

"A hen is only an egg's way of making another egg." (Samuel Butler)

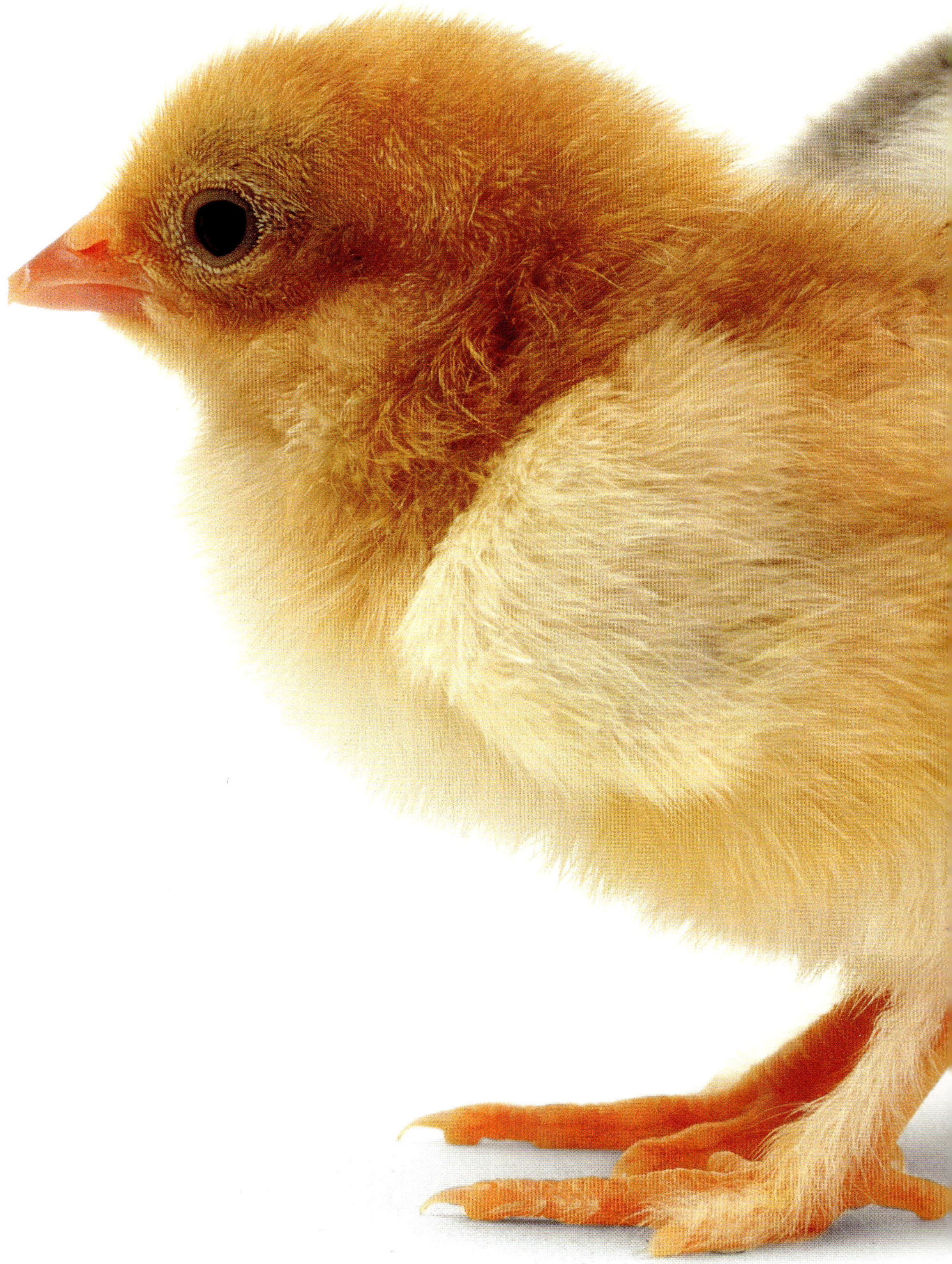

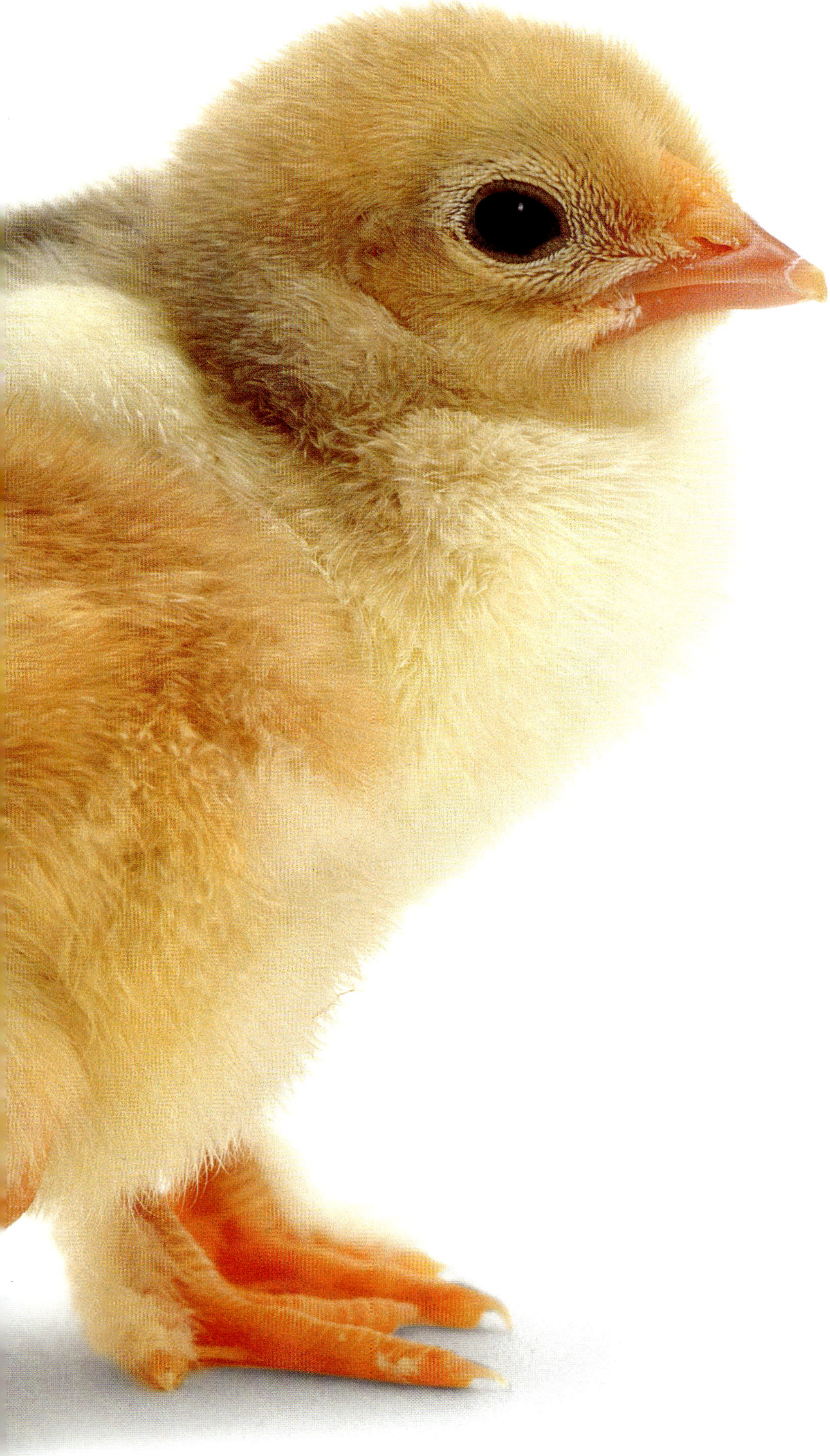

THE SIGN OF GEMINI

SINGING LESSON

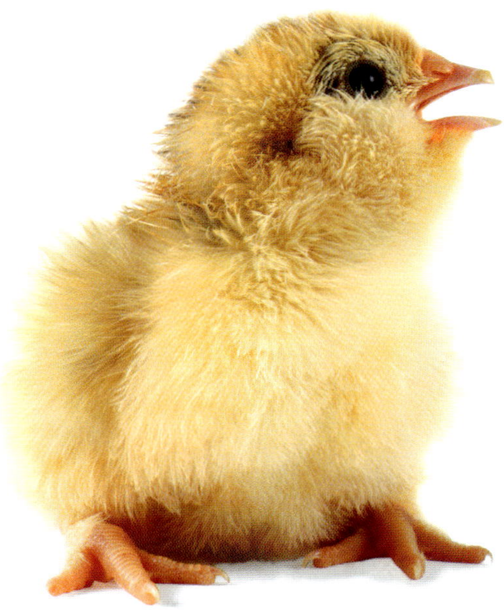

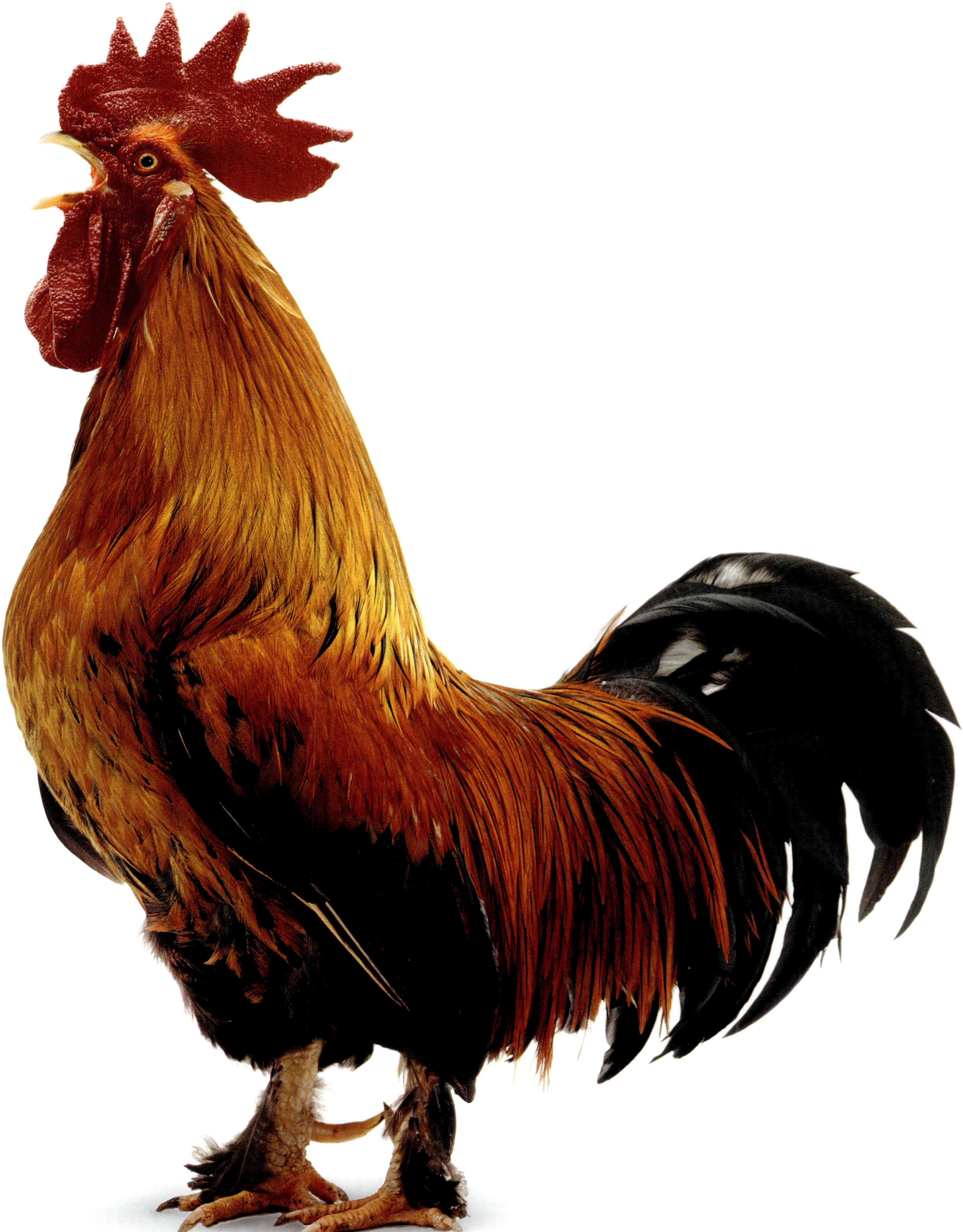

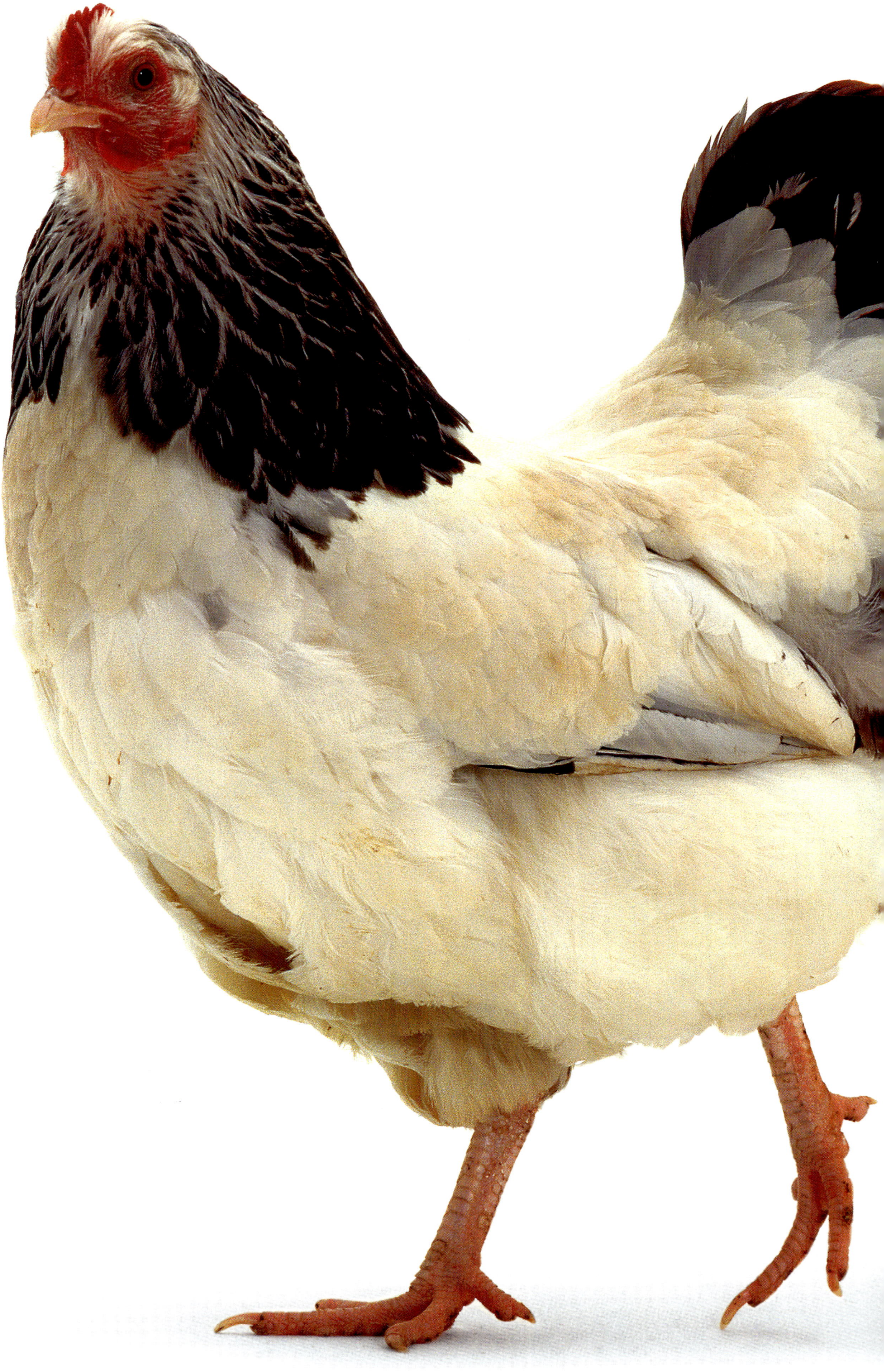

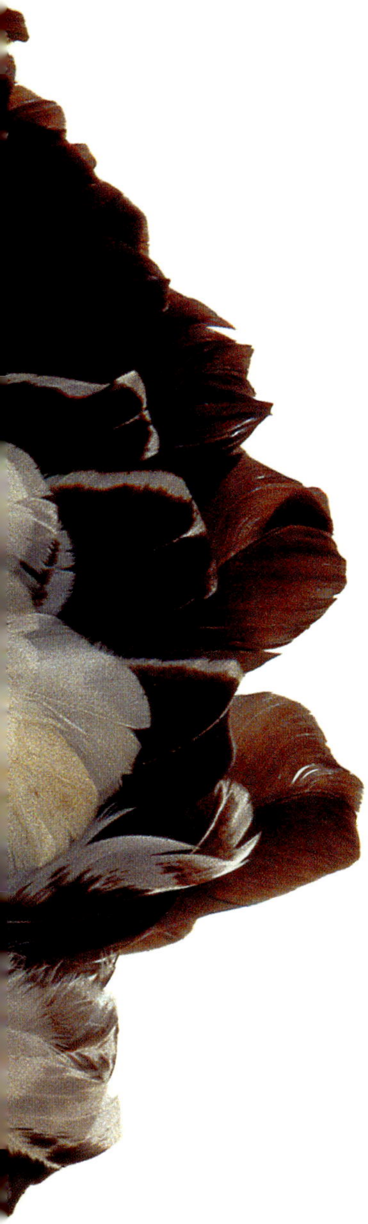

ELEMENTARY SCHOOL

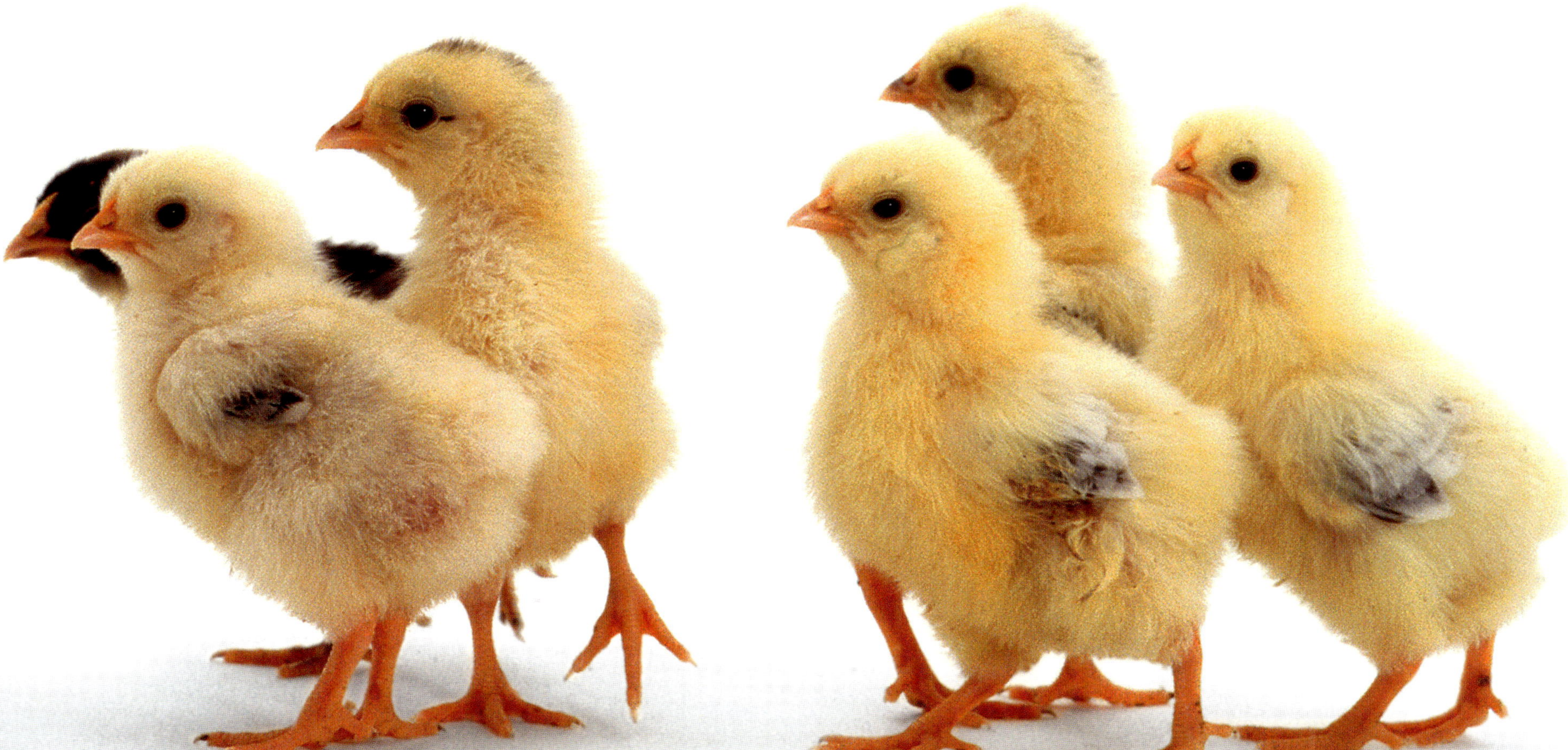

THE GREAT ESCAPE

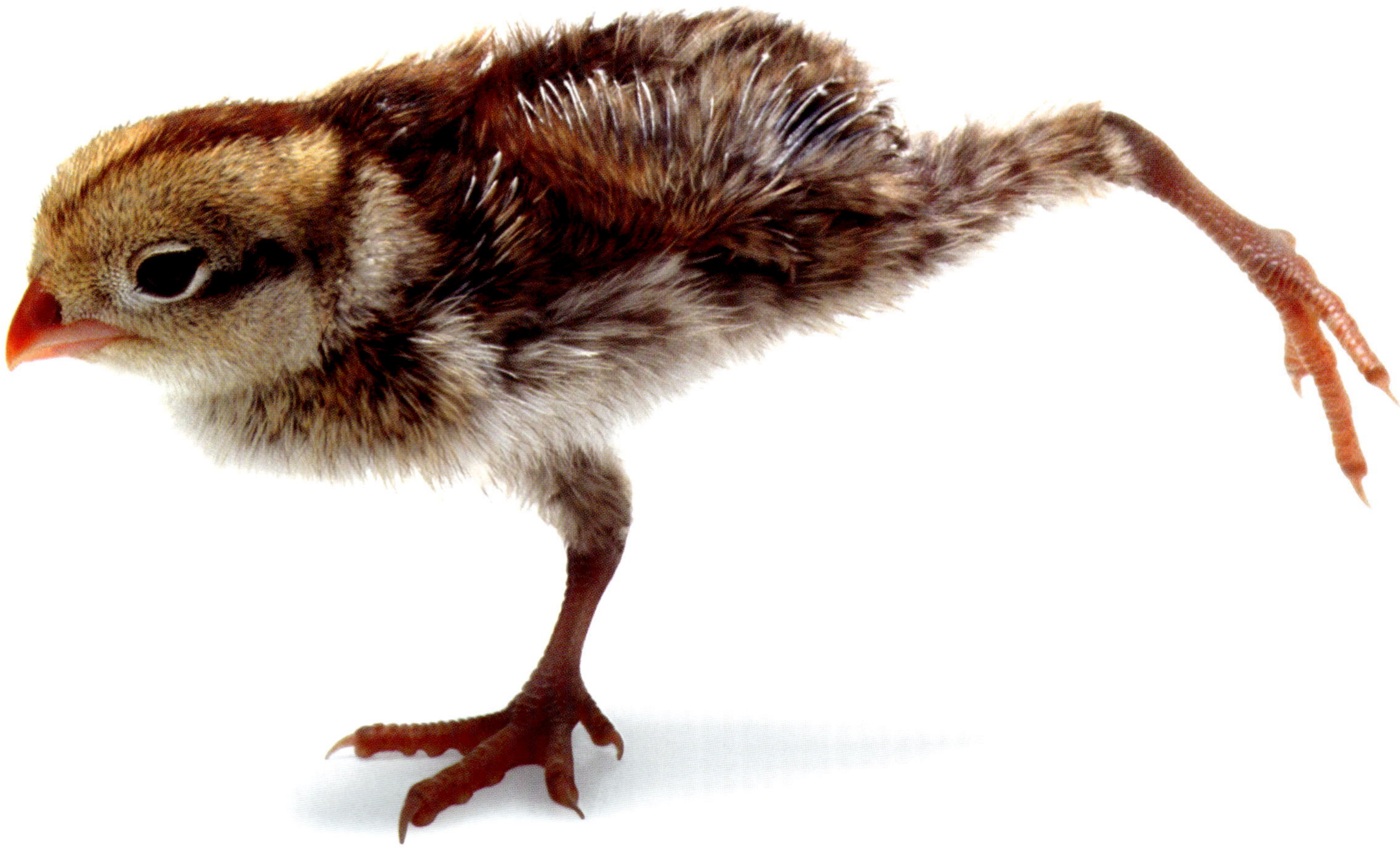

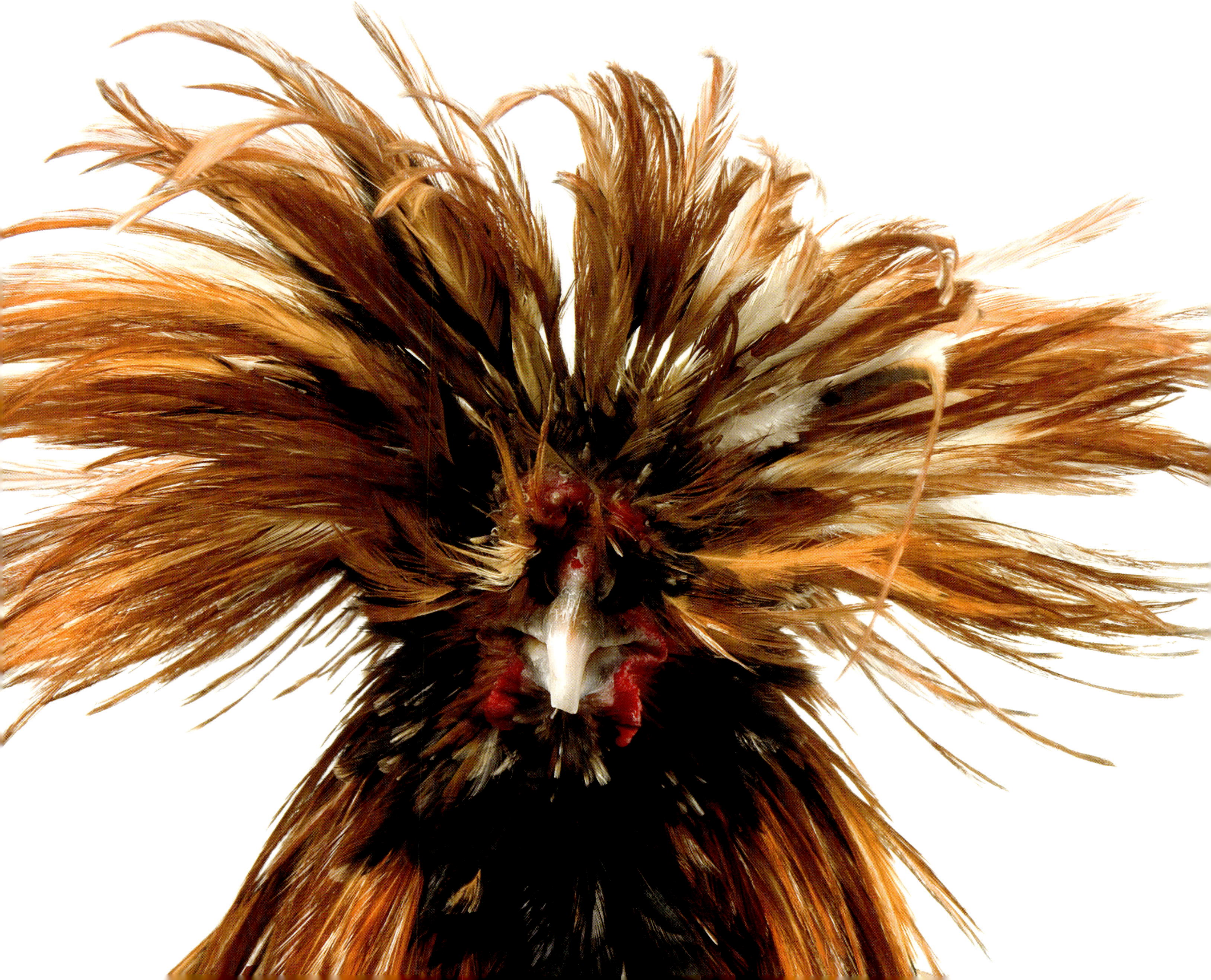

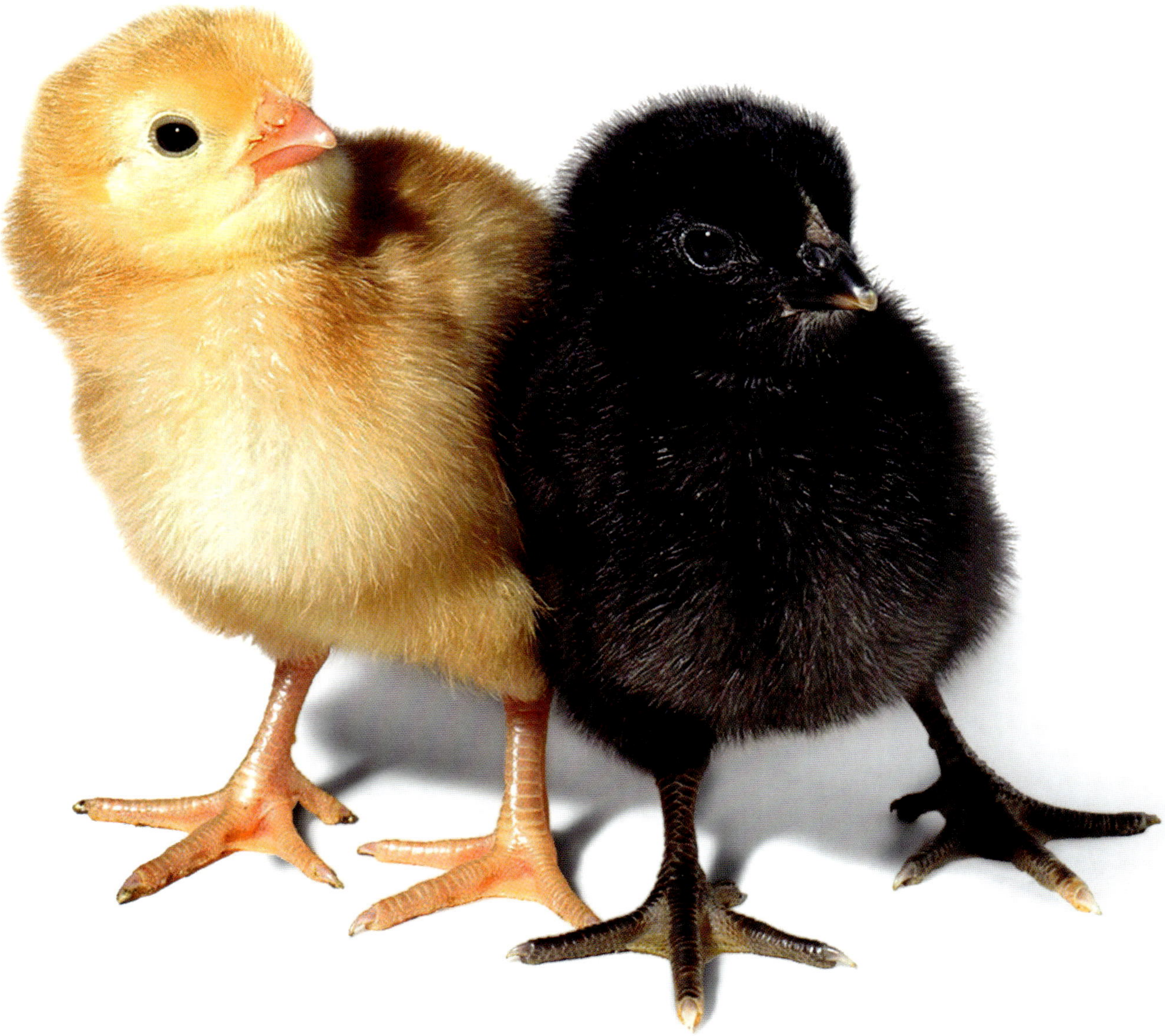

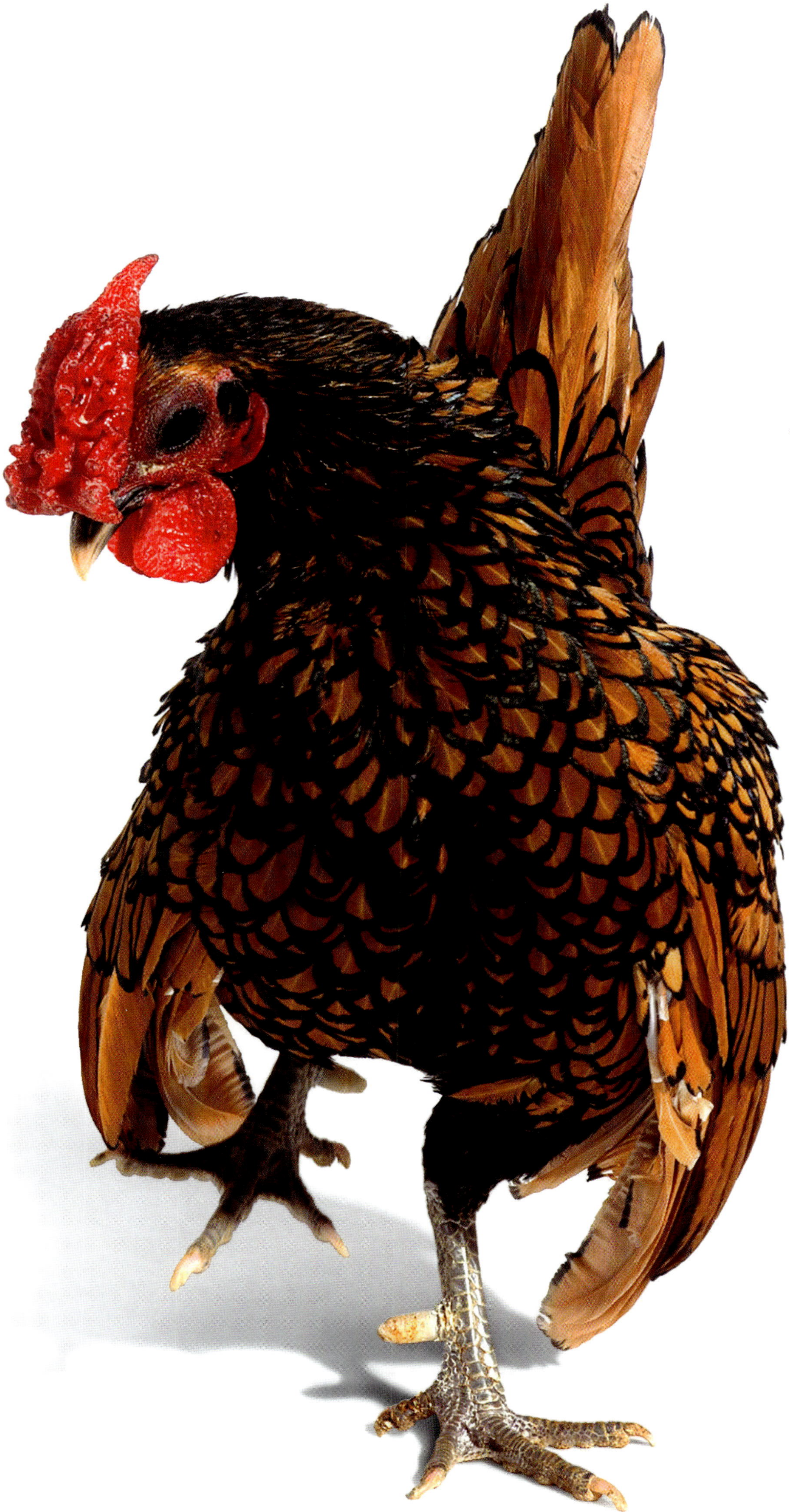

THE DAY OF RECKONING

AIR OF CONFRONTATION

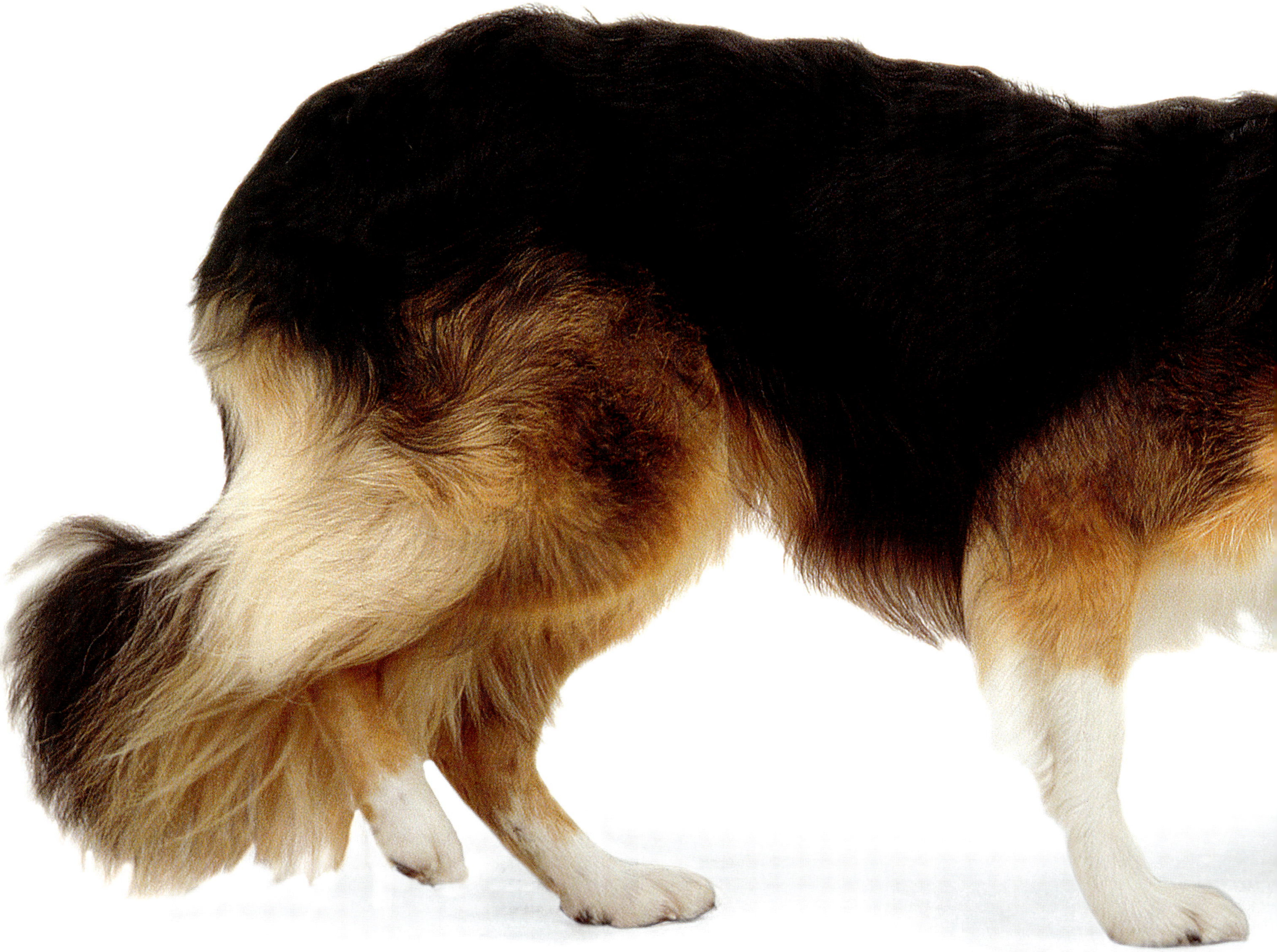

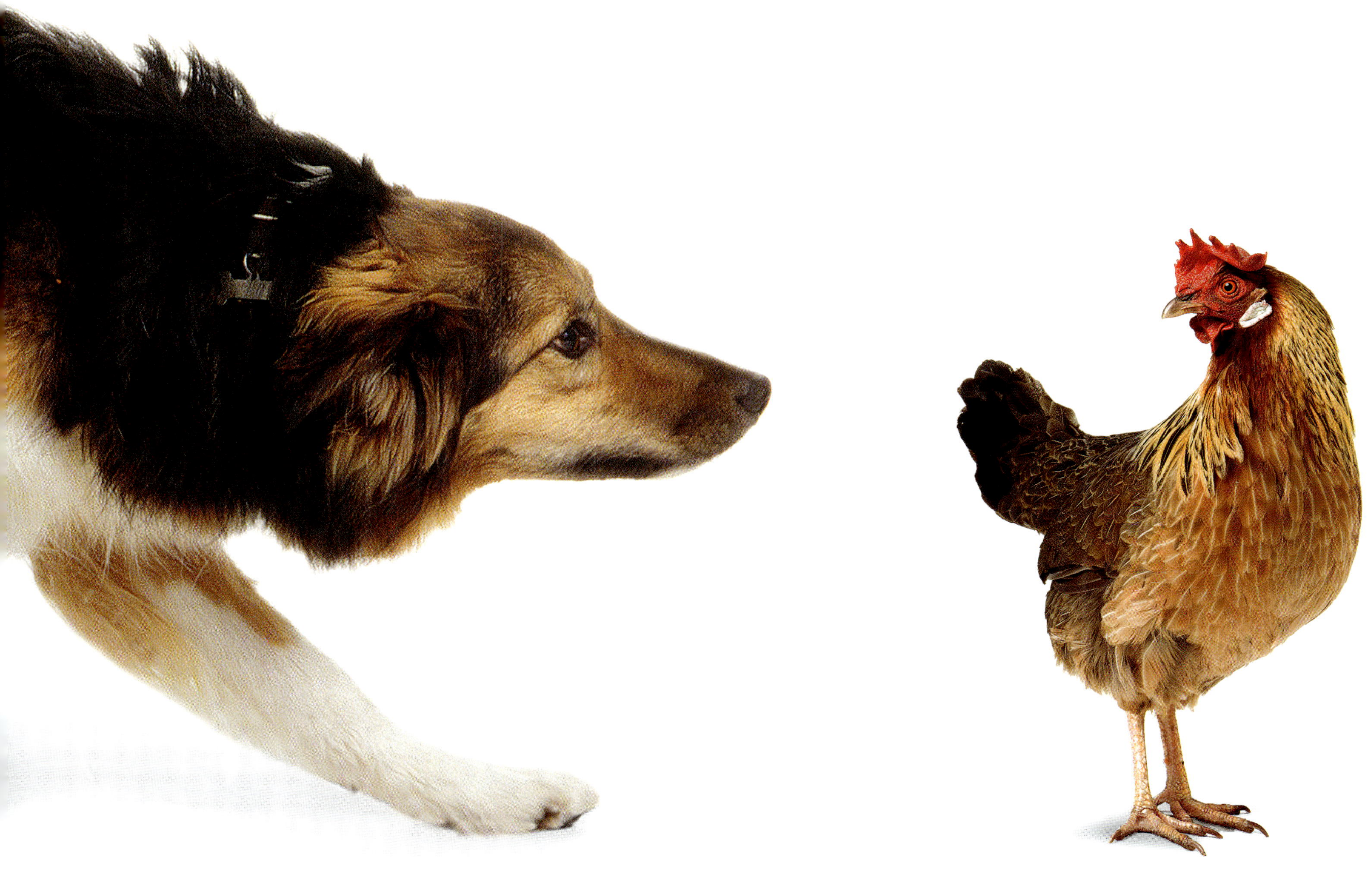

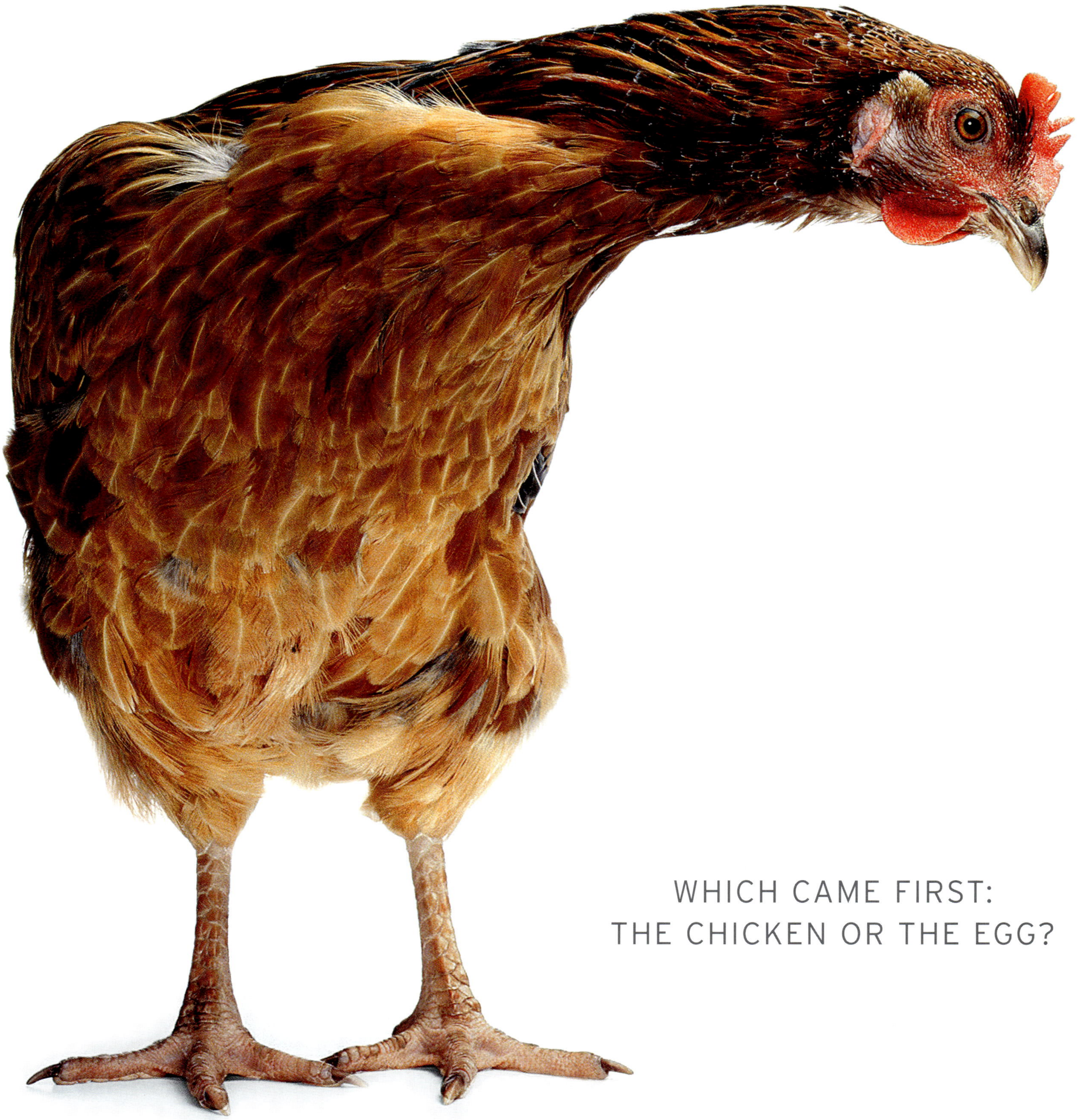

WHICH CAME FIRST:
THE CHICKEN OR THE EGG?

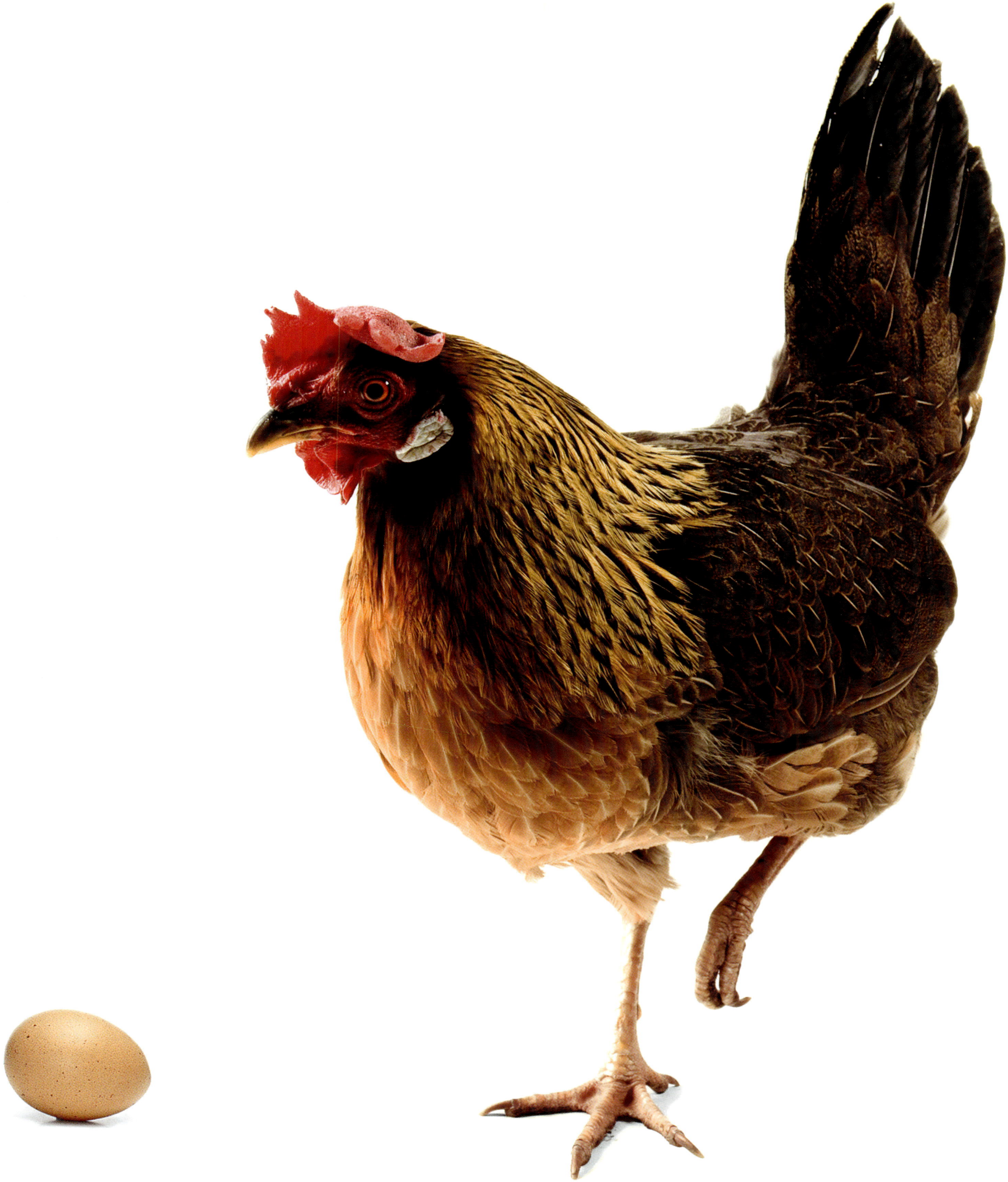

"The fox said: 'Hello, my hen friend.
I haven't seen you for a long time.
Somebody told me you went through a hard time,
I'm very worried about you. . .'"

(The Fox and the Hen – Aesop's Fables)

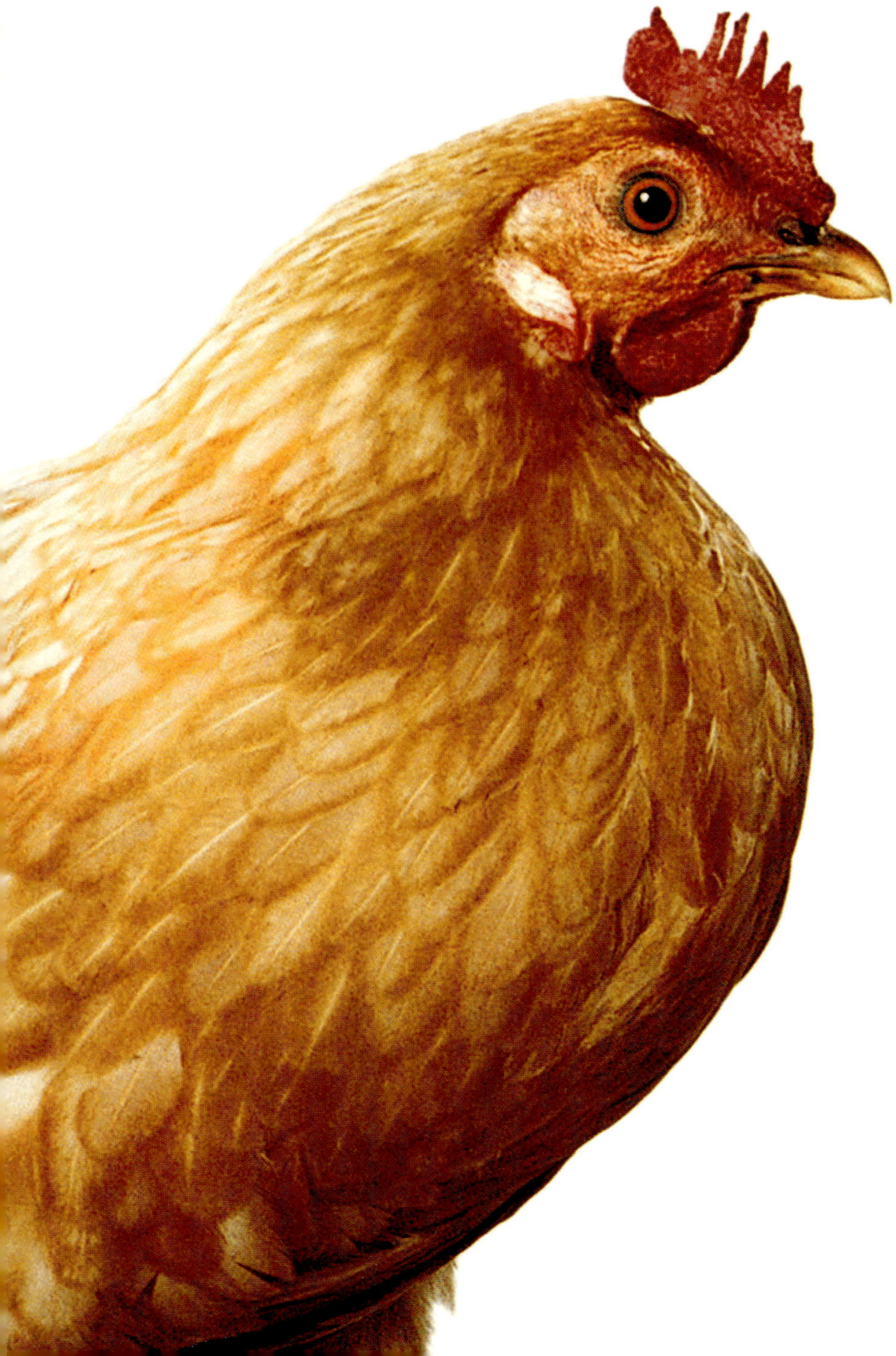

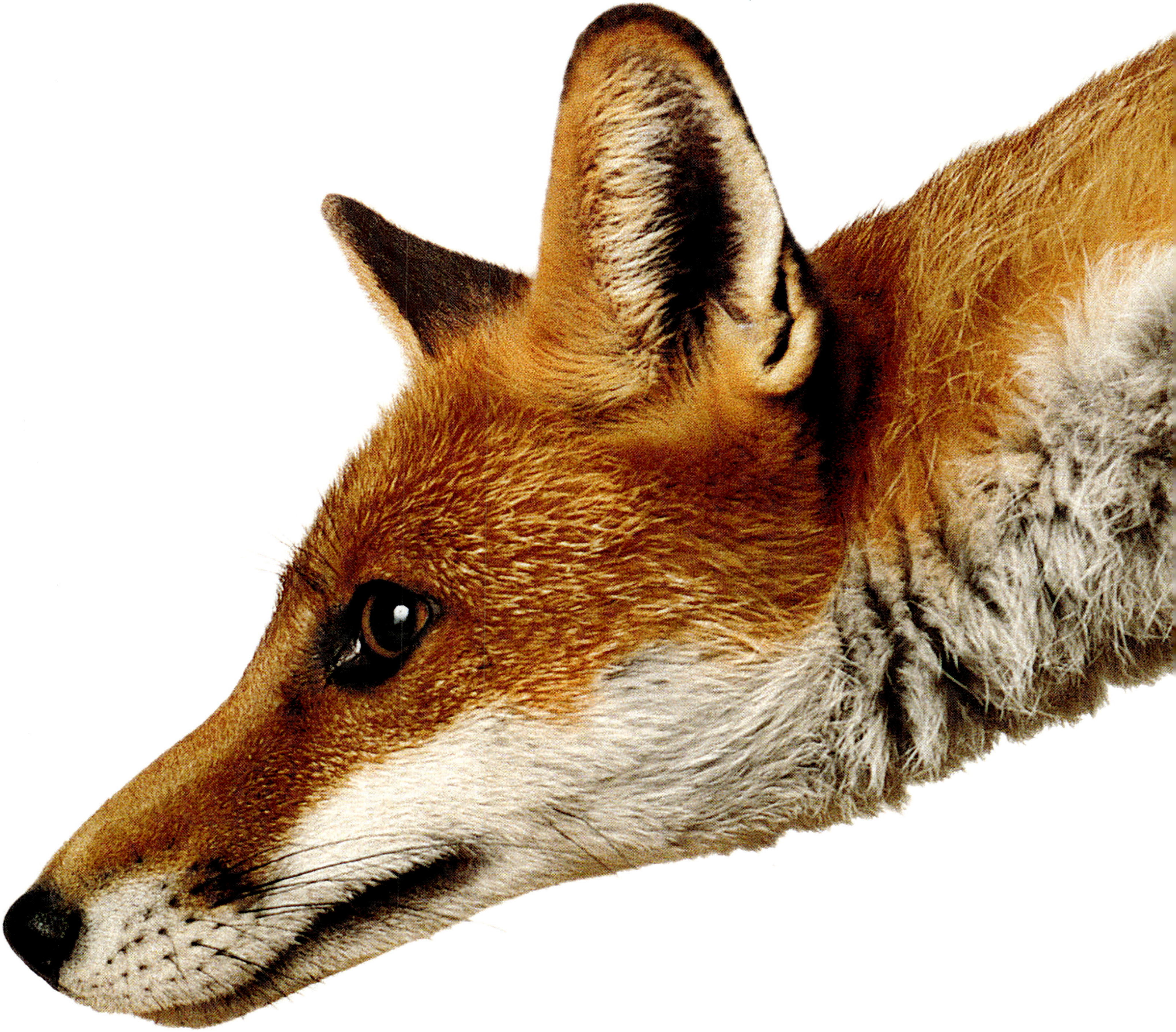

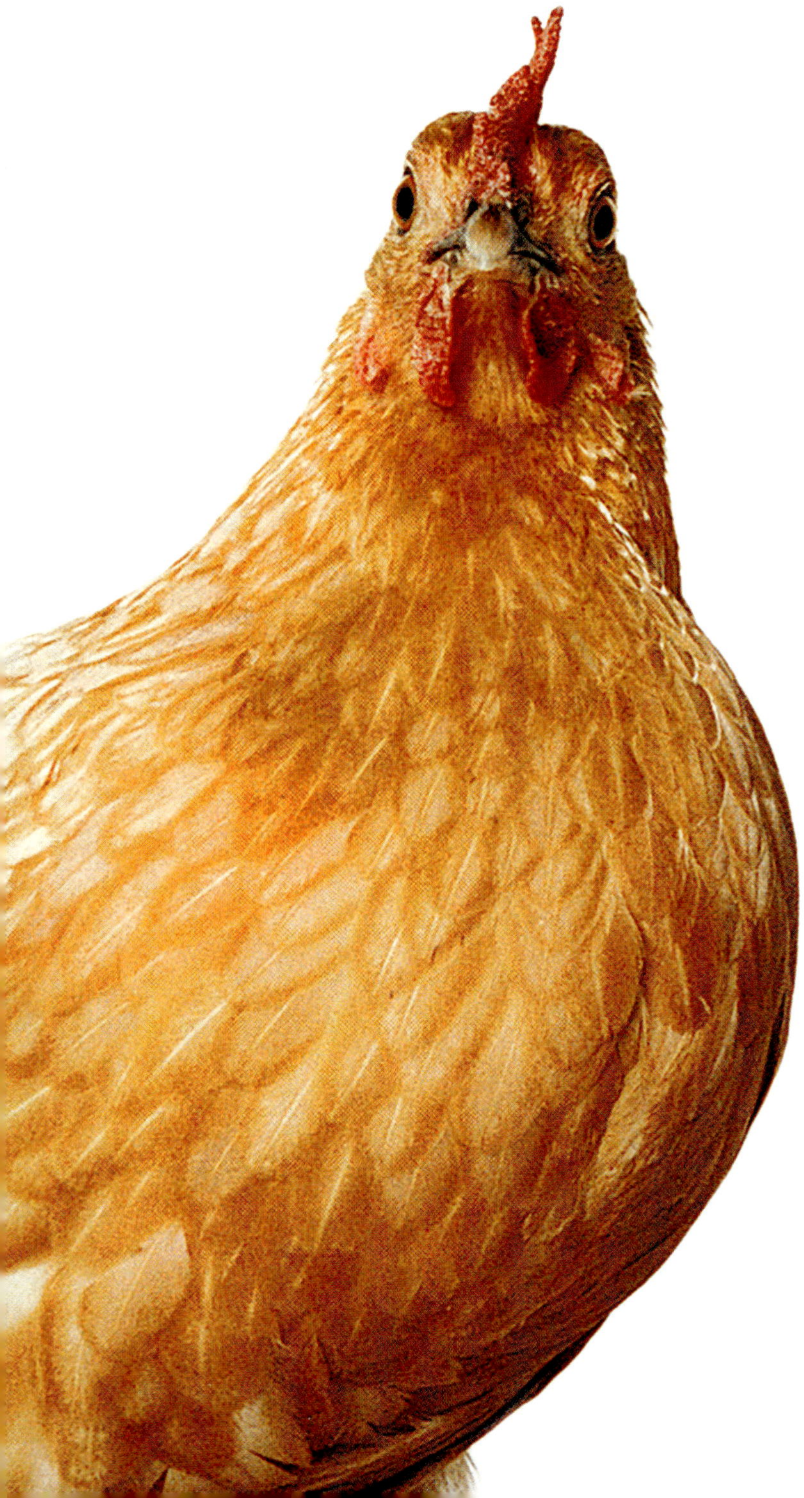

"The hen answered:
'Sorry, my friend fox, but my health is
such that if I were to come to you,
I should be frightened of dying.'"

(The Fox and the Hen - Aesop's Fables)

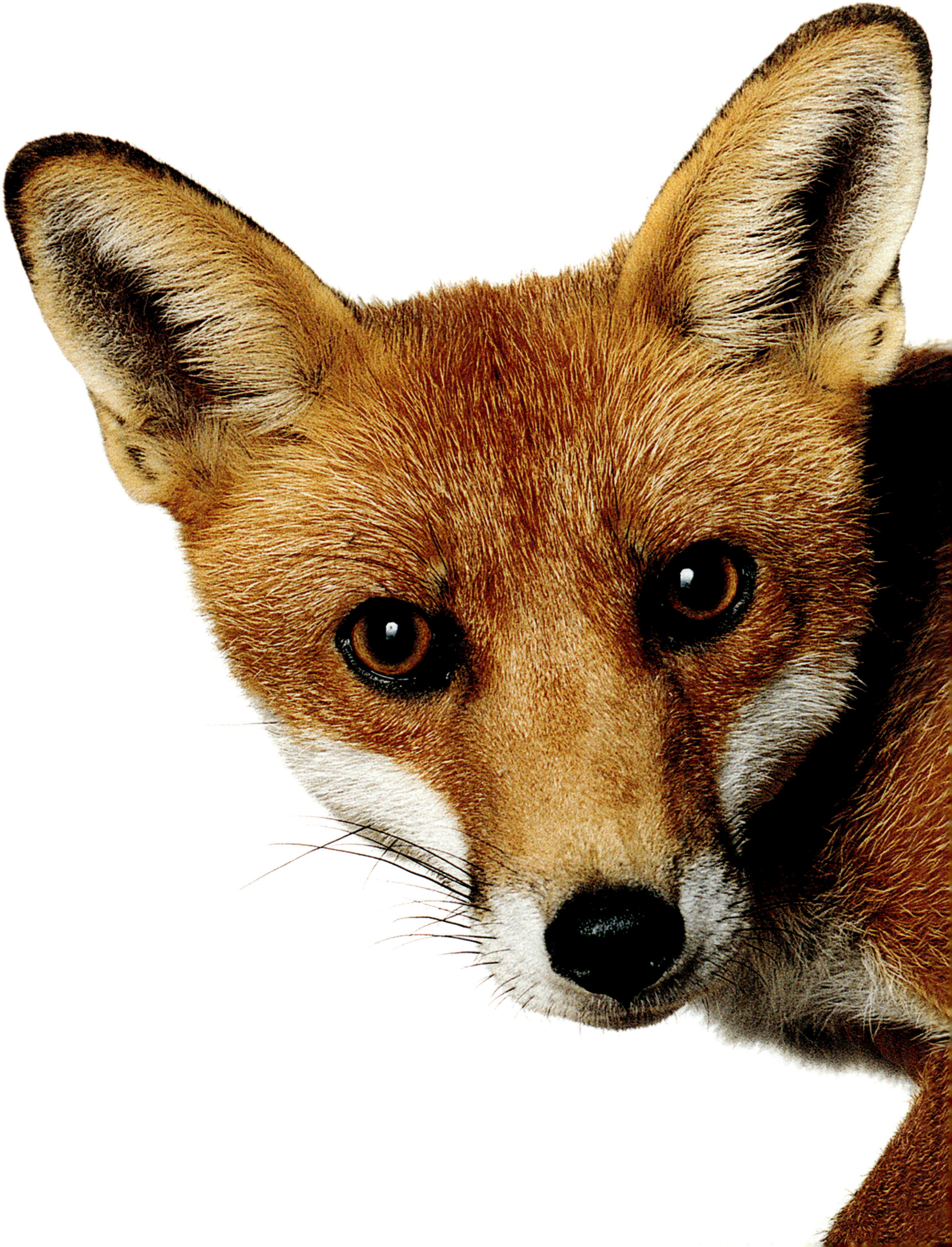

WHITE STAR PUBLISHERS

WS White Star Publishers® is a registered trademark
property of Edizioni White Star s.r.l.

Via Candido Sassone, 24
13100 Vercelli, Italy
www.whitestar.it

Translation: Olga Buongiorno

ISBN 978-88-544-0478-6
1 2 3 4 5 6   13 12 11 10 09

Printed in Italy

PHOTO CREDITS

Amann/Getty Images: pages 103, 176, 177, 179, 180-181, 182-183 * Geoff Brightling/Getty Images: pages 19, 43 * American Images Inc/Getty Images: pag. 118 * James Balog/Getty Images: pag. 149 * Jane Burton/Getty Images: front PLC left, pages 67, 100, 209 * Jane Burton/Warren Photographic: front PLC right and back PLC, pages 4, 8, 14, 15, 16, 17, 32-33, 38-39, 40, 42, 60 in basso, 60-61, 62-63, 68, 88-89, 90, 90-91, 98-99, 101, 104-105, 106, 107, 114, 115, 116, 119, 120-121, 121, 122-123, 128, 129, 130, 131, 132-133, 135, 136-137, 146-147, 163, 190, 192, 193, 194-195, 199, 205, 206, 210-211, 216-217, 219, 220-221, 225, 239 * Gordon Clayton/Getty Images: pages 66, 69, 184, 189 * John Daniels/Getty Images: pages 7, 21 * Digital Vision/Getty Images: pages 160, 161, 234-235, 236-237 * Digital Zoo/Corbis: page 3 * Digital Zoo/Getty Images: pages 30, 31, 34, 35, 86, 97, 166-167 * Dorling Kindersley/Getty Images: pages 80, 151, 152, 170, 191, 196, 204, 212-213, 213, 218, 223, 235 in basso * Michael Duva/Getty Images: pages 12, 55, 58-59, 71, 168 * Bob Elsdale/Getty Images: pages 20, 56, 57, 94-95, 102, 140 * Don Farrall/Getty Images: pages 26, 38, 99 in basso * Tim Flach/Getty Images: pages 46, 72, 74, 155, 188 * Neil Fletcher/Getty Images: pag. 150 * Fotototo/Blickwinkel: pages 111, 112-113 * Martin Gallagher/Corbis: pag. 173 * Tim Graham/Getty Images: pages 45 * Frank Greenaway/Getty Images: pag. 1 * GK Hart/Vikki Hart/Getty Images: pages 6, 11, 27, 28, 29, 36-37, 78-79, 125, 126-127, 198, 202-203, 207, 226 * Imagemore Co. Ltd/Getty Images: pages 82-83, 84-85, 92, 93 * Imagestate/Tips Images: pages 174-175, 186 * Mike Kemp/getty Images: pag. 162 * Dave King/Getty Images: pages 18, 22, 23, 24-25, 117, 142, 197 * Cyril Laubscher/Getty Images: pages 222, 224 * Jens Lucking/Getty Images: pages 76-77 * Ryan McVay/Getty Images: pag. 171 * Hans Neleman/Getty Images: pag. 145 * PIER/Getty Images: pages 232-233 * Max Poddler/Getty Images: pages 64-65 * Purestock/Getty Images: pages 208, 228-229 * Stuart Redler/Getty Images: pag. 201 * Joel Sartore/Getty Images: pages 227, 231, 233 * Preston Schlebusch/Getty Images: pag. 158 * H. Schmidbauer/Blickwinkel: pag. 81 * Steve Shott/Getty Images: pages 41, 138-139 * Justin Sullivan/Getty Images: pag. 49 * Harald Sund/Getty Images: pag. 157 * Kim Taylor/Getty Images: pag. 148 * David Trood/Getty Images: pag. 109 * Freudenthal Verhagen/Getty Images: pages 214-215 * David Ward/Getty Images: pages 230-231 * Michael Winokur/Getty Images: pages 52, 53, 50, 51 * Jerry Young/Getty Images: pages 164-165, 234-235

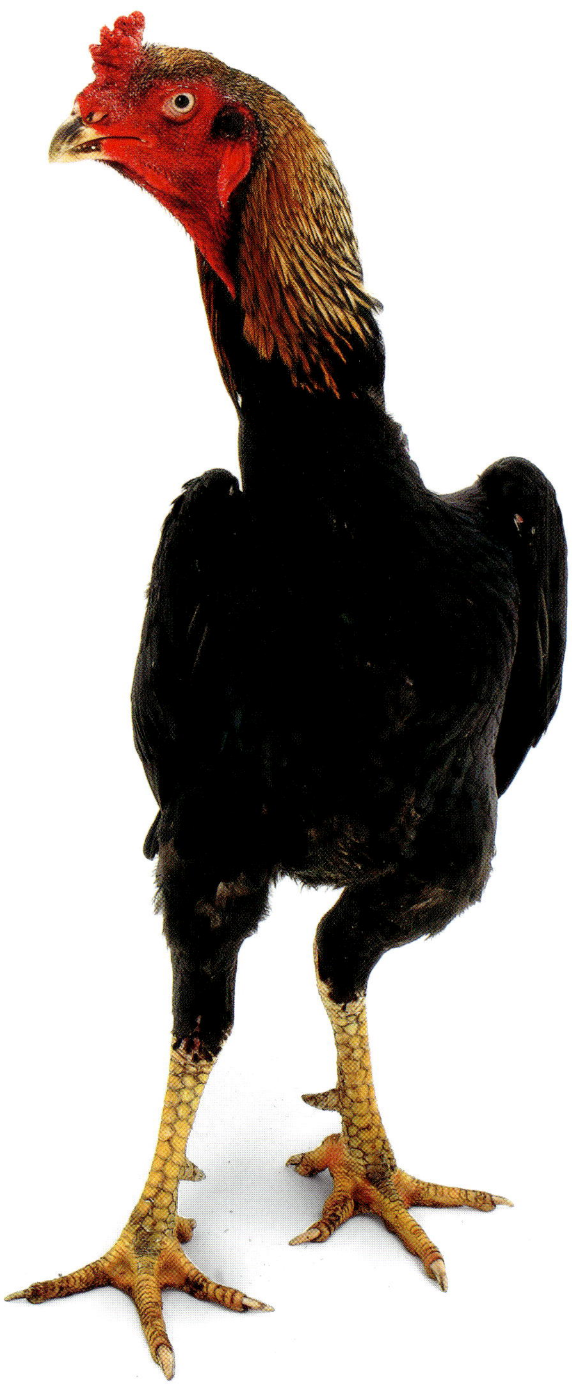

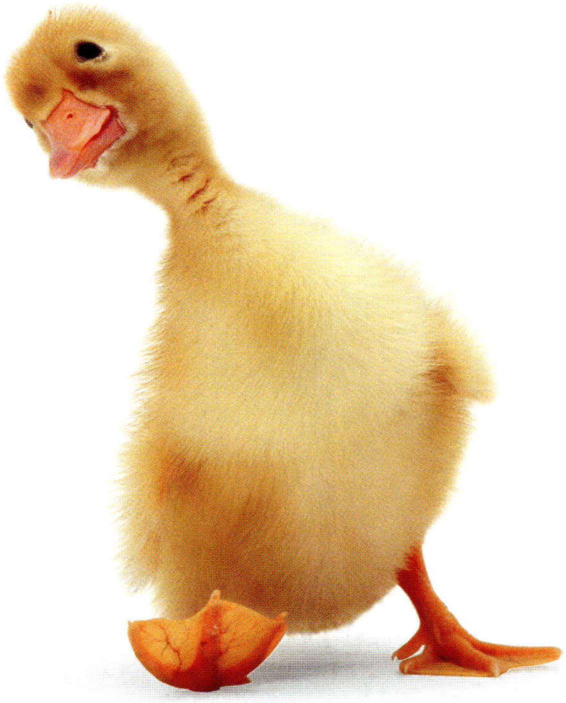

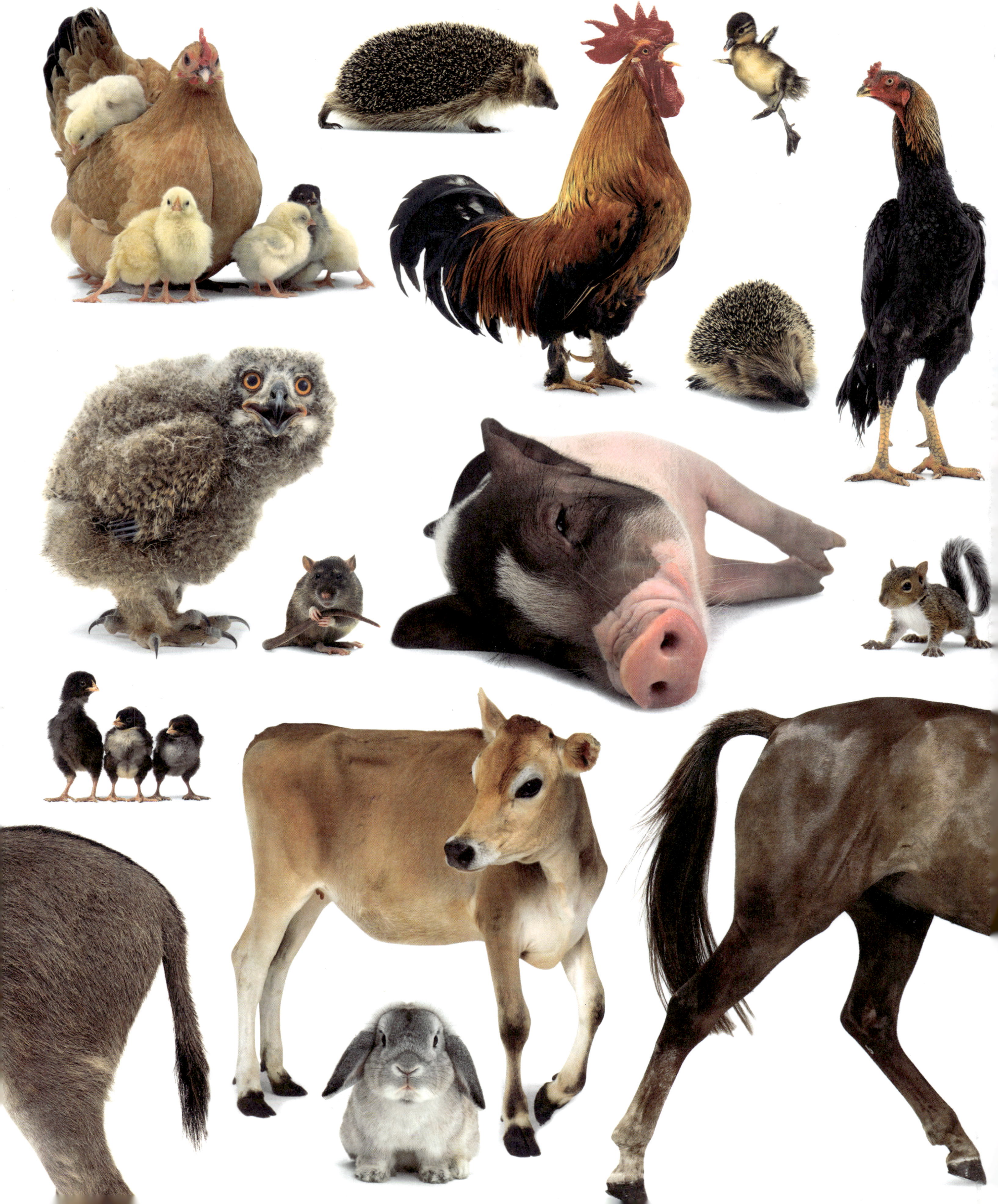